HIGH TEMPERATURE AIR COMBUSTION
FROM ENERGY CONSERVATION TO POLLUTION REDUCTION

Environmental and Energy Engineering Series

Series Editors

Ashwani K. Gupta
Department of Mechanical Engineering, University of Maryland, College Park, Maryland

David G. Lilley
School of Mechanical and Aerospace Engineering, Oklahoma State University, Stillwater, Oklahoma

Published Titles

Enclosure Fire Dynamics
Björn Karlsson and James G. Quintiere

Integrated Product and Process Design and Development
Edward B. Magrab

Advances in Chemical Propulsion: Science to Technology
Gabriel D. Roy

High Temperature Air Combustion: From Energy Conservation to Pollution Reduction
Hiroshi Tsuji, Ashwani K. Gupta, Toshiaki Hasegawa, Masashi Katsuki, Ken Kishimoto, and Mitsunobu Morita

HIGH TEMPERATURE AIR COMBUSTION

FROM ENERGY CONSERVATION TO POLLUTION REDUCTION

HIROSHI TSUJI
ASHWANI K. GUPTA
TOSHIAKI HASEGAWA
MASASHI KATSUKI
KEN KISHIMOTO
MITSUNOBU MORITA

CRC Press is an imprint of the
Taylor & Francis Group, an **informa** business

CRC Press
Taylor & Francis Group
6000 Broken Sound Parkway NW, Suite 300
Boca Raton, FL 33487-2742

First issued in paperback 2019

© 2003 by Taylor & Francis Group, LLC
CRC Press is an imprint of Taylor & Francis Group, an Informa business

No claim to original U.S. Government works

ISBN-13: 978-0-8493-1036-2 (hbk)
ISBN-13: 978-0-367-39564-3 (pbk)

Library of Congress Card Number 2001043860

This book contains information obtained from authentic and highly regarded sources. Reasonable efforts have been made to publish reliable data and information, but the author and publisher cannot assume responsibility for the validity of all materials or the consequences of their use. The authors and publishers have attempted to trace the copyright holders of all material reproduced in this publication and apologize to copyright holders if permission to publish in this form has not been obtained. If any copyright material has not been acknowledged please write and let us know so we may rectify in any future reprint.

Except as permitted under U.S. Copyright Law, no part of this book may be reprinted, reproduced, transmitted, or utilized in any form by any electronic, mechanical, or other means, now known or hereafter invented, including photocopying, microfilming, and recording, or in any information storage or retrieval system, without written permission from the publishers.

For permission to photocopy or use material electronically from this work, please access www.copyright.com (http://www.copyright.com/) or contact the Copyright Clearance Center, Inc. (CCC), 222 Rosewood Drive, Danvers, MA 01923, 978-750-8400. CCC is a not-for-profit organization that provides licenses and registration for a variety of users. For organizations that have been granted a photocopy license by the CCC, a separate system of payment has been arranged.

Trademark Notice: Product or corporate names may be trademarks or registered trademarks, and are used only for identification and explanation without intent to infringe.

Library of Congress Cataloging-in-Publication Data

High temperature air combustion : from energy conservation to pollution reduction / authored by H. Tsuji ... [et al.].
    p. cm. -- (Environmental and energy engineering series)
    Includes bibliographical references and index.
    ISBN 0-8493-1036-9 (alk. paper)
    1. Combustion engineering. 2. Environmental protection. I. Tsuji, H. (Hiroshi) II. Series.
TJ254.5.H54 2001
621.402′3—dc21   2001043860

Visit the Taylor & Francis Web site at
http://www.taylorandfrancis.com

and the CRC Press Web site at
http://www.crcpress.com

Dedication

Ryoichi Tanaka
March 29, 1928–Oct. 17, 1997

This only book on the subject of "high temperature air combustion: from energy conservation to pollution reduction" is dedicated to the late Ryoichi Tanaka, President of Nippon Furnace Kogyo Kaisha Ltd, Yokohama, Japan from April 12, 1950 to the time of his death on October 17, 1997. Tanaka dedicated his entire career to promoting the development of advanced industrial furnaces that are today known as the most efficient and environmentally benign. He made pivotal contributions to the practical application of high temperature air combustion principles for use in advanced furnace design by introducing most efficient regenerators. He was a patient educator to professionals in the field, a good colleague, an outstanding mentor and friend to members of his staff, a fine, talented, and intellectual individual, and an extremely generous friend. He promoted the art and science of high temperature air combustion that continues to gain good recognition in the technical community. His dedicated efforts during the 1990s on advanced smart industrial furnace design are now providing the fruits of his efforts, and are being utilized worldwide, including Japan, Asia, Europe and the United States.

Tanaka was born on March 29, 1928 in Nagoya, Japan. After graduating from Tohoku University in 1953 in economics, he started the operation of his Nippon Furnace Kogyo Kaisha Company. He began to participate in the social and business activities related to his company at the very early stage of his career. He was appointed as a director of the Industrial Furnaces Association in 1959, a former organization of the Japan Industrial Furnace Manufacturers Association (JIFMA), and was then appointed as vice-chairman of JIFMA in 1966. He was also appointed as director of the Japan Industrial Furnace Manufacturers Pension Foundation and then president in 1991. He founded the Japanese Flame Research Committee (JFRC) for the International Flame Research Foundation (IFRF, Netherlands) in Japan and served as the chairman of JFRC starting in 1977. In 1992 he was installed as the president of JIFMA and was then appointed as director of the energy conservation center, Japan, in the same year. He was appointed as vice president of IFRF in 1994. He received many prestigious honors and awards, including the Blue Ribbon Medal Award in 1986 and the Social Achievement Award of the Japan Society of Mechanical Engineers in 1995.

This book, as well as the development of high performance industrial furnaces, was one of the goals and a vision of Tanaka to enhance industry–university exchange as well as international exchange of combustion technologies. Members of the technical community value his commitment to and accomplishments in this remarkable technology of significant energy savings, downsizing of the furnace, uniform thermal field, and pollution reduction, all of which occur simultaneously.

Foreword

This book represents the outcome of the collective efforts of many scientists and engineers from industry and academia. The Japan Industrial Furnace Manufacturers Association (JIFMA), under the sponsorship of the New Energy and Industrial Technology Development Organization (NEDO), initiated a new High Performance Industrial Furnaces development project during 1993 to 1999. The goal of this project was to demonstrate significant reduction in energy consumption in the industrial sector. This major undertaking by the Japanese government involved many Japanese companies and organizations and some academic institutions. These research and development efforts have resulted in numerous technical publications, reports, mass media publicity, and international recognition. In addition, several technological awards and special recognitions were given for the innovative findings from this project.

The issue of global environmental protection has been discussed throughout the world as a major task, especially since the 1992 environmental summit conference held in Brazil. In November 1999, our study on The Development and Practical Application of High Performance Industrial Furnaces won the highly honorable 9th Nikkei Global Environmental Technology Grand Prize Award. The reason this project was so honored is that its new technological advances and developments have demonstrated the possibility of reducing energy consumption by about 30%, NO_x emission by about 50%, and equipment size by about 25%. These technological achievements established a new landmark, which was difficult to envision by many professionals and colleagues throughout the world. These results were highly appreciated by high government officials in Japan. We have very high expectations from this project. We believe that our technology will not only be utilized in several industries in Japan, but also be gradually transferred to many other countries. We also believe that this innovative technology, originating in Japan, will provide a major contribution to environmental protection on a global scale.

We will make our best efforts to promote and widely spread the high performance industrial furnace technology on a wide scale. We will also make dedicated efforts to expand this technology further for use in other technological areas that utilize fuel as an energy source.

This book is the culmination of this 7-year project. The book also highlights some recent accomplishments made in Japan and abroad on some peripheral and application technologies in cooperation with several universities and research institutions. We are indebted to all those who have contributed in any capacity to the development of this technology. Some of the organizations, industries, universities, and institutions contributing to this technology development effort are cited here.

This book represents the outcome of the collective efforts of many scientists and engineers from the related industry and academia. I would like to take this opportunity to express sincere gratitude to all the participants, including the corporations that undertook subcommissioned assignments, the personnel from academia and industry, and those from the Ministry of Economy, Trade and Industry (METI, formerly MITI) and NEDO. I hope that this book will prove to be of value to worldwide members of the technical community from the industry, academia, research organizations, and government.

Finally, I wish to thank all members of the technical community whose diligent efforts led to the outstanding success of this major undertaking by the Japanese government. Their efforts will be remembered not only in Japan but also worldwide. I wish to acknowledge my predecessor, the late Ryoichi Tanaka, President of Nippon Furnace Kogyo Kaisha Ltd. (NFK), who served as the chairman of JIFMA for all the research and development efforts. He devoted his whole life to the improvements of industrial furnace technologies and struggled to develop high performance industrial furnaces and high temperature air combustion technologies until the end. We owe our gratitude to his lifelong dedicated efforts on evolutionary and revolutionary combustion and heat transfer technology developments.

Tadashi Tanigawa
Chairman, Japan Industrial Furnace
Manufacturers Association

August 2002

Preface

This book is a comprehensive and illustrated work on high temperature air combustion (here called HiTAC), which has revolutionized our paradigm on the use of all kinds of fossil, alternative, waste, and derived fuels for energy conversion and energy utilization in industry. Significant experimental knowledge and insights from many practical devices have resulted in the utilization of HiTAC technology for many applications. The traditional definition of flame is that which gives heat and light during chemical reaction between reactants. However, under certain conditions with some fuels, this definition of flame can be revised.

The text is oriented toward the person who wishes to gain a good understanding of the principles and practice of HiTAC. The text also allows one to apply this technology to achieve significant energy savings, to reduce the size of equipment and environmental pollution, including CO_2, for specific applications. Combustion technology utilizing preheated combustion air in excess of 1000°C has drawn significant worldwide attention for many applications. The basic concept is that the combination of maximum waste heat recovery by high cycle regenerator and controlled mixing of highly preheated combustion air with burned gases yields uniform and relatively low temperature flames. Indeed, the revolutionary HiTAC technology has been demonstrated to provide simultaneous reduction of CO_2 and nitric oxide emissions and to reduce energy consumption for a specific process or requirement. Specifically, HiTAC has been demonstrated to provide about 30% reduction in energy (and hence also CO_2 emission), 50% reduction of pollutants, and about 25% reduction in the physical size of the facility compared with the conventional type of furnace design. Furthermore, extremely low levels of nitric oxide emissions, far below the present regulations, have been demonstrated in several field trials.

This book describes the development of HiTAC technology and its practical application to different kinds of furnaces of importance in industry. Future potential applications of this technology are also presented. Recognition of the vast scope and importance of HiTAC technology has prompted CRC Press to include the present text in their series of books on Environmental and Energy Engineering.

Other texts in the series delve deeply into other specific areas. This book focuses on all aspects and applications of HiTAC; good characterization of the combustion phenomena with high temperature combustion air is of prime concern. Particular reference is made to the work published in this area during the last decade. Other valuable information may be found in various research reports and journals in Japan and from international symposia and journals.

Chapter 1 describes the innovation of HiTAC, as well as the historical background and evolution of this combustion technology. Chapter 2 discusses the combustion phenomena associated with high temperature air combustion. A comprehensive view is provided of the fundamental differences in the thermal, chemical, and fluid dynamic characteristics of the flame. HiTAC technology provides significantly higher flame stability at all fuel–air mixtures (including very lean fuel mixtures), higher heat transfer, and low heat loss from the stack (waste heat). The fundamentals of gas, liquid, and solid fuel flames are also presented from the point of view of HiTAC. Also included here are the significantly different flame features, flame stability, reduced emissions, and significant energy savings with HiTAC. The flame color is found to be much different from the usually observed blue or yellow. Under certain conditions bluish green and green color flame has been observed using typical hydrocarbon fuels. In contrast, flameless (or colorless) oxidation of the fuel has also been observed. These characteristics of flames have not been cited before in the literature. In Chapters 3 and 4 the models for simulating high temperature air combustion as well as the impact of HiTAC on industrial furnace performance are presented. Chapter 5 provides the design guidelines for high performance industrial furnaces. General and optimal design guidelines for various kinds of furnaces, such as reheating furnaces, heat treatment furnaces, and melting furnaces, are presented from the point of view of higher heat transfer, reduced size, reduced pollution, and higher performance. Experience and field trials on different kinds of practical furnaces are also presented. In Chapter 6, potential applications of HiTAC to other energy-using sectors are presented. Some of the examples include the conversion of coals, biomass, and solid waste fuels to cleaner fuels, fuel reforming, stationary gas turbine engines, internal combustion engines, and many other advanced energy-to-power conversion systems. Reference data from several high performance industrial furnaces are also included as an appendix to the book.

This book is the first to be published on high temperature air combustion, including fundamental aspects, its practical use in furnaces and boilers, potential applications in other energy conversion systems, and projected developments and trends. We hope that our readers will be stimulated by the new developments in equipment for energy saving and low pollution for industry and commerce. The authors of any specialized text must select, abstract, and reframe the material that they find most suitable for exemplifying the principles and techniques. In this book we have selected the work of several prominent researchers. Nevertheless, a special attribute of this book is the strong practical emphasis of the portrayal of those concepts, which may be difficult to understand and apply. We have tried to strike the best balance among the physical, practical, and mathematical aspects, and to produce a text that appeals to students and practicing engineers in applying the latest available knowledge to solve their practical energy conservation needs and environmental pollution reduction.

The book is intended as a basis for engineers and researchers in the area of energy conversion using fuels, and also as a textbook for senior year undergraduate and graduate students. The scope of the book is to provide a solid foundation for those who intend to utilize HiTAC technology for their specific application for energy conservation and pollution reduction.

We wish to acknowledge the work of all those who contributed and collaborated on HiTAC research and development activities and of those who assisted in the preparation of this book. We are particularly grateful to our numerous colleagues in Japan and throughout the world who provided us with information for inclusion here. Specific acknowledgment to authors and sources is made in the text and in the lists of references. Special thanks are due to all the industries, institutions, and organizations listed in the Acknowledgments as well as the authors who have contributed to this book. Their help, support, encouragement, and friendship were most welcome. We wish to acknowledge the late Ryoichi Tanaka, President of NFK, for his vision and leadership on the development of HiTAC technology. All members of the technical community worldwide will remember him for his commitment and devotion to developing advanced furnaces. We are grateful for his lifelong dedicated efforts on the evolutionary and revolutionary burner and furnace technology developments. Finally, we most gratefully wish to thank CRC Press for their special cooperation in careful preparation of the book.

Hiroshi Tsuji
Ashwani K. Gupta
Toshiaki Hasegawa
Masashi Katsuki
Ken Kishimoto
Mitsunobu Morita

August 2002

The Authors

Hiroshi Tsuji was promoted to Professor of Combustion Science at the University of Tokyo in 1962. Since 1984 he is Emeritus Professor of the University of Tokyo. He became a member of the Japan Academy in 1992. He has made an honorable achievement through study on the fundamental aspects of flames, using a porous cylinder burner, called the "Tsuji burner". Professor Tsuji was awarded the Bernard Lewis Gold Medal (the Combustion Institute) in 1988, the JSME Thermal Engineering Achievement Prize in 1990, the Japan Academy Prize in 1990, and the Second Order of the Sacred Treasure in 1993.

Ashwani Gupta is Professor of Mechanical Engineering at the University of Maryland, College Park. He received his Ph.D. in 1973 and D.Sc. in 1986, both from Sheffield University, UK. He has co-authored 2 books, edited 10 books, and published over 300 archival papers. He is a Fellow of AIAA, ASME, and the Institute of Energy, UK. He received the AIAA Energy Systems award in 1990 and Propellants and Combustion award in 1999, and the ASME George Westinghouse Gold Medal award in 1998. He also received several best paper awards from AIAA and ASME. He is an associate editor for *AIAA J. Propulsion and Power* and co-editor of the Environmental and Energy Engineering series of books published by CRC Press.

Toshiaki Hasegawa is Director and General Manager of Combustion Equipment Division at Nippon Furnace Kogyo, Yokohama. He graduated from the Metallurgical Engineering Department at Iwate University and participated in a combustion engineering course at Waseda University in 1977. He joined NFK in 1974 and has been working on the development of technologies for high intensity combustion and low NOx combustion for industrial furnaces and boilers. In 1999, he was awarded the Grand Technology Prize of Ohkouchi Memorial Foundation following R&D on high temperature air combustion heating system. In 2000 he received the Best Paper Award from the Combustion Society of Japan.

Masashi Katsuki is Professor of Osaka University, Department of Mechanophysics, School of Engineering. He was the chairman of the Thermal Engineering Division of the Japan Society of Mechanical Engineers in 2001, and the Vice President of the Combustion Society of Japan from 1999 to 2000. He is a board member of the Japan Society of Mechanical Engineers, the Combustion Society of Japan, and the High Temperature Society of Japan. He was awarded Best Paper Awards from the High Temperature Society of Japan in 1990 and 1995, Gas Turbine Society of Japan in 1996, Japan Society of Mechanical Engineers in 1997, and the Combustion Society of Japan in 2000.

Ken Kishimoto is Professor of Mechanical Engineering, Graduate School of Technology at Kokushikan University. He is a graduate of Waseda University. He obtained his Masters degree from Waseda University in combustion engineering. He is experienced in the fields of combustion dynamics, and heat and energy engineering. He has published a number of papers on electrical augmented combustion, combustion driven oscillation, combustion noise, pulse combustion, and high temperature air combustion.

Mitsunobu Morita is General Manager of Technology and Senior Researcher in High Performance Industrial Furnace Development Center, Japan Industrial Furnace Manufacturers Association. He graduated from the University of Tokyo in Mechanical Engineering in 1967. He obtained his Masters degree from the University of Tokyo in 1969, and Doctorate degree from Nagoya University in 2001. He has worked in the field of thermal engineering and process engineering for 30 years after joining Nippon Steel Corporation in 1969. He is experiencee in the development and applications of surface combustion in sintering furnace, new hot strip mills reheating furnace design and construction, and reduction burner with strong recirculation.

Acknowledgments

The authors sincerely acknowledge the New Energy Development Organization (NEDO), and the Japan Industrial Furnace Manufacturers Association (JIFMA) for supporting this work. One of the authors, A. K. Gupta, would like to acknowledge the support provided to him by the National Science Foundation (NSF), National Aeronautics and Space Administration (NASA), and NFK. He would personally like to acknowledge the support and friendship provided by late President Ryoichi Tanaka of NFK and to Dr. R. H. Woodward Wesche for his valuable comments in the entire manuscript. The support and contributions provided by the following individuals from academia, industry, and research are gratefully acknowledged.

- Prof. Hiroshi Taniguchi, Hokkai Gakuen University
- Prof. Masatake Sadakata, University of Tokyo
- Prof. Emeritus Toshisuke Hirano, University of Tokyo
- Prof. Masanobu Hasatani, Nagoya University
- The late Prof. Kazutomo Ohtake, Toyohashi University of Technology
- Dr. Kiyoshi Sugita, Nippon Steel Corporation
- Hiroshi Mitsukawa, Tetsuo Ohmura, Toru Yamauchi, Shinichi Nishioka, Yoshio Takagi, New Energy and Industrial Technology Development Organization (NEDO)
- Prof. Akira Yoshida, Tokyo Denki University
- Prof. Yoshio Yoshizawa, Tokyo Institute of Technology
- Prof. Kuniyuki Kitagawa, Nagoya University
- Prof. Nilson Kuniyoshi, Okayama Prefectural University
- Prof. Ichiro Naruse, Toyohashi University of Technology
- Prof. Atsushi Makino, Shizuoka University
- Prof. Hideyuki Aoki, Tohoku University
- Prof. Roman Weber, Technische Universität Clausthal, Germany
- Shinichiro Fukushima, Shunichi Sugiyama, Toshikazu Akiyama, Yutaka Suzukawa, Toshio Ishii, Yoshimichi Hino, Tatsuya Shimada, NKK Corporation
- Susumu Mochida, Makoto Miyata, Jun Sudoh, Takeshi Tada, Kiyobumi Kurita, Tadahiro Araake, Nippon Furnace Kogyo Kaisha, Ltd
- Jyunichi Hayashi, Dr. Hideki Murakami, Toshiaki Saitou, Masataka Hase, Atsushi Hida, Nippon Steel Corporation
- Kouji Murakami, Morihiko Imada, Hisamichi Otani, Masao Uede, Chugai Ro Co., Ltd
- Kenjiro Sato, Yoshiyuki Tomita, Daido Steel Co., Ltd
- Osamu Takeuchi, Kazumi Mori, Toru Yoshida, Ishikawajima-Harima Heavy Industries Co., Ltd

Katsumi Hazama, Takayuki Inami, Sanken Sangyo Co., Ltd
Hirokiyo Shimomura, Jyunichi Ninomiya, Rozai Kogyo Kaisha, Ltd
Dr. Hideo Tai, Masayuki Tamura, Dr. Ryoichi Toriumi, Tokyo Gas Co., Ltd
Masahiko Nakahara, Yamatake Corporation
Hirotaro Kouhata, Yokogawa Electric Corporation
Eizo Maeda, Kawasaki Refractories Co., Ltd
Yasuo Moriguchi, Cosmo Oil Co., Ltd
Katsuyoshi Kobayashi, Taketo Sasaki, Tsutomu Yasuda, JIFMA

Table of Contents

Chapter 1 Introduction 1
1.1 Historical Background of High Temperature Air Combustion 1
    1.1.1 Environment and Energy Conservation 1
    1.1.2 Reduction of Pollutant Emissions and Energy Crisis 2
    1.1.3 Panorama of High Temperature Air Combustion Technology 4
1.2 Innovation of High Temperature Air Combustion 6
    1.2.1 Fundamentals of Combustion 6
        1.2.1.1 Heat Recirculating Combustion 6
        1.2.1.2 Definition of High Temperature Air 10
        1.2.1.3 Heat Recirculation and Exhaust Gas Recirculation 10
    1.2.2 Principle of Combustion Control for CO_2 and NO_x Reduction 13
        1.2.2.1 Carbon Dioxide 13
        1.2.2.2 Nitric Oxides 15
    1.2.3 Heat Transfer in High Temperature Air Combustion 17
        1.2.3.1 Convection Heat Transfer of High Temperature Air
                Combustion 18
        1.2.3.2 Radiant Heat Transfer of High Temperature
                Air Combustion 20
        1.2.3.3 Effect of Wall as Wavelength Conversion Body in High
                Temperature Air Combustion 21
    1.2.4 Thermodynamics of High Temperature Air Combustion 23
References 28

Chapter 2 Combustion Phenomena of High Temperature Air Combustion 29
2.1 Introduction 29
2.2 Flame Features 30
    2.2.1 Flame Stability 30
        2.2.1.1 Temperature Profiles 32
        2.2.1.2 Influence on NO_x Emissions 34
    2.2.2 Thermal Field Behavior 34
        2.2.2.1 350 kW-Scale Combustion Test 34
        2.2.2.2 Cold Flow Model Test 34
        2.2.2.3 Temperature Profiles 36
        2.2.2.4 Flow Patterns 38
    2.2.3 Flame Structure, Radicals, and Species 39
        2.2.3.1 Experimental Furnace for Optical Measuring 39
        2.2.3.2 Combustion Conditions 39
        2.2.3.3 Optical Measurement Results 42

		2.2.3.4	Summary 48
	2.2.4	Flame with Heat and Combustion Products Recirculation 49
		2.2.4.1	Improved Heating Method 49
			2.2.4.1.1 Heat and Combustion Product Recirculation 49
		2.2.4.2	Heat Balance in the System 50
			2.2.4.2.1 Gross Heat Input 50
			2.2.4.2.2 Heat Transfer in Furnace 51
			2.2.4.2.3 Heat Output 53
			2.2.4.2.4 Equation Arrangement 53
		2.2.4.3	Calculation Results 53
			2.2.4.3.1 Effect of Gas Recirculation 53
			2.2.4.3.2 Heat and Gas Recirculation 54
			2.2.4.3.3 Thermal Efficiency 57
		2.2.4.4	Discussion 57
		2.2.4.5	Summary 60
2.3	Fundamentals of Gaseous Fuel Flames 60
	2.3.1	Extinction Limit and No_x in Laminar Diffusion Flame 60
		2.3.1.1	Experimental Apparatus 61
		2.3.1.2	Velocity Field and Temperature Field 62
		2.3.1.3	Extinction and Re-ignition Temperatures of Laminar Diffusion Flame 64
		2.3.1.4	Distributions of Temperature and Concentrations of Species 66
		2.3.1.5	Effect of Flame Temperature on NO_x Formation 68
		2.3.1.6	Relationship between Flame Temperature and the Critical Velocity Gradient 69
		2.3.1.7	Summary 70
	2.3.2	Burning Velocity 71
		2.3.2.1	Simulation Model 71
		2.3.2.2	Simulation Results and Discussion 72
			2.3.2.2.1 Preheated but Not Diluted Premixed Flames 72
			2.3.2.2.2 Preheated and Diluted Premixed Flames 73
			2.3.2.2.3 Fuel Flux 74
			2.3.2.2.4 NO Formation 75
		2.3.2.3	Summary 78
	2.3.3	Mixing in Furnace 79
		2.3.3.1	Jet Mixing 79
		2.3.3.2	Unmixedness 83
		2.3.3.3	Well-Stirred Reactor 85
	2.3.4	Pollutant Formation 86
		2.3.4.1	Nitric Oxides 86
	2.3.5	Pollutant Formation and Emission 90
		2.3.5.1	Calculation Method 91
		2.3.5.2	Results and Discussion 91

| | | | Ignition of ∅ = 5 Mixture | 91 |
| | | | Ignition of ∅ = 2 Mixture | 98 |

- Summary 100
- 2.3.6 Radiation 100
- 2.4 Fundamentals of Liquid Fuel Flames 107
  - 2.4.1 Liquid Fuel Flame Characteristics and Stability 107
    - 2.4.1.1 Experimental Apparatus 107
      - 2.4.1.1.1 Spraying Device 107
      - 2.4.1.1.2 Combustion Device 108
      - 2.4.1.1.3 Spray Nozzle 108
    - 2.4.1.2 Experimental Method 109
      - 2.4.1.2.1 Air Preheating 109
      - 2.4.1.2.2 Spray Pressure 111
      - 2.4.1.2.3 Spraying Method 111
      - 2.4.1.2.4 Measurement of Flame 112
    - 2.4.1.3 Experimental Results 112
      - 2.4.1.3.1 Temperature of Blowout 112
      - 2.4.1.3.2 Flame Form and Flame Color 113
    - 2.4.1.4 Discussions 114
      - 2.4.1.4.1 Blowout of Flame 114
      - 2.4.1.4.2 Changes in Flame Form and Flame Color 115
      - 2.4.1.4.3 Spray Combustion in the High Temperature Preheated Diluted Air 117
    - 2.4.1.5 Summary 117
  - 2.4.2 Emissions in Liquid Fuel Flame 117
    - 2.4.2.1 Emissions on Liquid Fuel Combustion 117
- 2.5 Fundamentals of Solid Fuel Flames 118
  - 2.5.1 Solid Fuel Flame Characteristics 118
  - 2.5.2 Combustion Process of Coal 121
    - 2.5.2.1 Properties of Coal 122
    - 2.5.2.2 Combustion Phenomena around Particles 123
    - 2.5.2.3 Combustion Phenomena inside a Particle 126
    - 2.5.2.4 Final Stage of Combustion 126
    - 2.5.2.5 Combustion Behavior of Coal at Synthetic Air Condition of High Temperature 127
    - 2.5.2.6 Summary 130
  - 2.5.3 Emissions in Solid Fuel Flames 130
    - 2.5.3.1 The Furnace Setup 131
    - 2.5.3.2 Fuel Properties (Natural Gas/Coal) 133
    - 2.5.3.3 Experimental Program 133
    - 2.5.3.4 In-Flame Measurements 135
      - 2.5.3.4.1 Heat and Mass Balance 136
      - 2.5.3.4.2 Gas Composition 136
      - 2.5.3.4.3 Temperature Measurements 138
      - 2.5.3.4.4 Velocity Measurements 139
      - 2.5.3.4.5 Burnout 141

		2.5.3.4.6	Solid Concentration 142
		2.5.3.4.7	Total Radiative Heat Flux 145
		2.5.3.4.8	Total Radiance 146
	2.5.3.5	Input/Output Measurements 148
		2.5.3.5.1	Coal Gun Position 150
		2.5.3.5.2	Coal Transport Air Mass Flow 152
		2.5.3.5.3	Precombustor NO_x Level 155
	2.5.3.6	Summary 156
2.5.4	Combustion Rate of Solid Carbon 157
	2.5.4.1	Combustion Field and Solid Carbon Specimens 158
	2.5.4.2	Experimental Results 159
	2.5.4.3	Combustion Rate in Room Temperature Airflow 159
	2.5.4.4	Combustion Rate in High Temperature Airflow 160
	2.5.4.5	Dynamic Analysis of Reactive Gas 161
		2.5.4.5.1	Combustion Rate 161
	2.5.4.6	Lower Limit of Oxygen Concentration 163
	2.5.4.7	Surface Temperature When a CO Flame Is Formed 166
	2.5.4.8	Combustion Rate in High Temperature Airflow 166
	2.5.4.9	Summary 168
References 168

Chapter 3 Simulation Models for High Temperature Air Combustion 171

3.1 Present State of Combustion Simulation in Furnaces 171
    3.1.1 Introduction 171
    3.1.2 Problems of Existing Combustion Models 172
        3.1.2.1 Arrhenius Type One-Step Global Reaction Model 172
        3.1.2.2 Mixing-Is-Reacted Model 173
        3.1.2.3 Eddy-Break-Up Model 174
        3.1.2.4 Problems in Temperature Calculation 176
3.2 Combustion Model for High Temperature Air Combustion 176
    3.2.1 Characteristics of High Temperature Air Combustion 176
    3.2.2 Proposed Improvements 177
    3.2.3 Temperature Correction for Thermal Dissociation 178
    3.2.4 Reaction Model for High Temperature Air Combustion 182
        3.2.4.1 One-Step Global Reaction Model (Coffee) 182
        3.2.4.2 Four-Step Reaction Model (Jones and Lindstedt) 183
        3.2.4.3 Four-Step Reaction Model (Srivatsa) 184
    3.2.5 Comparison of Reaction Models 185
        3.2.5.1 Comparison of Flame Lifted Height by
                   Different Reaction Models 186
        3.2.5.2 Comparison of Maximum Flame Temperature by
                   Different Reaction Models 188
        3.2.5.3 Influence of Jet Velocity on Flame Lift Height 188
3.3 Heat Transfer Model for High Temperature Air Combustion 190
    3.3.1 Heat Transfer Models 190

		3.3.1.1	Gray Model 190

- 3.3.1.1 Gray Model 190
- 3.3.1.2 Weighted-Sum-of-Gray-Gases Model 192
- 3.3.1.3 Nongray Models 194
- 3.3.2 Radiative Heat Transfer Using Nongray Property of Radiation 195
- 3.4 Examples of Practical Application 197
  - 3.4.1 Nitric Oxide Emission 198
    - 3.4.1.1 Thermal NO 198
    - 3.4.1.2 Prompt NO 199
    - 3.4.1.3 NO Reduction Mechanism (Reburning) 199
    - 3.4.1.4 Results and Discussion 201
  - 3.4.2 Transient Behavior of Furnaces 202
    - 3.4.2.1 Fluid Dynamics Model 202
    - 3.4.2.2 Radiation Heat Transfer Model 203
    - 3.4.2.3 Combustion Model 204
    - 3.4.2.4 Temperature Distribution during Fuel Changeover 205
    - 3.4.2.5 Comparison with Measured Temperatures by Suction Pyrometer 206
    - 3.4.2.6 Calculation on Wide Regenerative Furnace 207
- References 208

Chapter 4   Practical Combustion Methods Used in Industries 211

- 4.1 Historical Transition of Industrial Furnace Technologies 211
  - 4.1.1 Energy Technologies Discussed at COP3 211
  - 4.1.2 Conventional Technologies of Energy Saving and Combustion Control for Industrial Furnaces 215
  - 4.1.3 Development of High Performance Industrial Furnaces 219
- 4.2 Energy Conservation 230
  - 4.2.1 Basic Approach 230
  - 4.2.2 Effect of Improvement 230
- 4.3 Pollution Reduction 235
  - 4.3.1 Basic Concept of Low NO_x Combustion 235
  - 4.3.2 Results of the Test 237
  - 4.3.3 Pollution Reduction 238
- References 241

Chapter 5   Design Guidelines for High Performance Industrial Furnaces 243

- 5.1 Flowchart on General Design 243
  - 5.1.1 Design Concept of a High Performance Industrial Furnace 243
  - 5.1.2 Optimal Design for Furnace Length and Height 243
  - 5.1.3 Optimal Design for Other Furnace Configuration 248
    - 5.1.3.1 Pitch and Capacity of Burner 248
    - 5.1.3.2 Partition Wall 248
    - 5.1.3.3 Analytical Study of the Effect of a Partition Wall 249
    - 5.1.3.4 Lower Part of Furnace 251

- 5.1.3.5 Furnace Width and Maximum Combustion Capacity.. 262
- 5.2 Heat Balance and Performance Estimation with Simulation Program........ 263
  - 5.2.1 Outline of Simulation Program 263
  - 5.2.2 Basic Functions of the Simulator 265
    - 5.2.2.1 Estimation Method of Fuel Flow Volume and Exhaust Gas Temperatures Using Heat Balance.............. 266
    - 5.2.2.2 Calculation Method of the Internal Temperature of the Semifinished Steel.. 266
  - 5.2.3 Calculation of Preheated Air Temperatures and Exhaust Gas Temperatures after Heat Exchange... 269
  - 5.2.4 Radiation Heat from the Furnace Body and Heat Loss by Cooling Water 269
  - 5.2.5 Outlines of System Operation Method and Simulation Result........ 271
  - 5.2.6 Comparison of Calculation and Measurement................................. 271
  - 5.2.7 Effect of Fuel Calorific Value on the Fuel Consumption of Reheating Furnaces 272
- 5.3 Combustion Control System.. 280
  - 5.3.1 Basic Combustion Control System for Stable Operation 284
  - 5.3.2 Signal Processing Method 287
  - 5.3.3 Disturbance Suppression Control of Door Open and Close 292
  - 5.3.4 Future Trends of Combustion Control Technology Using High Temperature Air Combustion.. 295
- 5.4 Application Design of High Performance Furnace.. 296
  - 5.4.1 Reheating Furnace.. 296
    - 5.4.1.1 Specifications and Performance of Facility 297
    - 5.4.1.2 Detailed Specifications of Facility 301
    - 5.4.1.3 Attachments 302
  - 5.4.2 Billet Reheating.. 305
  - 5.4.3 Heat Treatment Furnace... 307
    - 5.4.3.1 Heat Balance and Evaluation Method of Furnace Performance 308
    - 5.4.3.2 Furnace Scale-Up for Commercial Production................. 312
    - 5.4.3.3 Test Design of Heat Treatment Furnace 314
  - 5.4.4 Melting Furnace 320
    - 5.4.4.1 Energy Savings and Exhaust Gas Regulation................... 320
    - 5.4.4.2 Size Reduction.. 322
    - 5.4.4.3 Method of Improving the Heat Transfer Efficiency inside the Furnace... 324
    - 5.4.4.4 A Design Example of High Performance Aluminum-Melting Furnace 325
- 5.5 Field Trials and Experiences Obtained through Field Test Demonstration Project.. 327
  - 5.5.1 Outline of the Field Test Project 328
  - 5.5.2 Applications for the Field Test in Fiscal Years 1998 and 1999...... 328
  - 5.5.3 Characteristic Aspects of the 1998 Field Test Project 333

	5.5.4	Effects of Modifications in the Field Tests 334
	5.5.5	Summary 337
References 339

Chapter 6 Potential Applications of High Temperature Air Combustion Technology to Other Systems 341

6.1 Introduction 341
6.2 Combustion of Wastes and Solid Fuels 344
      6.2.1 Formation of Dioxins and Furans 350
            6.2.1.1 Refuse (or Waste) Derived Fuel 350
            6.2.1.2 Applied Technology for RDF 351
            6.2.1.3 Changes in the Calorific Value of Municipal Wastes 351
            6.2.1.4 Problems with Waste Derived Fuel Production and Combustion 352
6.3 Burning of Coals and Lowgrade Coals 353
6.4 Volatile Organic Compounds 354
6.5 Ash Melting 354
6.6 Compact Boilers 355
6.7 Gas Turbine Combustion, Micro Gas Turbines, and Independent Power Production 355
6.8 Paints, Oily Wastes, and Heavy Fuel Oils 356
6.9 Fuel Cells 356
      Example 1 357
6.10 High Temperature Air Combustion Using Pure Oxygen 358
6.11 Summary 359
References 359

Appendix A Results of Investigations on the Current State of Japanese Industrial Furnaces 361

A.1 Introduction 361
A.2 Items and Methods of Investigation 361
A.3 Results of Investigation 362
      A.3.1 Results of the Questionnaire with Users 362
      A.3.2 Results of Interview with Users 362
      A.3.3 Results of Estimate of Number of Installed Industrial Furnaces and Energy Consumption 364
A.4 Evaluation Based on Results of Investigation 368
      A.4.1 Evaluation of Estimated Number of Industrial Furnaces 368
      A.4.2 Evaluation of the Presumed Values of Energy Consumption of Industrial Furnaces 371
      A.4.3 Consideration of the Results of Interviews — Efficiency of Industrial Furnaces 372
A.5 Effect of Energy Saving by Development of High Performance Industrial Furnaces 373

	A.5.1	Assumptions of Calculations	373
	A.5.2	Results of the Calculation	376
A.6	Summary		377
References			378

Appendix B Constants and Conversion Factors 379

B.1 Universal Constants and Conversion Factors 379
B.2 Nondimensional Parameters 381
B.3 Nomenclature 382

Index 387

1 Introduction

1.1 HISTORICAL BACKGROUND OF HIGH TEMPERATURE AIR COMBUSTION

Global energy consumption in recent years has continued to increase not only in developed countries but also in developing countries, primarily as a result of rapid industrialization and improvement in the standard of living. This increased energy consumption has led to increased emissions of carbon dioxide and nitrogen oxide into the environment. Because energy and environmental issues have become of prime concern, it is now a matter of great urgency to deal with environmental preservation on a global scale and over a longer time duration. Under these circumstances, combustion technology utilizing preheated air in excess of 1000°C has drawn increased attention in many application areas. This combustion technology enables one to contribute greatly to the simultaneous reduction of carbon dioxide and oxides of nitrogen emissions. This high temperature air combustion (HiTAC) has achieved approximately 30% reduction in energy (and hence also carbon dioxide emission) and 25% reduction in the physical size of facilities as compared with the traditional type of furnace. Furthermore, HiTAC technology has demonstrated extremely low levels of emissions of nitric oxide, which are far below the present regulatory standards. This book describes the development of this attractive and innovative HiTAC technology and its practical applications to different kinds of furnaces for many industries.

1.1.1 ENVIRONMENT AND ENERGY CONSERVATION

In 1992, the United Nations Conference on the Environment and Development (the so-called Earth Summit) was held in Brazil with the objective of establishing international initiatives for the preservation of the global environment. In December 1997, the Kyoto Protocol was held on Global Climate Changes as the Third Conference of Parties (COP3) to the United Nations Framework Convention. At this conference it was stipulated that developed countries should reduce their total emissions of greenhouse gases by at least 5% from the level of 1990 between the years 2008 and 2012. Therefore, the reduction of carbon dioxide emissions became an urgent issue, particularly from those industries that consume large amounts of fossil fuel.

In many developed countries, including Japan, both energy consumption and industrial activities are quite vigorous. It has become an important responsibility for all developed countries, including Japan, to endeavor to achieve both environmental preservation and industrial growth by developing efficient and environmentally friendly energy utilization technologies, which harmonize with the conservation of resources and energy saving. Against this background, the Ministry of International

Trade and Industry (MITI) in Japan embarked on the *Development of High Performance Industrial Furnaces and the Like Project*. The project, aimed at developing an outstanding energy-saving technology for the 21st century, focused on significant reduction in energy consumption and simultaneous contribution to environmental preservation.

The *Development of High Performance Industrial Furnaces Project* has been implemented since 1993 as one of the activities of New Energy and Industrial Technology Development Organization (NEDO). The objective of this project has been to advocate the establishment of a basic concept for innovative technology. The research and development on the above-mentioned advanced combustion technology have been mainly conducted by the industrial sector during the 7-year history of the project. The fundamental research related to the technology has been carried out by the academic sector through cooperative exchange of information with the industrial sector. This research cooperation between academia and industry has played an important role in promoting the unique research and development of the project.

1.1.2 REDUCTION OF POLLUTANT EMISSIONS AND ENERGY CRISIS

Before the 1960s, the combustion technology had made magnificent progress. But, from the 1960s into the early 1970s, two new serious problems have emerged that result from combustion technologies.

The first was the problem of pollutant emission generated by combustion, such as sulfur oxide (SO_x), nitrogen oxide (NO_x), and smog. The other was the requirement for efficient use of energy caused by the oil crisis.

In developed countries the problem of air pollution has been caused by the combustion exhaust emissions brought by the rapid expansion of industrial production and the remarkable prevalence of motorization in the private sector. It has become a serious social problem. Legal regulations for the restriction of pollutant emissions have been enforced. To cope with this problem, each industry worked intensively on the development of combustion methods and combustion units with low levels of pollutant emission. During the development process of new combustion technology, a particular focus was on low NO_x burners and engines. Intensive research was conducted on the mechanisms controlling the generation and reduction of NO_x. It is also widely acknowledged that the achievements of Japanese industry in the development of new combustion technologies on low NO_x emission have been superb and well recognized by the technical community. This is supported by the many patents taken by industry.

With the oil crisis of 1973, the problem of the energy crisis was suddenly highlighted. Every energy-consuming sector has placed emphasis on the development of a combustion system with low fuel consumption. Furthermore, attention was focused on such issues as the effective combustion of alcohol, various low-grade fuels, etc., which had not been used before as principal fuels. An ultra-lean mixture is defined as a fuel/air lean premixed gas mixture with the fuel concentration near to the lower flammability limit. Because combustion of this ultra-lean mixture had not been possible under usual conditions, a new method to accelerate the reaction by using some type of mechanism has been sought. The evaluations of the combustion of ultra-lean mixtures have varied depending on the positions of the evaluators. However, research on the combustion of ultra-lean mixtures at the combustion limit has been a challenge to combustion engineers and attractive to researchers in the field. Because ultra-lean mixtures do not burn under normal pressure and temperature conditions, it becomes necessary for the mixture to accelerate the reactions by methods such as: (1) catalytic oxidation using a catalyst, (2) injection of active particles such as high energy radicals, (3) preheating, or (4) increase of pressure. Among these, except for oxidation with catalyst, interest has focused on a method of preheating the ultra-lean mixture using a heat recirculation method. The heat recirculation method mentioned here is proposed as a method of preheating the unburned mixture without using external energy. The method provides a means of recirculating the heat from the high temperature side (burned gas) back to the unburned mixture side using an appropriate heat exchange method. The preheating gives additional enthalpy to the unburned mixture without dilution by the combustion products. The combustion system utilizing heat recirculation, in fact, has been commonly used in industrial combustion units for many years. The methods of recovering waste heat and preheating combustion air using various types of heat exchangers have been employed primarily to improve thermal efficiency and stable combustion. Here, it should be pointed out that the high temperature air combustion technology is an extension of the above means. Therefore, from an industrial perspective, the heat recirculation combustion method used for the combustion of ultra-lean mixtures is not, in principle, a totally new method of combustion.

In 1971, Weinberg[1,2] of Imperial College perceived the fact that preheating the mixture using the heat recirculation method can attain stable combustion of an ultra-lean mixture and expand the flammablilty limits of the mixture. He proposed the concept of additional enthalpy combustion for ultra-lean mixtures. A special feature of this method is that a heat source for preheating is not necessary to maintain combustion except at the time of start-up. Furthermore, the temperature of the final exhaust gas does not necessarily become high. The novelty of Weinberg's idea lies in maintaining sustained combustion without any assistance from an external heat source, for example, by applying the heat recirculation method used in conventional industrial furnaces for ultra-lean mixtures. In this case, the success of applying the heat recirculation method depends on (1) whether or not a heat exchange method is appropriate and (2) how much of the heat loss from a combustion unit can be reduced. Although various heat exchange methods have been proposed, the major heat recirculation methods studied to date can be broadly classified into (1) indirect (external) heat recirculation methods and (2) direct (internal) heat recirculation methods.

The indirect (external) heat recirculation methods are methods of circulating heat at the exterior of a flame zone without essentially changing the structure of the flame as represented by the double spiral type burner proposed by Lloyd and Weinberg,[1,2] namely, a so-called Swiss Roll burner. The indirect (external) heat recirculation methods can further be classified into methods that mainly utilize conduction only and methods that also actively utilize radiation for the heat feedback. For the latter methods, a combustor was proposed wherein a porous solid wall is positioned to enclose the combustion chamber, because the porous solid wall has the function of converting the sensible heat of gas into radiation energy at a high rate of conversion efficiency. When burned gases behind the flame pass through the porous solid wall, the sensible heat is converted into radiation energy and the unburned mixture is preheated using this radiation energy.

In contrast, the direct (internal) heat recirculation methods are methods of feeding back heat directly from the side of burned gases to the side of unburned gases by inserting, for example, a porous metal with high thermal conductivity into the flame zone, changing the internal structure of the flame and forming an additional enthalpy flame. Many research activities using several of the above-mentioned methods have been conducted in the past and some remarkable research results have been reported.

1.1.3 PANORAMA OF HIGH TEMPERATURE AIR COMBUSTION TECHNOLOGY

A recuperator was used in large-scale industrial furnaces as a waste heat recovery unit to realize high thermal efficiency and energy conservation. The recovered waste heat was used to preheat the combustion air, which was then fed to a burner. The preheated air resulted in energy conservation and good combustion performance. However, the disadvantages included incorporating a large-scale heat recovery system for waste heat. Furthermore, the temperature of the preheated air was only about 600 to 700°C, at best. To obtain further substantial energy savings, it was found necessary to recover combustion waste heat thoroughly and feed it back to the unburned side effectively. An effective means for achieving this is the development of a regenerative burner having better function for effective waste heat recovery. This has been the topic of active research and development since the beginning of the 1980s.

At British Gas and later at Hotwork International (United Kingdom), throughout the 1970s and the 1980s, substantial efforts were allocated to the development of both recuperative and regenerative burners.[3] Recuperative burners could offer only modest fuel savings, since the combustion air could be preheated to temperatures typically not higher than 600°C. The burners equipped with regenerators (beds packed with ceramic balls) offered much higher preheating levels typically up to a 1000°C with a cycle time of 30 to 40 s. Such substantial air preheating was possible only when the furnace exit gases, entering the regenerator, were at temperatures typically of 1300 to 1400°C.

Further progress in the regenerator design was made in Japan, at Nippon Furnace (NFK), at the beginning of 1990s. New honeycomb-type regenerators were shown to be more compact and possessed smaller thermal inertia. The honeycomb regenerators operated at a very small temperature difference (typically of 50 to 100°C) between the furnace exit temperature and the combustion air temperature. They provided possibilities for achieving combustion air preheating up to 1200°C, thus further improving furnace efficiency.

The technology carried through in the *Development of High Performance Industrial Furnace Project* is high-cycle alternating regenerative combustion technology. It employs high temperature air, preheated to temperatures in excess of 1000°C, using a heat-storage-type heat exchanger that also has short switching time. It has been proved, through extensive research and development efforts to date, that not only carbon dioxide but also nitrogen oxide can be substantially reduced. This type of technology, utilizing high temperature air, has attracted significant attention for the past several decades. For example, since the 1960s, fundamental research has been implemented in the field of special combustion systems, such as a supersonic combustion ram-jet engine, called SCRAM-jet. SCRAM-jet is now attracting attention in the aerospace field, and substantial results have been obtained. However, the technological concept intending to apply high temperature air combustion to conventional industrial furnaces is innovative, and one can say that it is a novel combustion technology because it can contribute to further reduction in energy consumption and promote environmental preservation.

In general, by implementing regenerative combustion, the heat generated by combustion is not uselessly discharged but can be effectively recovered. Therefore, remarkable energy savings (about 30%) and prevention of global warming have been realized. In contrast, it is known that NO_x emission increases with an increase in temperature of the air used for combustion; this has even been shown with numerical simulation on a flame formed in the field of a simple flow. This fact has also been experimentally confirmed in conventional combustion systems, and is now widely acknowledged among researchers and engineers in the field of combustion technology. From the point of view of reducing the emission of pollutants (especially NO_x), many researchers and engineers had doubts on the possibility of utilizing high temperature air combustion. They believed that the technology has not been developed in either the industrial sector or the academic sector. Thus, in the industrial sector, finding a way to achieve a balance between energy savings and reducing emission of a pollutant has been a challenge for many years.

In the meantime, entering the 1990s, unexpectedly reduced values of NO_x were measured in an experimental furnace when high-cycle alternating regenerative combustion technology was applied. The NO_x values were reduced further when the velocity of the airflow injected into the furnace was increased. A concentration of very low NO_x, at most 80 ppm, was confirmed experimentally when the air of a temperature of 1350°C was injected with the velocity of 90 m/s. In addition, because of the temperature rise of combustion air, combustion in a low oxygen concentration atmosphere became possible. The flame was transparent and colorless. This colorless and transparent flame had not been observed previously. In contrast, most flames exhibit a local high temperature as seen in all existing industrial furnaces. Also observed was the fact that combustion proceeds across a wide region in the furnace. The temperature in the furnace is close to the limit of its operation, and the temperature distribution in the furnace is almost uniform. These characteristics revealed that an almost ideal heating furnace can be constructed. Thus, this high temperature air combustion technology has suddenly become the focus of a great deal of attention as a novel combustion technology not only in the industrial sector but also in the academic sector.

The development of this high temperature air combustion technology, especially the development of low NO_x burner technology for natural gas combustion, has been conducted with leadership from the industrial sector. At the beginning of the 1990s, the work at Tokyo Gas led to the development of an advanced fuel direct injection (FDI) concept.[4] The concept utilizes the idea of discrete injection of fuel gas into hot combustion products. During the same time at NFK a lot of efforts were made to reduce NO_x emission using the process of additional enthalpy.[5] The company tested several configurations and designs for air and fuel gas ports. The best results, corresponding to low NO_x emissions, were achieved when the injection ports for fuel and air were positioned apart.

Developments in other countries traveled different routes. Rather than developing new combustion processes, the work focused on burner designs. The flameless oxidation (FLOX) burner, developed in Germany in 1991 to 1992, utilizes a number of air jets, which entrain combustion products before mixing with the fuel.[6] In 1991, the IFRF designed a series of natural gas burners in SCALING 400 studies.[7] Substantial NO_x reduction was achieved when either 80 or 100% of the fuel gas was provided through the individual injectors located on the burner circumference. Furthermore, in this case, NO_x emissions were no longer dependent on preheating of the combustion air. When the burner was fired with 80 or 100% fuel staging, the combustion mode resembled the FDI system. During the tests, it was observed that the radiation heat flux of the staged flame were approximately 20% higher than those of the baseline (unstaged) flame. This, at first, appeared as a surprising result. However, it was later confirmed in other experiments in which, despite the lower flame temperatures with the staged flames, the higher flame volume caused an increase of the heat flux. The most recent developments include a Canadian Gas Research Institute (CGRI)[8] burner that also features discrete injections of fuel gas and air.

Based on the observations of combustion regime in high performance industrial furnaces, which will be shown in later chapters, it could be concluded that the key for high temperature air combustion technology (HiTAC) is the dilution of fuel and combustion air by burned gases in the furnace. Specifically, combustion in an atmosphere of low oxygen concentration is the key. This situation can be realized by the recirculation flow induced by high velocity air injection. These clarifications in terms of the true HiTAC mechanism have provided a framework within which the technology could be utilized for specific applications for energy conservation and pollution reduction.

1.2 INNOVATION OF HIGH TEMPERATURE AIR COMBUSTION

1.2.1 Fundamentals of Combustion

1.2.1.1 Heat Recirculating Combustion

Preheating of combustible mixture by recycled heat from flue gases has been considered an effective technology not only for combustion of low calorific fuels but also for fuel conservation. It is called heat-recirculating combustion in which reactants are heated prior to the flame zone by heat transfer from burned products without mixing of the two streams.[9] The temperature histories of premixed combustion in a one-dimensional adiabatic system are schematically compared in Figure 1.1[1,2] for

Introduction

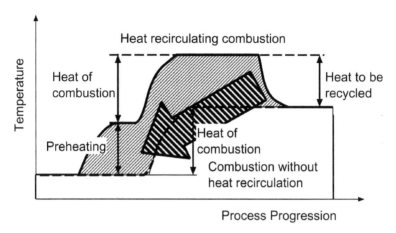

FIGURE 1.1 Temperature history of heat-recirculating combustion of premixed reactants in one-dimensional adiabatic system.

the cases with and without heat recirculation. The maximum temperature in heat-recirculating combustion is determined by the amount of recycled heat that is independent of the equivalent ratio of the mixture or the calorific value of the fuel used. This is certainly true for premixed combustion in adiabatic condition. Accordingly, it has been held that heat-recirculating combustion brings a temperature rise throughout combustion processes in proportion to the amount of recycled heat.

At normal ambient temperature, an ordinary hydrocarbon gaseous fuel mixed with atmospheric air exhibits a combustible domain around the stoichiometry, and an increase of temperature of the mixture expands the combustible limits significantly, as illustrated in Figure 1.2.[10] A large increase in the temperature may cause

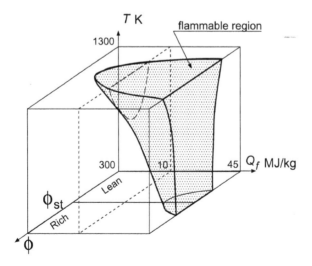

FIGURE 1.2 Flammable domain expressed by calorific value of fuel, Q_f, initial temperature of mixture, T_i, and mixture equivalence ratio, ϕ.

auto-ignition. In contrast, if a fuel of low calorific value is adopted, the combustible domain disappears at ambient temperature and reappears when the mixture is preheated over a certain temperature level, as shown in the figure exhibiting conceptual trends.

Most of the previous research in this field has been aimed at burning ultra-lean mixtures or low calorific value fuels produced in chemical processes or vented from coal mines. In those cases the resultant maximum flame temperature with heat recirculation was not crucial to the tolerance of materials used in the system because of the low calorific value. Further, scientific studies on heat recirculating combustion have been mostly carried out on relatively small-scale premixed flames. However, for large-scale industrial use, diffusion or non-premixed combustion is more common because of its controllability and safety. Heat recirculating combustion in diffusion or non-premixed combustion can be achieved by heating combustion air with the recycled heat from burned products. The temperature of combustion air in an adiabatic system can theoretically be raised to almost the same temperature as the exhaust stream by regenerative heat exchangers. In practice, the regenerative combustion system for implementing high temperature air combustion is the system shown in Figure 1.3, where a heat recirculating method (by use of honeycomb-type regenerators) is applied to a heating furnace. A pair of burners, operating alternately, is used as a unit and the flow path for air ejection in each burner is filled with ceramic

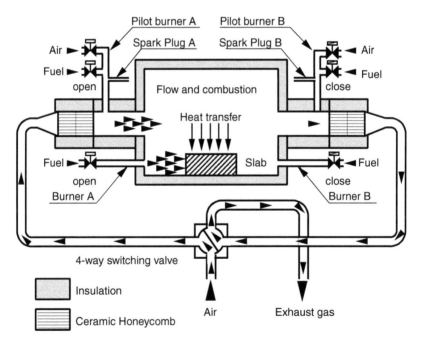

FIGURE 1.3 Schematic diagram of HiTAC furnace operated with high frequency alternating flow regenerators.

honeycombs. Each flow path also acts as a flow path for discharging high temperature burned gases. The high temperature burned gases generated by the one burner are introduced into the flow path of the other side, and the sensible heat of the burned gas is stored in the ceramic honeycomb for a while. Then, the operation of the burner is switched to the other burner, and combustion air is introduced through the ceramic honeycomb by reversing the flow direction. Preheated air at a temperature of about 1273 K is easily obtained and used for combustion. The high-cycle operation, typically 30 to 60 s, is adopted to reduce the heat loss escaping with the waste gas.

The principal merit of heat-recirculating combustion is fuel saving, which is achieved by efficient recovery of waste heat in exhaust gases. Higher preheating temperature assures less rejection of heat with the exhaust, which results in more fuel saving. Therefore, heat-recirculating combustion is surely an attractive technology for future design of any industrial furnaces as far as energy conservation and pollution reduction are concerned. However, it was believed that diffusion flames inevitably harbor near-stoichiometric flame temperature somewhere within their structure, borrowing the words of Weinberg,[9] which tends to generate increased levels of nitric oxides even in great excess air ratio. If stricter air quality regulations, particularly regarding nitric oxides, are applied to furnaces, reduction of nitric oxides in non-premixed combustion is the first issue to be solved for the future utilization of heat-recirculating combustion in a variety of furnaces.

Regarding emissions from combustion systems, it has been generally held that the emission of nitric oxides increases with the temperature rise of combustion air when preheated air is used. Therefore, any practical trade-off between thermal efficiency and emission control has always been a critical issue for designers and engineers. Numerous efforts to strike a balance of furnace between fuel saving and reduction of nitric oxides emission have been made during the last decade. In practice, direct injection of fuel into a furnace,[4,11] high momentum ejection of staging air,[12,13] and mixing control were found to be effective to some extent in reducing nitric oxide emissions in regenerative combustion. Therefore, the practical extent of heat recirculation in industrial furnaces has been specified taking account of the trade-off between energy conservation and tolerance of materials or air quality regulations.

In the process of developing a high-cycle regenerative furnace, extremely low nitric oxide emission was reported from an experimental furnace operated with high temperature combustion air of 1400 K.[5] Because it was difficult to interpret the results, based on existing knowledge of nitric oxide formation mechanisms, this motivated an extensive, collaborative study of high temperature air combustion between industry and academia. Eventually, practical developments and applications of the concept in industry have achieved great progress in energy saving as well as reduction of nitric oxide emission. The details of the technology are explained throughout this book. The basic concept is the combination of maximum waste heat recovery by high-cycle regenerator and controlled mixing of highly preheated combustion air with burned gases to yield relatively low temperature flames.

1.2.1.2 Definition of High Temperature Air

The term high temperature air is used throughout this book. The temperature of the air indicated often varies depending on the situation where it is used. What is the definition of high temperature air when furnace combustion is considered? Imagine that a gaseous fuel at ambient temperature is injected into an air stream. When the fuel mixes with combustion air, some heat is necessary to initiate combustion, and a recirculating flow of combustion products behind a flame holder or a pilot flame is frequently utilized for stabilizing flames in furnaces. However, if combustion air is sufficiently heated prior to mixing, combustion takes place somewhere downstream in the furnace following the mixing of two reactants, even if the flame in the near field of the fuel jet is blown off by a strong shear motion.

Although the temperature level of preheated air does not seem important when discussing preheated air combustion, the fact described above is significant in realizing advanced low NO_x combustion technology, which will be explained later. Therefore, the auto-ignition temperature of a gaseous fuel with air as the limit of high temperature, that is, the air temperature at which a gaseous fuel is ignited automatically in it and in which continuous combustion is sustained, should be called high. Although the definition of high temperature is not given by a fixed value, it is now possible to give high temperature air combustion (HiTAC) a clear meaning. Following the above definition, preheated air combustion (PAC) is defined as combustion with the air of preheated temperature below the auto-ignition limit that has long been utilized in industry.

Once the combustion air is preheated to higher than the PAC limit, a method to stabilize a flame is not necessary for furnace combustion. This auto-ignition temperature of a gaseous fuel varies depending on the kind of fuel and concentration of oxygen of the diluted air. Figure 1.4 shows auto-ignition and combustible limits for propane in preheated air or diluted air with inert gas. auto-ignition and combustion occur even in an atmosphere of oxygen content as low as 3% when it is preheated above 1200 K.

1.2.1.3 Heat Recirculation and Exhaust Gas Recirculation

Global excess air ratio or equivalent ratio is one of the combustion parameters that characterize the operating condition of furnaces. The temperature of combustion products in adiabatic circumstances can be easily defined by the ratio. However, heat subtraction by the material being heated in the furnace and the heat loss from practical systems are influencing factors in defining gas temperature in the furnace. Therefore, the reduced temperature level of burned gas as well as its recycling flow rate largely affects flame temperature with gas recirculation.

Exhaust gas recycling, whether it is internal or external, is an effective method to reduce flame temperatures, and thereby nitric oxides emission. Combustion with normal ambient air usually becomes unstable when the exhaust gas recycling rate, defined as the mass ratio between exhaust gas and fresh reactants, exceeds 30%. As is shown in Figure 1.4, however, a stable combustion domain appears for high rates of exhaust gas recycling, if combustion air is preheated over the auto-ignition

Introduction

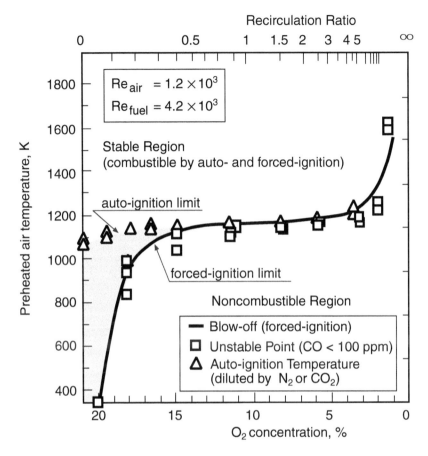

FIGURE 1.4 auto-ignition limits and blow-off limits of propane in a preheated air or a diluted air with nitrogen.

temperature of the fuel. Actually, very diluted air, whose oxygen concentration is as low as 3%, can sustain combustion when it is preheated up to 1200 K.

Hasegawa et al.[14] discussed the individual and multiple influences of heat and gas recycling. In Chapter 2, Figure 2.29 shows contours of the maximum flame temperature on the combined effect of preheated air temperature and recycling rate of burned gases, where R is the gas recycling rate. A combination of highly preheated air and high recycling rate of burned product generates relatively low maximum flame temperature. One can understand that the stoichiometric flame temperature in very diluted air, where mass fraction of oxygen is far below the value in normal atmospheric air, is not as high as is usually expected. This is the key for HiTAC when it is applied to practical combustion systems. Keeping the global equivalence ratio constant, the flame temperature in the furnace can be varied or regulated by combining the preheated air temperature and the recycling rate of burned gases.

The concept of the HiTAC is illustrated in Figure 1.5, compared with that of a conventional furnace combustion. Extremely high temperature flames are usually

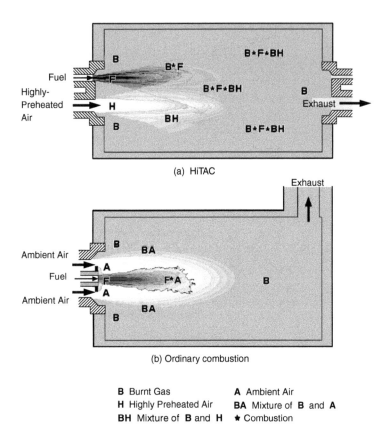

FIGURE 1.5 Mixing and combustion in furnace.

generated in furnaces, if direct combustion between fuel and high temperature fresh air occurs. As a result of the modified furnace geometry, not only extinction of base flames occurs by the shear motion of high velocity inlet air but also the dilution of air with burned gas (BH) must occur prior to combustion by separating fuel and air inlets. Note that those are the conditions in which ordinary combustion cannot be sustained with ambient temperature air. In addition, the fuel injected separately into the furnace also entrains burned gas in the furnace, and some changes in the fuel, such as pyrolysis, decomposition, and vaporization of liquid fuel, if any, during this preparation period. Weak combustion reactions may occur between fuel and entrained product (B*F) and the main combustion follows in the mixing zone of fuel and diluted air with a large amount of burned gas (B*F*BH). The change of flame due to a low concentration of oxygen caused by the high rate of recycling of burned gas probably yields a broadened reaction zone, where relatively slow reactions may be taking place. In established combustion without preheated air, direct combustion between fuel and fresh air (F*A) occurs in the near-field of the burner. Thereafter, some combustion in diluted condition with burned gases may follow in the downstream portion of the flame because of the entrainment of recirculated burned gas by the incoming combustion air. Combustion (F*A) in the vicinity of the burner shows the maximum temperature in the furnace, and most of the nitric oxides emitted from the furnace are formed there. However, combustion in this region is essential to sustain the combustion in the furnace, and whole flame cannot exist if extinction occurs in this portion.

Figure 1.6a shows conceptual temperature histories along streamlines passing through and by the flame zone. The former experiences near-stoichiometric flame temperature, which is slightly below the theoretical adiabatic temperature, T_{ad}, and the latter rises only as mixing progresses. Turbulent mixing between the two produces large temperature fluctuations that can usually be observed in ordinary turbulent flames. If combustion air is preheated, it forces up all temperature profiles to some extent, with the same degree of fluctuations exceeding the theoretical adiabatic temperature of ordinary combustion T_{ad}, as shown in Figure 1.6b. This is a well-known feature of the preheated air combustion (PAC) achieved by use of a recuperator. However, once the preheating temperature exceeds the auto-ignition temperature of the fuel, HiTAC becomes possible. Then, as shown in Figure 1.6c, the temperature history along the streamline passing through the reaction zone shows a relatively mild temperature rise due to slow heat release in a low oxygen concentration atmosphere compared with those in previous cases. The other extreme indicates a temperature profile along the streamline outside the reaction zone, and it rises only with the progress of mixing between preheated combustion air and burned gas recirculating in the furnace. Accordingly, temperature fluctuations generated between these two are very small compared with the previous cases. In spite of the use of highly preheated air, the mean temperature as well as the instantaneous peak temperature is considerably lower in HiTAC than in ordinary combustion.

1.2.2 Principle of Combustion Control for CO_2 and NO_x Reduction

1.2.2.1 Carbon Dioxide

As long as hydrocarbon fuels are used, CO_2 will be emitted in proportion to the carbon content of the fuel, unless it is artificially fixed or removed. The only way for combustion engineers to suppress CO_2 emission from combustion devices is energy conservation by raising the thermal efficiency of the device. In that sense, HiTAC is one of the most attractive technologies for use in the furnace industry. The basic concept is easy to understand. If all of the heat loss and waste from a heating furnace can be eliminated, for example, all the heat generated from fuel will be transferred to the material being heated in the furnace. Therefore, the high efficiency of waste heat recovery is one of the most promising measures available to suppress CO_2 emission, that is, to reduce greenhouse gases. The higher the efficiency of the regenerator, the less CO_2 emitted.

Although available heat in exhaust gases can be transferred efficiently to the incoming cold combustion air using an infinitely long heat exchanger, the actual heat transfer rate is limited by the geometry of heat exchangers. Accordingly, the maximum obtainable temperature of combustion air depends not only on the tolerance of materials used, such as heat-resistance alloys and refractory, but also on heat losses of the system.

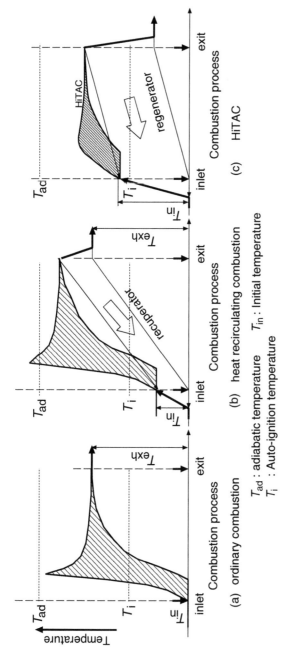

FIGURE 1.6 Conceptual temperature histories and fluctuation intensity.

Regenerators for heat-recirculating combustion appeared first in Europe and studies on heat-recirculating combustion were reviewed by Weinberg.[9] The typical regenerative heat exchanger for preheating combustion air was a bed packed with ceramic balls.

The size varies depending on the cycle time between 30 s to several minutes, which determines the amount of heat storage. A typical regenerator at that time produced preheated air exceeding 1273 K, generally within 300 K below the furnace temperature.[15] However, if the hot gas flow escaping through a shortcut of minimum pressure drop occurs in the bed, it gives rise to uneven temperature distribution in a cross section, resulting in inefficiency of the regenerator.

In contrast, the volume of a honeycomb-type regenerator of necessary heat capacity can be minimized because of its large surface area-to-volume ratio. As a result, direct installation of a regenerator into a burner becomes possible, forming a thermal dam at the exit of the furnace as illustrated in Figure 1.3. Temperature distribution in a ceramic honeycomb is quasi one-dimensional, because ceramic honeycombs assure the uniformity of temperature in a cross section. There is an example in which combustion air of 1570 K was actually obtained by this type of regenerative system for the mean furnace gas temperature of 1623 K, that is, an approximately 50 K difference. Almost constant temperature of combustion air, less than 50 K variation during a cycle, was realized by the use of regenerators with small heat capacity, and about 40% fuel saving, and hence CO_2 reduction, was achieved.[16]

1.2.2.2 Nitric Oxides

During the last quarter of a century, extensive experiments and detailed chemical kinetic calculations have been carried out to clarify the formation mechanisms of nitric oxides. As a result, the temperature rise in combustion air has been recognized as one of the influencing factors on nitric oxides emission from combustion systems because it often causes higher flame temperatures, where most nitric oxides are rapidly formed. The influence of inlet air temperature on nitric oxides emission from a prototype furnace is shown in Figure 1.7, demonstrating an exponential increase of nitric oxide emission with temperature rise of combustion air. These characteristics have been widely taken among combustion engineers as common knowledge of nitric oxides emission from combustion devices.

Clearly a large amount of nitric oxide is formed with the increase in flame temperature. Consider what will happen when preheated air higher than the auto-ignition temperature of a fuel is used. If reaction between fuel and preheated normal air at near-stoichiometric ratio occurs, the flame temperature must be extremely high. Therefore, a large quantity of nitric oxides may be emitted when non-premixed combustion takes place with high temperature pure air in furnaces. If this happens, the reduction of nitric oxides in HiTAC seems unpromising. There is one more important dependency factor that needs to be discussed. This is the oxygen concentration in the reaction zone where local combustion reactions take place. Combustion always occurs in near-stoichiometric mixture even though the mixture is non-premixed. However, when referring to stoichiometric ratio, few

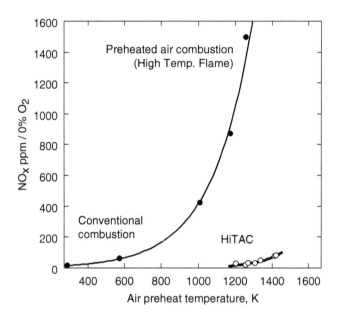

FIGURE 1.7 Preheated air temperature on NO_x emission.

people refer to the actual concentration of oxygen because utilization of atmospheric air as the oxidizer in practical combustion systems is far too common to be noted. Thus, we must pay special attention to the influence of oxygen concentration on nitric oxide formation.

Suppose that we use highly diluted air with inert gas such as nitrogen for combustion. Figure 1.8 shows numerically predicted turbulent diffusion flames between methane and air or their dilution together with their temperature and species concentration profiles across the flame. The ordinary methane–air flame has a typical thin reaction zone called a flamelet, where rapid combustion reactions produce a steep concentration gradient in methane and oxygen as well as a sharp and high-temperature rise due to the heat release in the flamelet. In contrast, the diffusion flame between diluted methane (40%) in nitrogen and diluted air in nitrogen, in which actual oxygen concentration is 8%, shows a broadened reaction zone associated with a lower peak temperature in spite of the preheating of both flows up to 1273 K. The unexpected low flame temperature is caused by a combination of high preheating of reactants and the dilution by a large amount of inert gas. Because the local reaction rate becomes small in diluted circumstances, increased volume of the reaction zone results from burning fuel at the same rate as ordinary combustion, hence the same total heat release rate in the furnace. Therefore, the reaction and heat release zone of combustion in the dilution with plenty of burned gases may become widely distributed compared with that of ordinary combustion and yield a mild temperature rise locally. These facts must be taken into account when attempting to suppress nitric oxides emission from practical HiTAC systems.

Introduction

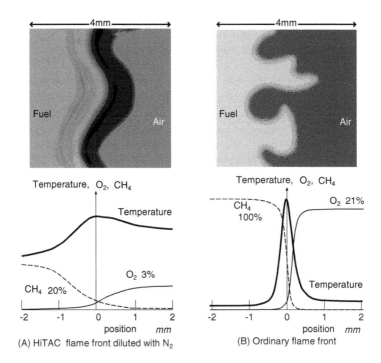

FIGURE 1.8 Thickness of reaction zone and temperature rise.

1.2.3 HEAT TRANSFER IN HIGH TEMPERATURE AIR COMBUSTION

Heat transfer in furnaces depends not only on the internal temperature distribution but also on the physical properties of combustion gases and of furnace walls. Although the heat transfer rate is augmented with the increase of furnace temperature, there is a maximum operating temperature due to the maximum temperature limit of furnace materials used, such as fire bricks. Also, temperature fluctuations in turbulent combustion, because a peak temperature at an instant, sometimes deteriorates the surface of the material being heated or the insulation on the wall. Therefore, we must achieve effective heat transfer under these limitations.

An earlier section explained that temperature fluctuations in HiTAC are much smaller than in ordinary combustion. This fact allows us to raise the operating temperature because a small instantaneous peak temperature does not exceed the limit. So, it is possible to increase the combustion load for the same size furnace. Or, we can reduce the furnace size by adopting HiTAC for the same combustion rate.

If combustion air can be preheated to a high temperature at the entry of a furnace using the recovered sensible heat, it can save some quantity of energy in heating the materials to the specified temperature. The heat capacity of the material being heated does not change because of the operating systems or because of the heat input to the furnace. However, the efficiency of heating, the heating time, and the uniformity of temperature rise of the material being heated do change depending on the local heat release rate in the furnace as well as on the resultant temperature distribution. We consider the heat transfer in furnaces from this point of view.

To realize the effective heat transfer in HiTAC furnaces, discussion is required on each mode of heat transfer (conduction, convection, and radiation). Among these, conduction from flames to the materials being heated is not so important when we consider the heat transfer in furnaces, although conduction is important when we discuss the depth of heating in the material. Thus, it is enough if conduction is taken into consideration only when the loss through furnace walls is discussed.

1.2.3.1 Convection Heat Transfer of High Temperature Air Combustion

The heat transfer in a conventional industrial furnace can be described in the schematic diagram shown in Figure 1.9. The convection heat transfer comprises a very small proportion of the total heat transfer rate from combustion gas to the material being heated. The heat convection rate is expressed by the product of heat transfer coefficient, contact area, and the temperature difference between the solid surface and its adjacent gas. The characteristic length of convection heat transfer depends on the boundary layer thickness. If there is a large deviation in the spatial distribution of temperature in combustion gases, the local convection heat transfer will generate nonuniform heat flux onto the material, and hence the temperature distribution on the surface. The redistribution of heat will follow by way of radiation and conduction, which will take somewhat longer.

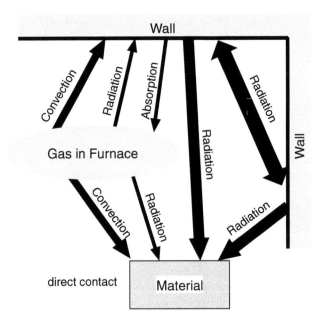

FIGURE 1.9 Heat transfer in furnaces.

Introduction

We may consider that a similar mechanism as shown in Figure 1.9 is still valid in HiTAC furnaces. However, from the viewpoint of the material being heated and the furnace walls, it is not necessary to consider the heat conduction in a solid due to the uniform heating resulting from the small deviation in temperature distribution. That is a characteristic feature of heating in HiTAC furnaces. Furthermore, as is described in a later section, the characteristic length of radiation heat transfer is large and the energy, which travels as an electromagnetic wave, can be transferred from frontside to backside of a furnace directly. Therefore, because wide wall surfaces are usually maintained at a nearly uniform temperature in HiTAC, heating by radiation dominates the heat transfer in the furnace. The direct convection heat transfer to the material being heated plays only a supplementary role.

Nevertheless, convection heat transfer in HiTAC is important as an initiator of heat transfer in the furnace. Since flames in HiTAC are not like conventional luminous flames but are blurred with low luminosity, as shown on the cover photograph of this book, we cannot expect effective radiation heat transfer from flames to materials being heated. However, the convection heat transfer from gas to walls works as the starting point for heat transfer in the furnace.

A simulation dealing with reaction and heat transfer in three dimensions is described in Section 3.4 in this book; we show another example here. This simulation was carried out assuming that the flame temperature was constant, and each element in the furnace, such as a wall surface, flame zone, and material being heated, was treated as a distributed constant. The dependence of radiation heat transfer on wavelength was considered. The flame gas absorptivity was expressed using the band absorption of CO_2 and H_2O by taking the respective mole fractions into account. Wall surface was regarded as a selective absorption surface, assuming typical refractory. As a result, it acted as a low absorption body for far-infrared radiation having a wavelength longer than 4 μm, and as a blackbody for a wavelength shorter than 2 μm. The furnace was a small rectangular furnace (length 2.5 × width 1.2 × height 0.8 m, Al_2O_3 wall thickness 0.3 m), in which a flame was positioned in the center and oriented longitudinally and the material being heated was placed 0.3 m above the bottom surface.

Some of the results are shown in Figure 1.10. As the emissivity and absorptivity of flame become smaller, the ratio of the radiation heat transfer from flame to wall to the convection heat transfer to the wall becomes smaller. On the contrary, the contribution of radiation heat transfer in the total heat transfer to the material being heated increases. This is in contrast to our expectation when the flame is in low luminosity. This is a result of the increase of radiation heat transfer from walls, which are heated by the convection, to the materials being heated. Therefore, the thermal field in a furnace is retained in equilibrium by radiation heat transfer. This has a long characteristic transfer distance, when the flame luminosity is low. It means that the field is dominated by radiant heat, and the necessary heating time is shortened by the change of luminous flames into nonluminous ones.

The increase of radiant heat transfer between walls means the rapid redistribution of the heat, transferred by local convection, to other walls. Therefore, the radiant heat from a wall covers most of the heat flux to the material being heated, although

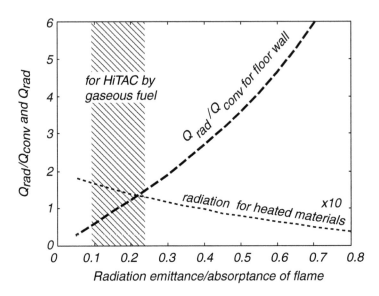

FIGURE 1.10 Radiation and convection heat transfer in furnaces.

the direct radiation heat transfer from flame to the material being heated is small compared with the heat transfer in a conventional combustion furnace. The increase from the effect described above can be considered as the effective increase in the exchange area, or configuration coefficient, of radiation heat transfer.

As explained above, in HiTAC where a flame is nonluminous, the convection heat transfer to the wall triggers the radiation heat transfer to the material being heated. The precise convection heat transfer coefficient depends mainly on the orientation and velocity of the flow. At the same time, it is more effective to agitate the flow and to avoid the development of a boundary layer on the solid wall surface by disposing refractory as a spacer. The material being heated against the flow is more effective than constructing the furnace with simple plane walls. Furthermore, the obstacles in the flow often work as heat accumulators and can be used as wavelength conversion bodies, which are discussed later, although it is difficult to estimate the convection heat transfer coefficient in a complicated flow.

1.2.3.2 Radiant Heat Transfer of High Temperature Air Combustion

The uniformity of temperature distribution comes from the dispersed reaction zone accompanied by a relatively mild temperature rise because combustion air or fuel is diluted with burned product before combustion. Accordingly, the reaction rate decreases, resulting in a long flame or a reaction zone if it is invisible. Generally, the length of a turbulent jet flame is kept nearly constant, because the momentum increase of a fuel jet generates enhanced mixing by entraining larger amounts of surrounding air. However, under low oxygen concentration atmosphere, the flame length will increase by the decreased reaction rate. Example calculations predict a

Introduction

flame two and a half times as long when the oxygen concentration in the air is reduced to 5% with the addition of nitrogen and the flame volume also increased dramatically.

A large volume reaction zone recognized by a mild temperature rise with low luminosity is a typical feature of HiTAC, which is clearly different from ordinary combustion burning with luminous flames. When discussing the heat transfer in HiTAC furnaces, we must take this low luminosity flame into account. Specifically, the primary heating mode in HiTAC furnaces is radiation heat transfer to the materials being heated from the walls, which have been heated by the convection heat transfer of the nonluminous combustion gas. Therefore, despite this undesirable property of HiTAC flame, the utilization of the furnace wall effectively increases the direct exchange area for the material being heated. This is another most important role of the radiation heat transfer in HiTAC furnaces.

1.2.3.3 Effect of Wall as Wavelength Conversion Body in High Temperature Air Combustion

A typical material for furnace walls has physical properties acting as a selective absorption body for radiation energy. They function as an absorber for short wavelengths (<2 μm) and as a reflector for long wavelengths (>5 μm) for typical properties of the material frequently used in furnaces. Although nonmetallic heat insulation materials often possess the properties mentioned above, their emissivity lies in the range of 0.8 to 0.95 and increases with an increase in temperature. The emissivity generally depends on the surface roughness as well as on the fine-scale temperature distribution of the material surface. In contrast, iron or copper as a material being heated has emissivity of 0.85 to 0.95 for iron oxide and 0.55 to 0.65 for copper oxide, respectively, since its surface is an oxide. Accordingly, high emissivity material being heated, which is placed in high temperature surroundings, acts as an absorber for short wavelength and a weak reflector for long wavelengths. On the other hand, regarding energy exchange between gas and walls, the short wavelength energy radiated from combustion gas is mostly absorbed by lower temperature walls which have high absorptivity in short wavelength. Because gas is considered a transparent medium in long wavelengths except for band absorption spectra of H_2O or CO_2, the net exchange of radiant energy between walls balances each other because of the small deviation in wall temperature. Therefore, radiant energy is transferred from walls to materials being heated according to their respective temperatures. This is because both materials have high emissivity at long wavelength. These characteristics in terms of wavelength produce a desirable function of wall working as a wavelength conversion body in the radiation heat transfer.

The mechanism of wavelength conversion in the vicinity of 1200°C is schematically shown in Figure 1.11. The radiation energy from combustion gas is emitted mainly from CO_2 and H_2O and sometimes from hydrocarbon contents, such as CH_4, as band emissions are usually distributed in the wavelength range over 2 μm, as shown in Figure 1.11a. Then, it is absorbed by the walls having the absorptivity indicated by the dotted line in Figure 1.11b, and it is converted into heat recognized

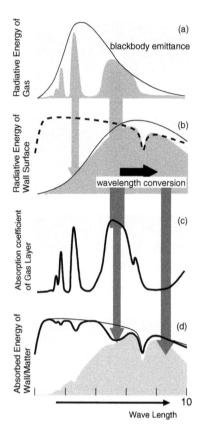

FIGURE 1.11 Radiation energy exchange by wavelength conversion.

as a temperature rise. The radiant intensity from the wall is determined by the product of its emissivity and the emissive power of a blackbody at the same temperature as the wall shown in Figure 1.11b. Although the combustion gas absorbs a part of the radiation emitted from the walls selectively as shown in Figure 1.11c, most of the radiation energy converted into different wavelengths reaches the opposite wall as shown in Figure 1.11d. Since the radiation energy absorbed in the combustion gas is small (2 to 5% at most in the example case), and the average transmission distance is about 1.4 m, the exchange of heat in the furnace is dominated by the radiation heat transfer between walls.

Only a fraction of the energy radiated in a short wavelength from a flame in HiTAC is trapped by semi-transparent combustion gas due to its low absorptivity. Once it reaches an opposing wall, it is converted into radiation with a long wavelength spectrum, and the radiation energy is retained in the furnace, except for a portion that is reflected from the furnace, and contributes to the heating of surrounding materials.

1.2.4 THERMODYNAMICS OF HIGH TEMPERATURE AIR COMBUSTION

It is quite reasonable to use exergy loss in precise evaluation of the usefulness of a thermal system aiming at the transformation of heat energy into power as well as to define its theoretical limit for thermal efficiency. However, when we utilize heat in heating furnaces, we usually use the specific units of fuel consumption instead of thermal efficiency units, because there are always nonequilibrium processes represented by, for example, heat conduction. Nevertheless, a thermodynamic analysis can be applied if we assume a quasi-equilibrium change for the heating process, and such thermodynamic discussion can be useful in evaluating the ability of energy saving of the process.

The available energy of fuel E_{av} is the difference between Gibb's energy of reactant (fuel and air) at ambient condition and that of burned product at the same condition:

$$E_{av} \equiv Ex_I = G_I - G_E$$

Exergy, Ex, and Gibb's free energy, G, are expressed by using temperature, T, enthalpy, H, and entropy, S:

$$Ex = H - H_0 - T_0(S - S_0) \quad (1.1)$$

$$G = H - TS \quad (1.2)$$

where subscript 0 denotes the ambient condition. Of course, there must be an exergy change of mixing between fuel and air, because the initial value of exergy was estimated from the separated fuel and air. However, because it varies depending on the equivalence ratio and because the exergy change due to mixing is relatively small as compared with that of combustion, it will be ignored here to simplify the explanation.

Thermodynamic analysis is conveniently made using H-T diagram shown in Figure 1.12, where the ordinate is enthalpy of unit mole (specific enthalpy) and the abscissa is temperature. The Hg and Hp curves are the relation of enthalpy of unburned mixture and its preheat temperature and the relation of enthalpy of chemically equilibrated burned product and its temperature, respectively.

Ordinary combustion of non-preheated mixture is ideally considered to be an isenthalpic change, and it is exemplified by the change from the initial point I on the Hg-curve to the theoretical burned point F. In general cases associated with a heat loss, the final point F_c is somewhat lower than point F. If the mixture is preheated to the point H, the burned point will be F_r, the temperature that increases with preheating. However, the Hp-curve increases exponentially in the high temperature range because of the increase of apparent specific heat caused by thermal dissociation in product, which makes the heat of combustion less. Since it becomes zero at the crosspoint of the two curves, L, which corresponds to the adiabatic limit temperature,

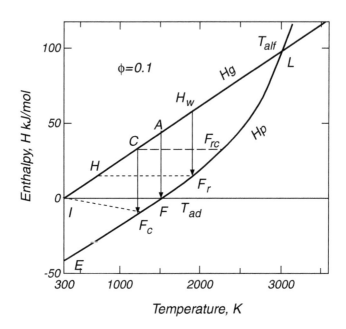

FIGURE 1.12 Enthalpy-temperature diagram for H_2–O_2 mixture ($\phi = 0.1$).

T_{alf}, higher temperature above this point is unrealistic. Accordingly, isenthalpic combustion, no matter how it is preheated or not, is generally associated with a temperature rise until it reaches the adiabatic limit temperature, T_{alf}.

In contrast, if the mixture preheated to the point C is introduced into a large quantity of burned product such as a well-stirred reactor and the heat subtraction in proportion to the heat release rate is assumed, we can consider an isothermal combustion burning at the preheated temperature. In this case, the condition at the reactor exit is indicated by the point F_c on the Hp-curve, and no temperature rise occurs during the combustion process. Further, the same quantity of heat subtracted from the reactor can be utilized effectively for heating. Therefore, we can consider that the idealized extreme of high temperature air combustion is an isothermal combustion.

Figure 1.13 shows the relationship between the adiabatic flame temperature (broken line) and the adiabatic limit temperature (solid line) in terms of fuel concentration in a hydrogen–air mixture. The change in adiabatic limit temperature seems much less compared with that in the adiabatic flame temperature, which strongly depends on the equivalence ratio of the mixture. Therefore, we understand that a considerably high temperature can be realized by preheating even in lean combustion.

Now, we introduce the thermodynamically ideal engine into the analysis of combustion process in furnaces to evaluate the effective energy in various types of combustion explained above, where the effective heat is taken as the heat transformable to the work. An ordinary furnace operated without heat recirculation, modeled in Figure 1.14a and Figure 1.14b, shows the heat recirculation furnace equipped

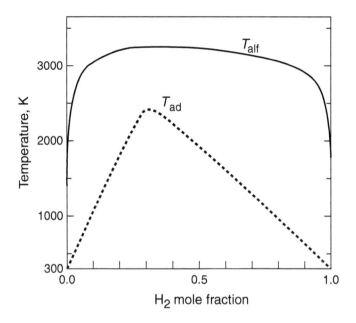

FIGURE 1.13 Adiabatic flame temperature and adiabatic limit flame temperature of H_2–air mixture.

with an ideal regenerator is better. The combustion process of both cases is isenthalpic combustion. Figure 1.14c, in contrast, includes the isothermal process of high temperature air combustion. We define the thermodynamic efficiency as the ratio between the work and the enthalpy difference at inlet and exit, as shown by the following equation:

$$\eta = W/(H_I - H_E)$$

Figure 1.15 shows the relationship between the thermodynamic efficiency and the maximum temperature during the combustion process. The point η_{ad}, indicated by an open circle on any curve, corresponds to the efficiency of the ordinary combustion without any heat loss, expressed by *I-F-E* in Figure 1.12. The lower broken branch of the curve shows the drop in efficiency of the ordinary combustion with a heat loss, *I-F_c-E* in Figure 1.12, and the upper real branch of the curve corresponds to the efficiency increase due to preheating by heat recirculation, *I-H-F_r-E* in Figure 1.12. It is reasonable that we can get higher efficiency if the maximum temperature is raised by preheating, *I-C-F_{rc}-E*. However, it should be noted here that the combination of higher preheating and a leaner mixture produces the higher thermodynamic efficiency, when we keep the maximum temperature constant.

The relationship between the thermodynamic efficiency of isothermal combustion and the maximum temperature during the combustion process, *I-C-F_c-E* is shown in Figure 1.16. The same curve of isenthalpic combustion for $\phi = 0.1$ in Figure 1.15 is also drawn in the figure for the convenience of comparison. The

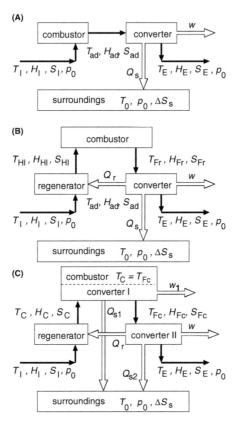

FIGURE 1.14 Steady flow models of heat engine: (A) ordinary combustion, (B) with a regenerator, (C) with an isothermal combustor/converter.

efficiency of isothermal combustion is always higher than that of the isenthalpic one; particularly, the former still remains considerably high efficiency when the latter starts to fall. Actually, the efficiency of isothermal combustion reaches as high as that of Carnot cycle in the temperature range when the thermal dissociation in burned product is negligible. It is clear now that the contribution of heat-recirculating combustion in raising thermodynamic efficiency is significant even at low temperatures where the efficiency of isenthalpic combustion is not high, although we tend to think that raising the maximum temperature is the only way to get higher efficiency.

Introduction

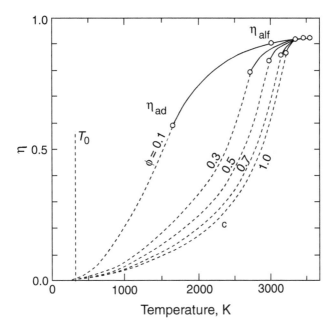

FIGURE 1.15 Thermodynamic efficiency of isenthalpic combustion for H_2–O_2 mixtures.

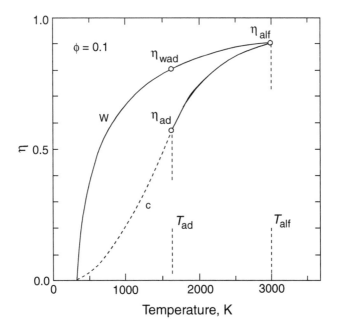

FIGURE 1.16 Thermodynamic efficiency of isothermal combustion for H_2–O_2 mixtures ($\phi = 0.1$).

REFERENCES

1. S. A. Lloyd and F. J. Weinberg. *Nature*, 251:47, 1974.
2. S. A. Lloyd and F. J. Weinberg. *Nature*, 257:367, 1975.
3. J. R. Cornforth. *Combustion Engineering and Gas Utilization*. 3rd edition, 1992.
4. T. Sugiyama, T. Nagata, and I. Nakamachi. *9th IFRF Members Conf.*, 1989.
5. T. Hasegawa and T. Hoshino. *JFRC Ann. Conf.*, 1992 (in Japanese).
6. J. A. Wunning and J. G. Wunning. *Prog. Energ. Combust. Sci.*, 29:81, 1997.
7. R. Weber et al. Technical Report F40/y/8, Ijmuiden, Sept. 1993.
8. S. H. A. Rahbar. *AFRC Spring Meeting*. Kingston, 1994.
9. F. J. Weinberg. *Combustion Science and Technology*, 121:3, 1996.
10. T. Niioka. Private communication, 1998.
11. M. Matsumoto et al. *11th IFRF Members Conf.* IFRF, 1995.
12. M. Flamme and H. Kremer. *Int. Gas Research Conf.* Orlando, FL, 1992.
13. M. Flamme. *4th Int. Conf. on Technologies and Combustion for a Clean Environment.* Lisbon, 1997.
14. T. Hasegawa, R. Tanaka, and K. Kishimoto. *AFRC/JFRC Int. Conf.*, p. N. 9C, 1994.
15. L. M. Dearden et al. *J. Inst. Energ.*, 69:23, 1996.
16. Y. Suzukawa et al. *J. Inst. Energ.*, 38:1061, 1997.

2 Combustion Phenomena of High Temperature Air Combustion

2.1 INTRODUCTION

Combustion with highly preheated air has been shown to provide significantly different combustion characteristics from that obtained with normal temperature or moderately preheated air. Fundamental studies, as well as applied studies, on combustion were carried out during the early 1990s at NFK under the leadership of the late President Ryoichi Tanaka. Evolution of the studies was accomplished largely through the support of NEDO via a research program involving several Japanese companies and universities. The goal of this program was to achieve 10 to 30% savings in energy, downsizing of the equipment by 10 to 20%, and reduction of pollutants below regulation levels. Most of the studies were on LPG (liquefied petroleum gas) and low calorific value gases, although studies have also been carried out on light oil and pulverized coal. The role of the fundamental studies was to provide insight into the thermal and chemical behavior of the flames, whereas that of the applied research was to develop optimum utilization of the technology for several specific applications. Fundamental understanding of these flames for a range of fuels and conditions is indispensable for wider application of HiTAC technology.

Issues of interest for the studies included precise quantification of the thermal field uniformity with regard to integral and microthermal time scales under various operational conditions, fuel–air mixing, flame chemistry, flow dynamics, and emissions using various fuels. In the absence of these, it is not certain if the mixture burns as a thickened reaction zone or as small flamelets. Studying alleviation of the hot spot in flame zones, as well as the reaction chemistry involved, helps determine the criteria for low NO_x emission. The species and radicals formed in HiTAC are significantly different from those formed in ordinary flames. How these radicals are formed is not known. An understanding of the reactions involved will be useful when meeting required operational conditions.

HiTAC results in high heat flux, but just how high remains unknown. The high heat flux may be due to species and radicals or soot. No quantitative information is available. The usual definition of flame is emissions of heat and light. Under certain conditions, however, flameless oxidation of the fuel has been observed. We must understand the features of flameless oxidation if we want to utilize this combustion behavior. Information on mixing, flow dynamics, and the role of the initial fuel puff during each cycle is not adequate. The roles of high temperature air to enhance stability limits and flame spectral emission are not fully understood. The effects on health of various pollutants emitted from HiTAC have not been examined. This is particularly important for waste and low-grade fuels which may result in the formation of dioxins and polycyclic aromatic hydrocarbon (PAH).

In this chapter, issues relating to the regenerative combustion, which provides a convenient means of obtaining highly preheated air for combustion, are omitted. The focus is on understanding the characteristics of HiTAC itself. Thus, discussion is primarily limited to combustion phenomena affected by variations of both temperature and oxygen concentration of combustion air.

2.2 FLAME FEATURES

2.2.1 FLAME STABILITY

The combustion limits of flame with hydrocarbon fuel were studied using the 1.3 kW scale test rig under various conditions of air preheating temperature and oxygen concentration. Stable flame with low oxygen concentration was achieved by diluting the air with either nitrogen or carbon dioxide. Air at desired oxygen concentration was preheated using the test facility regenerator prior to its entry into the test section.

A schematic diagram of the experimental facility, used for 1.3 kW scale combustion tests, is shown in Figure 2.1. The facility consists of a hot air generator, flow control unit for air, fuel and dilution gases, sequence controller, a flue gas analyzer, and a test section. The combustion air in the model combustor was preheated with the regenerator, which could preheat the air to temperatures greater than 1300°C. Low oxygen concentration diluted air was supplied to the regenerator. In the test procedure, a mixture of air and N_2 (or CO_2), after passing through the heat storage media of the regenerator, was allowed to heat to a temperature of about

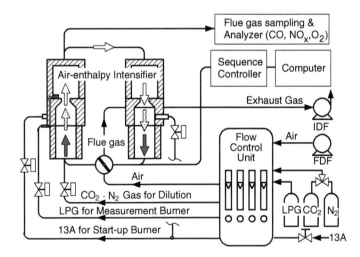

FIGURE 2.1 Schematic of the 1.3 kW-scale hot air combustion test rig.

Combustion Phenomena of High Temperature Air Combustion

1200°C. The total flow rate of the diluted air was maintained constant at 15.0 mm_N^3/h. By maintaining a fixed flow rate of diluted air, a similarity in the flow pattern between the diluted air and the fuel flow could be maintained over the temperature range of interest. It is recognized that complete similarity will require compensation for the temperature changes that occur during the test run. The switching device incorporated in the facility allowed the flow direction in the regenerator to be changed after a prescribed time duration. In the results, the switching device was allowed to change the flow direction after every 30 s. The LPG fuel flow rate was maintained constant at an energy release rate of 1.28 kW. The results have been obtained at an overall air ratio of 12.6 with 21.0% O_2 in air and 2.40 with 4.0% O_2 diluted air. These conditions represent air ratio values significantly larger than the theoretical value for stable combustion.

The results presented in Figure 1.4 show stable flame combustion conditions with low oxygen concentration air (less than 5%) at air temperatures above about 800°C. However, for air temperatures below 800°C the flame stability was significantly affected at less than 15% oxygen concentration in the combustion air. For example, a stable flame at 700°C could only be achieved at much higher oxygen concentration of about 16%.

The physical appearance of the flame (e.g., flame color, size, shape, and global flame structure) was found to significantly change when the concentration of oxygen was decreased from 21 to 3% (by volume) while maintaining the air temperature constant at about 1000°C. Figure 2.2 shows the result of flame intensity profiles

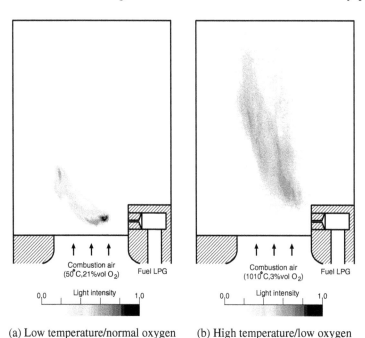

(a) Low temperature/normal oxygen   (b) High temperature/low oxygen

FIGURE 2.2 Visible light intensity profile of flame.

measured instantaneously with a charge-coupled device (CCD) camera. The global flame features observed under conditions of 21% O_2 and 50°C air temperature showed the flame to be short and compact having the expected flame luminosity and blue flame color; see Figure 2.2a.

In contrast, the flame observed with air having 3% O_2 and 1010°C temperature had extremely low luminosity and a very large volume; see Figure 2.2b. The color of the flame was found to be bluish green to green depending on the input conditions of the air. (See the color pictures in preface.) These results suggest that combustion under high temperature and low oxygen concentration air provide lower heat release rates per unit volume compared to the flames obtained with 21% oxygen in the combustion air. The low oxygen concentration and high temperature air flames had very good flame stability limits compared to the normal air combustion case.

2.2.1.1 Temperature Profiles

The mean temperature of the flame was measured with an R-type thermocouple (Pt-Pt/Rh 13%) with a wire diameter of 0.05 mm. The thermocouple was coated with a ceramic layer to alleviate catalytic effects of Pt in the flame. The temperature of diluted (low oxygen concentration) and preheated air was measured at a location of 105.0 mm upstream from the fuel nozzle using the R-type thermocouple. Then, 36 measurement positions were selected in the test section and are shown in Figure 2.3. The X-axis is along the direction of the fuel injection and the Y-axis is along the direction of airflow. The results obtained with 21% oxygen in air at 35°C are shown in Figure 2.3a, and

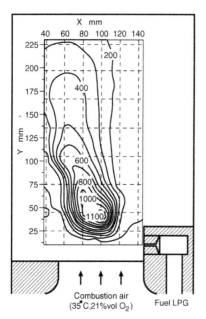
(a) Low temperature/normal air

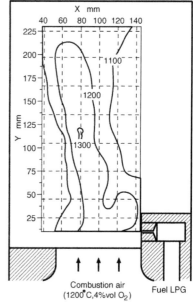
(b) High temperature/low oxygen conc. air

FIGURE 2.3 Temperature distribution.

those obtained with 4% oxygen in air at 1200°C are shown in Figure 2.3b. In both cases, the air flows vertically upward as shown by the three arrows in the figure and the gaseous fuel flows from right to left in the X-direction at a position of $Y = 0$ mm, as shown by a single horizontal arrow in Figure 2.3.

Temperature profiles were measured for the two different flames obtained with air at 35°C and 21% O_2 (as in the case of ordinary combustion) and preheated air at 1200°C with 4% O_2. The latter conditions represent highly preheated and diluted air combustion conditions. It can be seen from Figure 2.3a that the peak temperature is located near the vicinity of the fuel nozzle when air with 21% oxygen is used without preheating. However, in contrast, when low oxygen concentration air is preheated to very high temperatures, a far more uniform distribution of the thermal field can be observed; see temperature profiles shown in Figure 2.3b.

Temperature fluctuations were also measured at the spatial position where the mean temperature was found to be maximum. These temperature fluctuations were also measured using an R-type thermocouple (Pt-Pt/Rh 13%) with a wire diameter of 0.05 mm. It is recognized that this size of thermocouple would not capture the very high frequency turbulent fluctuations in the flow. The results showed that temperature fluctuations decrease drastically when the combustion air conditions were changed from 21% oxygen in air at 35°C to 4% oxygen in air at 1200°C. The results shown in Figure 2.4 clearly indicate the high levels of temperature fluctuations (root mean square value of about 197°C) associated with the ordinary combustion conditions; see Figure 2.4a. Temperature fluctuation levels of around 50 to 100°C are characteristic of many turbulent diffusion flames.

The results obtained here show a significant reduction in the rms value of temperature fluctuations (less than 5°C) with the flame having 4% oxygen in air at 1200°C; see Figure 2.4b. The characteristic features of the results obtained at two additional data points, measured upstream and downstream of the maximum flame temperature position, remained very similar to the results shown in Figure 2.4b. This suggests that turbulent mixing does not appear to be governed by the high temperature and low O_2 content air combustion process. This requires further clarification since the local velocity and fluctuation levels are not known.

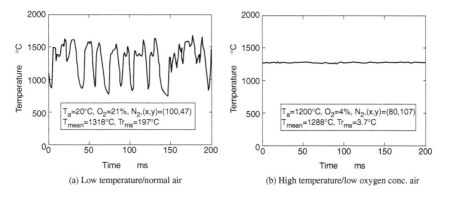

FIGURE 2.4 Temperature fluctuation.

2.2.1.2 Influence on NO_x Emissions

The influence of both air temperature and gas chemical composition on NO_x emissions was studied using LPG gas as the fuel. The CO emissions were found to be less than 100 ppm within the combustion stability limits shown in Figure 1.4. The results presented in Figure 2.5 show a significant influence of the dilution gas composition on NO_x emission levels. Much higher levels of NO_x are obtained with N_2 as the dilution gas compared to CO_2 at any degree of air preheating. This shows the importance of gas chemical composition for controlling NO_x in highly preheated air combustion. Thus, in the combustion process, the use of burned gases (natural products of combustion) can be more effective to control NO_x in addition to the amount of oxygen in the combustion air.

High temperature and low O_2 air combustion extended the flame size (both in length and width) as shown in Figure 2.2. The size of the flame has a direct effect on the distribution of temperatures in the flame. Therefore, in higher temperature air combustion it is possible to maintain the maximum combustion temperature, similar to that achieved in the case of conventional normal air combustion or lower than in the case of conventional combustion, depending upon the choice of air preheating temperature and air chemical composition.

2.2.2 THERMAL FIELD BEHAVIOR

2.2.2.1 350 kW-Scale Combustion Test

The primary objective of this study was to evaluate the effect of fuel–air nozzle configuration as well as location of the exhaust gas exit in a large-scale furnace on the combustion process. The physical dimensions of the 350-kW combustion test facility were 1.0 m wide, 1.0 m high, and 3.8 m long. A schematic diagram of this facility is shown in Figure 2.6. The combustion air in this facility was preheated with the high-cycle switched regenerator operating with a switching time of 30 s (i.e., same as that used for the 1.3-kW facility). Town gas (13A) was used as the fuel. The firing rate was maintained constant at 349 kW with a constant air ratio of 1.2.

Three configurations modes were examined in the rig to determine their effect on the distribution of temperatures in the furnace. The locations of 15 thermocouples for temperature measurements are indicated in Figure 2.7. Each thermocouple in the furnace was positioned midway between the ceiling and the bottom. However, provision existed to traverse the thermocouple within the furnace, if desired.

2.2.2.2 Cold Flow Model Test

Flow visualization for the three configurations modes was made using a cold water flow model as shown as Figure 2.8. This facility is a scaled down model test rig made from transparent acrylic material and is 1/10 scale in length. A light scattering technique was used for flow visualization. A sheet light beam formed using a 3W argon ion laser allowed illumination of the desired cross section of the flow. Small-size scatters were introduced to the flow to enhance the quality of flow pattern images obtained. The facility provided information on the resulting flow patterns obtained at various operational conditions.

Combustion Phenomena of High Temperature Air Combustion

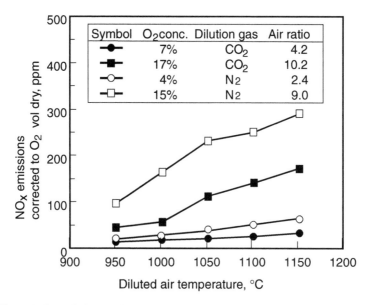

FIGURE 2.5 Effect of diluted air temperature on NO_x emission.

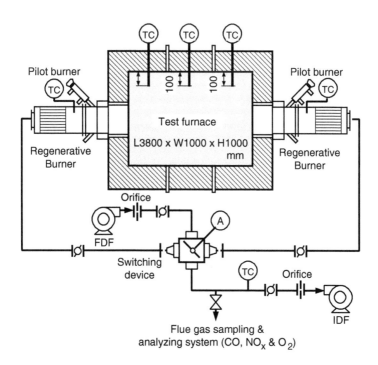

FIGURE 2.6 Schematic of 350 kW-scale test rig.

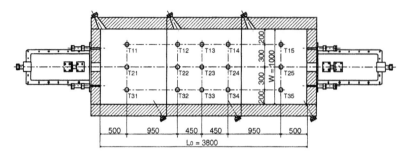

FIGURE 2.7 Locations of thermocouples for temperature distribution measurement.

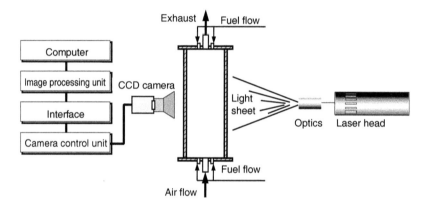

FIGURE 2.8 Schematic of cold flow model test.

2.2.2.3 Temperature Profiles

Three different fuel–air nozzle configurations shown in Figure 2.9 have been examined. Mode I represents an arrangement for ordinary combustion in which the mixing of fuel with air occurs in the air nozzle or near to the air throat. Mode II is a parallel flow type of arrangement for high temperature combustion in which fuel–air preparation and dilution processes with the products of internal combustion occur through the use of high-velocity turbulent airflow that creates a natural entrainment effect to the airflow. Mode III also represents an arrangement for the high temperature combustion process but with a counter flow type of arrangement. This arrangement is intended to enhance mixing of the fuel–air dilution process in the test furnace.

Temperature profiles averaged over one switching cycle were measured, and results for the three modes are shown in Figure 2.10. The maximum temperature difference in Mode I was 120°C and this temperature difference decreased to 80°C in Mode II. The minimum temperature difference of 20°C was obtained in Mode III, which suggests that high temperature and low oxygen concentration combustion occurred not only in the designed 1.3-kW scale test facility but also in the 350-kW scale test rig under the arrangements of Mode II and Mode III. In addition, NO_x

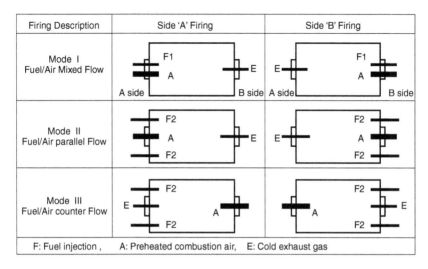

FIGURE 2.9 Fuel–air nozzle configurations.

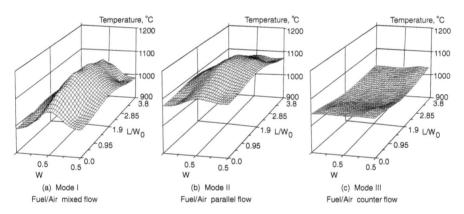

FIGURE 2.10 Temperature distribution averaged over one switching cycle in 350 kW-scale test rig.

and CO emissions for Mode I were 213 and 16 ppm, respectively, at 11% O_2 correction, whereas they were both 15 ppm in Mode II. In Mode III, NO_x and CO emission levels were 13 and 17 ppm, respectively, which is very similar to the Mode II performance.

In these combustion tests, the average temperature of the preheated air reached 1150°C when the average flue gas temperature was 1250°C in the test rig. This temperature was obtained by using the high-cycle switched regenerator with 30 s switching time. The air side temperature efficiency for this condition was 91%, and the efficiency of the waste heat recovery was 77%. These results demonstrated that significant energy saving can be attained by introducing the high-cycle regenerator.

2.2.2.4 Flow Patterns

Flow patterns of the two modes of the fuel–air nozzle configurations, i.e., Mode II and Mode III, were investigated to see what occurs in the mixing process in the furnace. Results shown in Figure 2.11 clearly show that each mode creates a very different flow pattern in the model furnace. Flow pattern distribution obtained for Mode II (Figure 2.11a) shows that fuel flow and airflow are issued into the furnace in a parallel fashion and that the mixing is moderate. In Mode II, proper preparation of the fuel–air dilution process occurs to cause the high temperature combustion phenomena. Mode III (Figure 2.11b) also demonstrates that sufficient fuel–air preparation occurs for the high temperature combustion and that the general features of the flow dynamics are very similar to those obtained for Mode II.

The observed flow patterns in Mode II and Mode III, therefore, suggest that high temperature air combustion conditions are strongly related to the mixing process and flow patterns in the entire furnace space. In contrast, ordinary combustion is highly sensitive to mixing for flame holding in the nozzle near field. The flames corresponding to Mode II and Mode III observed in the furnace had enlarged flame volume and the furnace space seemed to be mostly filled with low luminosity flames. We can treat the flames in Modes II and III as well-stirred reactors. Note that high temperature air combustion was not caused by the agitating momentum although it has similar features to that of the well-stirred reactors.

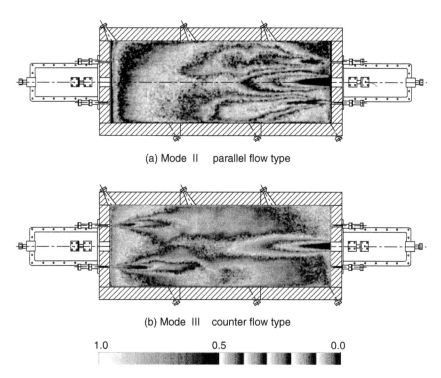

FIGURE 2.11 Measured flow patterns.

2.2.3 FLAME STRUCTURE, RADICALS, AND SPECIES

2.2.3.1 Experimental Furnace for Optical Measuring

The optical measuring method was applied to the combustion flame in the high temperature air formed in a small regeneration furnace with almost the same structure as the actual furnace. A regenerative test furnace was made in which openings for optical measuring were arranged at the upper, lower, and side parts of the furnace. Its external appearance is shown in Figure 2.12 and its specifications in Table 2.1. The inside of the furnace is a 1 m square regular cube. The fuel used was city gas 13A or LPG, and four burners were arranged, two pairs of which were on combustion alternating every 30 s. The layout drawing of holes for the preheated air and fuel nozzles is shown in Figure 2.13. The combustion rate when using the 13A fuel was 58 kW/pair at maximum, and the in-furnace temperature was raised to 1300°C at maximum. A quartz glass was set up for each opening at the side part in order to observe the inside of the furnace, and to each opening at the lower part of the furnace for irradiating the laser beam. Up to the in-furnace temperature of 850°C, the fuel and air were mixed within each burner to carry out concentrated combustion (F1 mode), and at a temperature higher than 850°C, distributed combustion (F2 mode) was practiced to enable operation of the test furnace.

2.2.3.2 Combustion Conditions

A schematic diagram of the laser induced fluorescence (LIF) measurement experiment is shown in Figure 2.14. The airflow velocity, the fuel jet velocity, and the fuel jet angle were changed to generate combustion flames with different concentrations of exhaust NO_x, which were the objects to be measured. Table 2.2 shows the conditions for measurement. All the conditions were unified on the basis of fuel 13A, air ratio 1.2 (airflow rate 52 m_N^3/h, fuel flow rate 4 m_N^3/h), and in-furnace

FIGURE 2.12 Combustion test furnace to be measured.

TABLE 2.1
Specifications of the Test Furnace

Furnace size	1000 mm in length, 1000 mm in width, 1000 mm in height
In-furnace temperature	Maximum, 1300°C
Combusting apparaus	Combusting volume: maximum 58 kW (50,000 kcal/h) pair
	Fuel: 13A, LPG
	Pressue: fuel 2000 mmAq
	air 800 mm Aq
Incidental facilities	Combusting air feeding fan
	Exhausting facility (IDF)
	Compressor

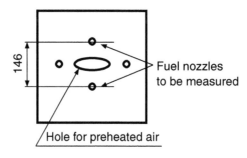

FIGURE 2.13 Burner nozzles.

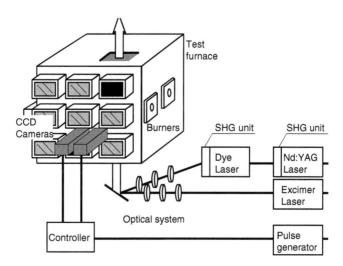

FIGURE 2.14 Laser induced fluorescence (LIF) measurement system.

TABLE 2.2
Combusting Conditions

	Preheated Air (m/s)	Fuel (m/s)	Fuel Injection Angle	Exhaust NO$_x$ (ppm; O$_2$ = 11%)
Type (a)	52	52		350
Type (b)	52	79	Parallel	110
Type (c)	52	40	Parallel	100
Type (d)	52	40	9 degree	148
Type (e)	74	40	Parallel	74

temperature 1100°C, and the flow velocities were adjusted according to the nozzle sizes of burners and air holes so that the air ratio was constant. Table 2.2 also shows the results obtained by measuring the concentration of exhaust NO$_x$ under various conditions (conversion rate of oxygen = 11%) using an NO$_x$ meter (Horiba Manufacturing Co. PG235).

Photographs of the flame under conditions (a) and (e) are shown in Figures 2.15a through c. The condition (a) produced the usual flame which combusts the concentrated fuel and air, and the resulting NO$_x$ concentration was 350 ppm, a high value. However, in the cases of conditions (b) through (e) of the distributed combustion flame, the flame elongated, and it became difficult to observe the flame with the eye

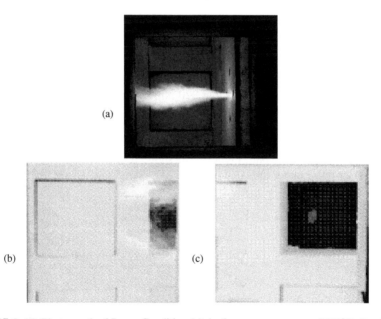

FIGURE 2.15 Photograph of flame. Condition (a), in-furnace temperature 1100°C, Condition (b), downstream of flame, and Condition (c), upstream of flame 1100°C.

in the vicinity of the burner. But, in the area between two fuel nozzles where the preheated airflow and the fuel jet flow contacted each other, a spontaneous emission light of a slightly blue color was observed, which was considered to be an incipient decomposition of the fuel.

Furthermore, the NO_x concentration in any of the conditions (b) through (e) was significantly lower, compared with the condition (a). But, the lowering of NO_x is affected more by both the airflow velocity and the fuel jet angle than the fuel jet velocity, resulting in a maximum difference in NO_x concentration of twice as much depending on conditions. Figures 2.16a through c show the results of the distribution of the in-furnace temperatures measured, when the maximum temperature was 1190°C among the temperatures measured by an R-type thermocouple at 25 points at a distance of 100 mm from the upper part of the in-furnace combustion chamber for the cases of conditions (a), (d), and (e).

From these results, the deviation between the maximum temperature and the minimum temperature at the 25 points measured is as follows:

$$\text{condition(a) } (117°C) > \text{condition(d) } (96°C) > \text{condition(e) } (69°C)$$

This shows a correlation between the exhaust NO_x concentration and the combustion conditions. In the case of condition (a), the flame at large becomes a luminous flame and its length is rather short, and further a steep section is observed in the temperature distribution. As for the distributed flame — conditions (d) and (e) — condition (e) gives a comparatively flat temperature distribution and none of the local portions at a higher temperature was observed.

2.2.3.3 Optical Measurement Results

Under the conditions shown in Table 2.2, LIF measurement was carried out in a range (shown in Figure 2.17) of about 220 mm in distance on the left side of each base portion of the burners. At the center of the base portion of the burners, holes for jetting the preheated air are arranged and the fuel nozzles are set up at the top and bottom portions of each hole. Here, the acetone steam was added only to the fuel nozzles at the bottom part. Figure 2.18a shows the results of the OH distribution for one laser pulse under the condition (a). As shown in the figure, the intensity of

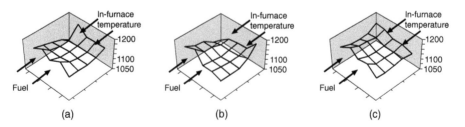

FIGURE 2.16 In-furnace temperature distribution. Condition (a) $T_{infmax} = 1190°C$, $\Delta T_{inf} = 117°C$, Condition (b) $T_{infmax} = 1190°C$, $\Delta T_{inf} = 96°C$, Condition (c) $T_{infmax} = 1190°C$, $\Delta T_{inf} = 69°C$ (T_{infmax}: maximum in-furnace temperature, ΔT_{inf}: temperature deviation).

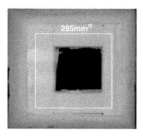

FIGURE 2.17 Range of LIF measurement (285 mm square).

the OH distribution per unit area is strong, to make the combustion reaction active. However, Figures 2.18b through e show the results of simultaneous measurement of both the OH distribution and the acetone distribution for one laser pulse, in the case of the distributed combustion flame. In either figure, the OH distributes outside the fuel jet diffusion is the state of a thin film in the vicinity of the burners, to form a diffusion flame. The OH distribution spreads out gradually at the lower portions of the flame stream.

As for the acetone distribution, since the wavelength range (340 to 450 nm) of the fluorescence signals is nearly the same, a distribution that is conjectured as a PAH and is visualized at the same time is observed at the downstream portion of flame. This phenomenon can be interpreted as follows. When simultaneous measurements are carried out in the vicinity of the fuel nozzles while the in-furnace temperature changes in the range between 900 and 1100°C under the condition of using LPG as the fuel and setting the air ratio at 1.2, the results obtained show that the signal intensity at the downstream portion of the flame increases as the temperature rises (see Figure 2.19). Based on this information, it can be considered that the fuel is jetted into the inside of the furnace to cause a thermal decomposition, resulting in the fuel being reformed to produce PAH, a precursor of soot. Furthermore, a flow of the preheated air is jetted between two flows of the fuel jet diffusion. When attention is paid to the OH distribution in this airflow jetted area, the distribution contacts the airflow at the lower portion of the flame stream in the case of condition (e), compared with conditions (b) through (d). As for the fuel jet diffusion, the acetone is added only to the nozzle set below the air nozzle. When comparing the signals from these two nozzles, it can be observed that the fuel is thermally decomposed at a distance of about 50 mm downstream from the nozzle. Under condition (d), where the NO_x concentration becomes maximum in the F2 mode, the PAH signals are taken into the preheated airflow in a shorter distance from the nozzle and the distribution steadily disappears. This phenomenon is different from those under other conditions. Furthermore, the OH distribution is intensified at the side contacting the airflow, and a feature is observed that the combustion reaction is carried out at a higher portion of the flame stream, compared with those under different conditions.

Entrainment of the in-furnace exhaust gas to the fuel jet diffusion is examined on the basis of the results obtained by measuring spontaneous emission spectra in the combustion flame. The spontaneous emission spectrum of the radical is proportional to the density of the radical in an excitation state. In an emission spectrum of the flame

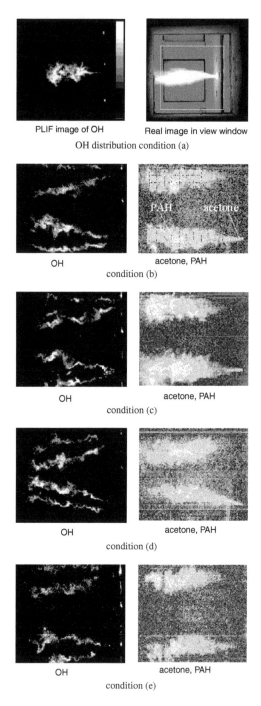

FIGURE 2.18 Distributions of OH and acetone (as fuel distributed domain).

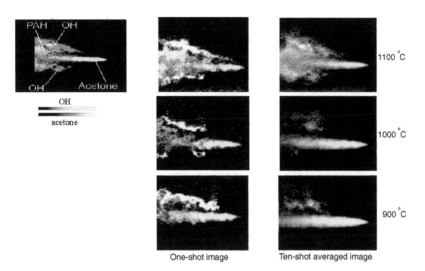

FIGURE 2.19 Results of simultaneous measuring of the combustion reaction domain (OH) and acetone (fuel distributed domain).

of hydrocarbon, the intensity of a CH radical or a C_2 radical, which is an intermediate product of combustion, varies according to the air ratio,[1,2] and the emission from the C_2 is made more intense at the visible side as the ratio of the combustible gas is higher.[1,2] Also, as for the diffused flame which is formed when LPG is used as the fuel in a high temperature airflow at 1100°C, the spectrum intensity of a C_2 Swan band (516.5 nm) against the intensity of CH A-X(0,0), 431.5 nm in the spontaneous emission spectrum of the flame increases when the oxygen concentration is decreased. This fact has been reported by the Basic Research Working Group, based on results obtained from the following experiment. When methane, a main component of city gas, is used in the experiment, the amount of change in the emission spectrum according to an equivalence ratio of the methane is small.[3] Therefore, LPG was applied as the fuel in the experiment, and the spontaneous emission of the flame was measured under the combustion conditions where the NO_x concentration differed. However, under condition (a) for the usual flame that makes the NO_x concentration maximum, the flame became luminous at its downstream portion, resulting in making the background larger due to black body radiation even in the case of the distributed flame. Three spectra were compared, which were measured at three points from the base portion of the flame contacting with the airflow at the central portion extending to the downstream portion of flame just before the luminous flame.

In the experiment, images of the flame were formed on the end face of an optical fiber of the spectrometer (Hamamatsu Photonics C7473) at the magnification of 1 to 1, using a spherical lens with a focal length of 200 mm, and the spectra were obtained. The measurement was carried out at an in-furnace temperature of 1100°C, the air ratio 1.2, and under the same combustion conditions as conditions (d) and (e) shown in Table 2.2. In this case, the exhaust NO_x concentration was 153 ppm (the conversion rate of $O_2 = 11\%$) under condition (d), and 60 ppm under condition (e). The measured results of the spontaneous emission spectra are shown in Figure 2.20. No significant

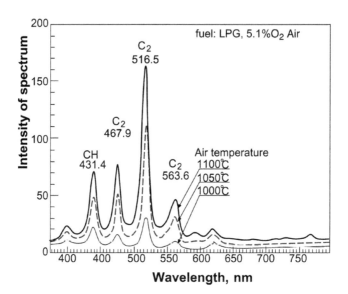

FIGURE 2.20 Chemical luminescence of HiTAC.

difference in the ratio of the spontaneous emission spectra of CH A-X(0,0) (431.5 nm), C_2 (516.5 nm) was observed at any point. As shown in Figures 2.18 and 2.19, the domain of combustion reaction is distributed outside the fuel jet diffusion in the state of a thin film in the vicinity of the burners, which is a typical shape of a diffused flame. When the effect of entraining the in-furnace gas is much higher, it can be considered that the fuel jet diffusion takes in the in-furnace gas before its ignition to contain some concentration of oxygen, resulting in a form of premixed combustion flame. It was reported by Weber and others[4] in the IFRF that, according to the results obtained by measuring the in-furnace flow, the EGR (exhaust gas recirculation) to the airflow is small in the vicinity of the burners, in the case of the distributed combustion flame when fuel jetting is made parallel to the airflow. But the EGR at 3 m down from the nozzle attains 400%. It may be conjectured that the existence of OH at the outer side of the fuel is caused by oxygen being supplied dominantly by its diffusion from the outside, and therefore a burned-gas recirculating effect in the vicinity of the burner is relatively small. In the case of condition (e), which showed the lowest NO_x, it can be also confirmed from results (shown in Figure 2.21) obtained by measuring the OH distribution at the lower portion of the flame on and after 385 mm from each burner, that the combustion range is distributed not in a thin film state but in a cubic state, compared with the state of distribution at the upper portion of the flame.

From these results, it can be judged that the reason no noticeable change is found in the spontaneous emission spectra is that the EGR is small at the upper stream portion of the flame. As for the cases of combustion conditions (d) and (e), it can be also assumed that since the difference in NO_x concentration between both conditions is relatively small, the difference in oxygen concentration in the jet diffusion due to the EGR is relatively small, resulting in showing none of the changes in the spectrum. In other words, it can be interpreted that, in the case of the distributed

Combustion Phenomena of High Temperature Air Combustion

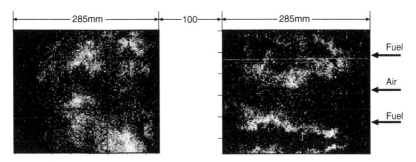

FIGURE 2.21 OH distribution, condition (e), upstream and downstream of flame.

combustion flame that produces a low NO_x combustion, the EGR in the vicinity of the burners is small, but the airflow is diluted by the EGR at the lower portion of the flame stream, to become an airflow with a low oxygen concentration, and low NO_x combustion is attained since no domain at a high temperature is locally formed.

Further, the authors carried out PIV (particle image velocimetry) measurement in the vicinity of the burners under condition (c) (fuel parallel jetting) and condition (d) (fuel oblique jetting) for distributed combustion, as shown in Table 2.2, with a view to measuring the difference of the in-furnace flow according to the fuel jetting conditions. A twofold wave of Nd:YAG lasers emitting from two machines was used as the light source, and the time difference between two laser pulses was set at 10 μs. Such laser beams as formed in a sheet-shape were radiated into the furnace, to obtain two images from their scattered light. At that time, tracer grains of SiO_2 (average grain diameter: 2.7 μm, apparent specific gravity: 0.45 g/cm^3) with enough heat resistance even in the flame were seeded into the 13A fuel, and the measurements were carried out in a combustion state at an in-furnace temperature of 950°C. The results obtained are shown in Figures 2.22 and 2.23.

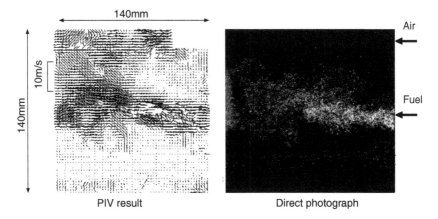

FIGURE 2.22 Results of measuring the in-furnace flow by PIV, condition (c), in-furnace temperature 950°C.

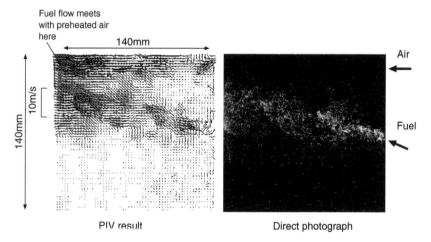

FIGURE 2.23 Results of measuring the in-furnace flow by PIV, condition (d), in-furnace temperature 950°C (F2 mode).

In either figure, the side image shows a laying of a velocity vector besides the measured photograph. Figure 2.23 shows a color bar at a maximum velocity 34 m/s. The velocity in terms of magnitude is the same in both Figures 2.22 and 2.23, but the position where the fuel jetting diffusion comes across the airflow to start combustion is much different. It has been found from the loci of velocity vectors that the combustion starting position is 140 mm downstream (divergent angle: 22°) under condition (c) and 92 mm downstream (divergent angle: 25°) under condition (d).

2.2.3.4 Summary

The optical measurement was carried out to investigate the characteristics of the flame having a similar structure to high temperature air combustion in an actual furnace using regenerative burners. In the case of a flame generated by the usual combustion condition, the combustion reaction takes effect suddenly with fresh airflow. Without dilution of airflow caused by entrained flue gas, a very high temperature results. This causes a high concentration NO_x emission as well. However, in the case of the distributed flame of F2 mode, the fuel is taken into the airflow to start combustion at an earlier stage when the fuel jetting is oblique to the airflow, compared with the case of fuel parallel jetting. This results in a high concentration NO_x emission as well. From these results, it has been confirmed that the concentration of NO_x emitted is lowered under these conditions such that a diffused mixing at the initial stage is controlled. Also the fuel is mixed with the airflow in the state of a low oxygen concentration in the wake where the self-EGR is enhanced.

2.2.4 FLAME WITH HEAT AND COMBUSTION PRODUCTS RECIRCULATION

A further improved heating method can be obtained by combining heat recirculation with hot combustion products recirculation in a furnace, which decreases the maximum flame temperature and creates a more uniform profile of combustion gas temperature in the furnace. In this section, the advantages of this heating method are described and are compared, by using a simple heat balance calculation, to the conventional additional enthalpy combustion method, which does not employ hot combustion products recirculation.

2.2.4.1 Improved Heating Method

2.2.4.1.1 Heat and Combustion Product Recirculation

To avoid localized overheating of the load, there has been a restriction of heat input and amount of heat recirculation for additional enthalpy combustion using fuel with high heating value. It is, however, possible to overcome this thermal restriction by controlling the following two parameters in the combustion process, which influence adiabatic flame temperature. They are (1) initial temperature of the air and fuel and (2) vitiation of the inlet air and fuel when combustion occurs under the condition of constant air ratio, constant fuel heating value, and atmospheric pressure in a furnace. Thus, if there is a way to control the flow rate of combustion products to vitiate the inlet air and fuel, it is possible to increase the amount of heat recirculation, particularly for high heating value fuels, over the restriction level without raising flame temperature. This heating method combining heat recirculation with high temperature combustion product recirculation in the furnace, schematically shown in Figure 2.24, can significantly improve the uniformity of the temperature profile formed in the furnace when compared to periodical alternate additional enthalpy combustion.

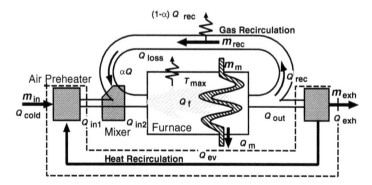

FIGURE 2.24 Schematic diagram of model furnace system with heat recirculation and gas recirculation.

The macroscopic mass balance and the heat balance of the system can be expressed by Equation 2.1:

$$\begin{cases} \dot{m}_{in} + \dot{m}_{rec} = \dot{m}_{exh} + \dot{m}_{rec} \\ Q_{cold} + Q_f + Q_{ev} + \alpha Q_{rec} = Q_m + Q_{loss} + Q_{exh} + Q_{ev} + Q_{rec} \end{cases} \quad (2.1)$$

where $\dot{m}$ = mass flow rate, Q = heat flow rate, and α = ratio of the recirculated heat to the total heat contained in the recirculated gas flow. The subscripts are in = input, rec = recirculation, exh = exhaust, $cold$ = initial state, f = fuel, ev = excess value, m = heating materials, and $loss$ = loss. Based on this equation, the characteristics of the improved heating method with heat and combustion product recirculation, particularly in the thermal field where radiation controls heat transfer, were studied by changing major parameters affecting furnace temperatures and thermal efficiencies of furnaces.

2.2.4.2 Heat Balance in the System

2.2.4.2.1 Gross Heat Input

The quantities of heat transferred from the hot flue gases to the fuel and air by a counter type of heat exchanger can be given by Equation 2.2:

$$\begin{cases} Q_{ev} = \xi_h (C\dot{m})_{in} (T_{out} - T_{cold}) \\ \xi_h = (LK/(C\dot{m})_{in})/(1 + LK/(C\dot{m})_{in}) \end{cases} \quad (2.2)$$

where
- T_{cold} = initial temperature of fuel and air
- T_{out} = flue gas temperature at the furnace outlet
- $(C\dot{m})_{in}$ = sum of heat capacity of fuel and air
- ξ_h = heat exchange coefficient
- L = length of heat exchanger
- K = gross heat transfer coefficient of heat exchanger

The temperature at the exit of the heat exchanger, T_{in1}, and the temperature after the mixer, T_{in2}, are given in Equations 2.3 and 2.4, respectively. The ratio of mass recirculation to the inlet mass is defined in Equation 2.5.

$$T_{in1} = T_{cold} + \frac{Q_{ev}}{(C\dot{m})_{in}} \quad (2.3)$$

$$T_{in2} = \frac{T_{in1} + \alpha Q_{rec}/(C\dot{m})_{im}}{1 + R} \quad (2.4)$$

Combustion Phenomena of High Temperature Air Combustion

$$R = \left(\dot{m}_{rec}/\dot{m}_{in}\right) \quad (2.5)$$

The maximum temperature in the furnace, T_{fmax}, is given by Equation 2.6 as it is assumed to be equal to the adiabatic flame temperature neglecting the effect of dissociation.

$$T_{fmax} = T_{in2} + \frac{Q_f}{(C\dot{m})_{in}(1+R)} \quad (2.6)$$

2.2.4.2.2 Heat Transfer in Furnace

A simple heat transfer model is introduced to express the relationship between combustion gas and the materials to be heated without taking into account the details of the actual heat transfer process taking place in a furnace. The amount of heat, Q_m, gained by the heated materials is expressed by Equation 2.7 as the sum of the radiation, Q_{rad}, and the convection heat transfer, Q_{conv}, from the combustion gas, using f representing heat transfer ratio Q_{rad}/Q_{conv} here.

$$Q_m = Q_{rad} + Q_{conv} = Q_{conv}(1+f) \quad (2.7)$$

Equation 2.8 is given by rewriting f as a function of temperature ratio, r_T, of combustion gas, T_g, and materials, T_m, and coefficient, ϕ_{CG}.

$$f(r_T) = \frac{\phi_{CG}\sigma\left(T_g^4 - T_m^4\right)}{h\left(T_g - T_m\right)} = C_r T_g^3\left(1 + r_T + r_T^2 + r_T^3\right) \quad (2.8)$$

where, $r_T = T_m/T_g$, $C_r = \phi_{CG}\sigma/h$

where σ = Stefan-Boltzmann constant, ϕ_{CG} = overall thermal absorption coefficient, and h = convection heat transfer coefficient.

In a furnace, $f(r_T)$ generally is affected by flow condition, the temperature profile of the combustion gas and of the heated surface configuration or furnace shape. The following three assumptions were made to simplify the heat balance calculation:

- C_r is constant, 0.17×10^9 defined under the condition of T_{gref} = 2000 K, T_{mref} = 600 K and f_{ref} = 2.
- f is a function of T_g, T_m and T_{out}.
- r_T varies with time proportionally when time is $0 \leq t \leq \tau$, and becomes constant when $\tau \leq t \leq 1$ as shown in Figure 2.25. In this study, 0.5 is used as a fixed value of τ.

According to the above assumption, time averaged heat transfer ratio, $\overline{f(r_T)}$ can be expressed by the equation 2.9.

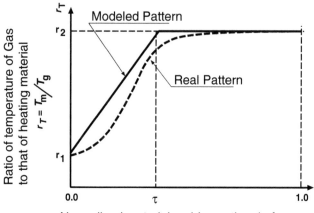

FIGURE 2.25 Assumed change pattern of the temperature ratio $r = T_m/T_g$.

$$\overline{f(r_T)} = \int_0^1 f(r_T(t))\, dt = C_r T_{out}^3 \int_0^1 \left(1 + r_T + r_T^2 + r_T^3\right) dt \tag{2.9}$$

The range of the integration of the Figure 2.25 is given when integral variable t is converted to r_T. The temperature of the combustion gas at the furnace entrance, T_{gin}, is assumed to be the maximum temperature of the gas, T_{fmax}, and the temperature at the furnace exit T_{gout} as T_{out}.

$$r_1 = \frac{T_{m_{in}}}{T_{gin}} = \frac{T_{m_{in}}}{T_{fmax}}, \quad r_2 = \frac{T_{m_{out}}}{T_{gout}} = \frac{T_{m_{out}}}{T_{fmax}}$$

The temperature of the heated material at the furnace exit can be given by using the heat quantity of the heated materials.

$$T_{m_{out}} = Tm_{in} + \frac{Q_m}{(C\dot{m})_m} \tag{2.10}$$

The amount of convection heat transfer gained by the heated materials is given by Equation 2.10 obtained by treating the convective heating process as a parallel flow type of heat exchanger.

$$Q_{conv} = (1 - \xi_m)(C\dot{m})_m (T_{mix} - T_{m_{in}})$$

$$T_{mix} = \{(C\dot{m})_{in} T_{fmax} + (C\dot{m})_m T_{m_{in}}\} / \{(C\dot{m})_{in} + (C\dot{m})_m\} \tag{2.11}$$

$$\xi_m = \exp(-b_m K_m L_m)$$

where K_m = overall coefficient of heat transfer, L_m = length of the heat exchanger, and b_m = the sum of reciprocal of the heat capacity

2.2.4.2.3 Heat Output

The amounts of heat flowing out of the furnace, Q_{out}, and flowing out of the counter type heat exchanger, Q_{exh}, is given by Equations 2.12 and 2.13, respectively.

$$\begin{cases} Q_{out} = Q_{in2} + Q_f - Q_m - Q_{loss} \\ T_{out} = T_{fmax} - \dfrac{Q_m + Q_{loss}}{(C\dot{m})_{in}(1+R)} \end{cases} \quad (2.12)$$

$$\begin{cases} E_{exh} = Q_{out} - Q_{ev} \\ T_{exh} = T_{out} - \dfrac{Q_{ev}}{(C\dot{m})_{in}} \end{cases} \quad (2.13)$$

2.2.4.2.4 Equation Arrangement

Through the above heat balance analysis, a relationship between combustion gas temperature at the furnace outlet, T_{out}, and other variables was obtained as Equation 2.12. Another relationship in terms of the quantity of the heat transferred from combustion gas to the heated materials in the furnace, Q_m, was also obtained as Equation 2.7 by estimating the convection and radiation heat transfer rate based on the same assumptions, which simplifies the actual heating process.

Equations 2.7 and 2.12 can be rewritten with the use of the basic variables shown in Equations 2.14 and 2.15, respectively. These equations can be calculated as two simultaneous equations numerically when specific values are given to those variables except for two arbitrary variables. It is necessary to provide initial conditions for heat input with fuel, Q_f, heat loss in furnace, Q_{loss}, and temperature of fuel and air, T_{cold}. For example, Q_m and T_{out} are calculated from these two equations by giving certain values to other variables, R, α, ξ_h, ξ_m, τ, $(C\dot{m})_{in}$, $(C\dot{m})_m$.

$$Q_m = F_1\{Q_m, T_{out}, R, \alpha, \xi_h, \xi_m, \tau, (C\dot{m})_{in}, (C\dot{m})_m\} \quad (2.14)$$

$$T_{out} = F_2\{Q_m, T_{out}, R, \alpha, \xi_h, \xi_m, \tau, (C\dot{m})_{in}, (C\dot{m})_m\} \quad (2.15)$$

2.2.4.3 Calculation Results

2.2.4.3.1 Effect of Gas Recirculation

Increasing the gas recirculation ratio of the model furnace caused the maximum flame temperature to decline and the flue gas temperature at the furnace outlet to rise gradually when there was no heat loss from the recirculating gas, i.e., $\alpha = 1.0$. Some heat loss (i.e., $\alpha < 1.0$) from the recirculating gas has a considerable effect in lowering both temperatures, T_{fmax} and T_{out}.

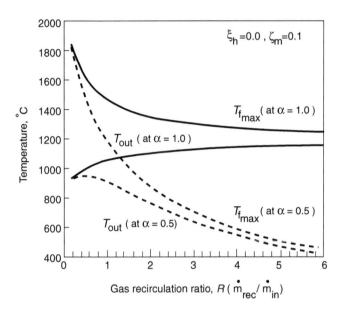

FIGURE 2.26 Combustion gas temperature in the model furnace affected by gas recirculation and its heat loss.

Figure 2.26 shows the change of these temperatures plotted as a function of the recirculation ratio R in the range of 0.0 to 6.0, and when the heat loss coefficient, α, is changed in two conditions, 0.5 and 1.0. Thermal efficiency was also calculated as shown in Figure 2.27 with the same condition as the case of Figure 2.26. A large amount of recirculation gas does not affect the thermal efficiency so strongly if the heat loss from the recirculating flue gas is relatively small, i.e., less than around 10% loss corresponding to α larger than 0.9.

Heat recirculation by means of preheating the combustion air or fuel results directly in increasing the maximum flame temperature. Figure 2.28 describes the effect of the temperature increase and existing restrictions due to the temperature limit of furnace materials and combustion stability. Available conditions in actual furnace operation may hence be restricted between the "over temperature zone" and "blow-off zone." If we take a specific temperature to distinguish those two zones, for example, 2150°C for the maximum and 1350°C for the minimum, heat recirculation brings about an expansion of the available operating range when shown as a function of gas recirculation ratio R.

2.2.4.3.2 Heat and Gas Recirculation

Equations 2.14 and 2.15 give the maximum flame temperature when the temperature at the heat exchanger exit, T_{in1}, and gas recirculation ratio, R, are given. The maximum flame temperature, T_{fmax}, was calculated and plotted as a function of T_{in1} and the gas recirculation ratio R with the condition of ξ_h less than 0.9 indicated as a heat exchanger limit. The contour lines of T_{fmax} from 1350 to 2150°C, divided into eight zones, is shown in Figure 2.29. It clearly demonstrates that increasing the gas recirculation ratio without any change of T_{fmax} is possible if T_{in1} can be sufficiently controlled.

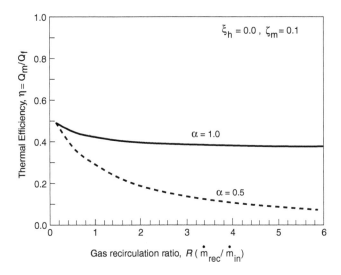

FIGURE 2.27 Thermal efficiency of the model furnace affected by gas recirculation and its heat loss.

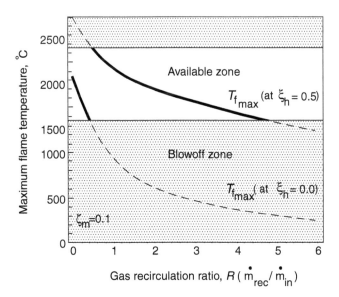

FIGURE 2.28 Maximum combustion gas temperature of the model furnace affected by gas recirculation and heat recirculation.

In addition to this mapping, $\Delta T (= T_{fmax} - T_{out})$ was also calculated and plotted as a function of T_{in1} and gas recirculation ratio, R, with the condition of ξ_h less than 0.9. The contour lines of ΔT from 100 to 1000°C divided into nine zones are shown in Figure 2.30. This mapping implies there is an area where the temperature difference

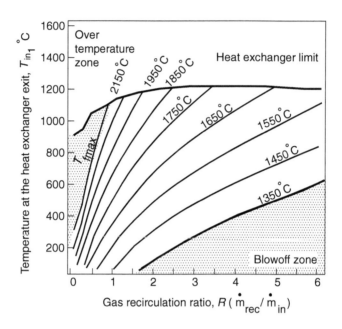

FIGURE 2.29 Contour map of the maximum combustion gas temperature of the model furnace affected by gas recirculation and heat recirculation.

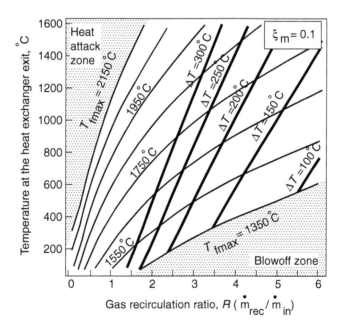

FIGURE 2.30 Contour map of the combustion gas temperature difference affected by gas recirculation and heat recirculation.

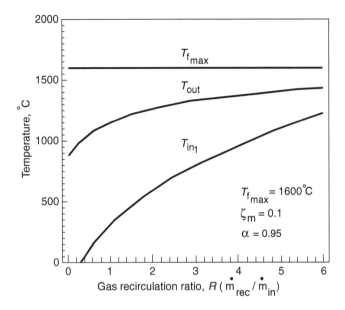

FIGURE 2.31 Superimposed contour map of the maximum combustion gas temperature and the temperature affected by gas recirculation and heat recirculation.

is remarkably small compared with conventional furnaces. Both T_{fmax} and ΔT distributions were then superimposed in Figure 2.31. Here the contour lines are shown of ΔT for every 50°C in the range from 100 to 300°C. The results show that an increase of both heat recirculation by air preheat and gas recirculation results in decrease of flame temperature and also decrease of the temperature difference between the flame and the flame gas.

2.2.4.3.3 Thermal Efficiency

The operation to increase heat and gas recirculation while maintaining the maximum flame temperature can be performed as shown in Figure 2.32, where T_{fmax}, T_{out} and T_{in1} are indicated in Figure 2.31. T_{in1} has to be increased to compensate the T_{fmax} decay resulting from the effect of dilution due to the combustion gas recirculation. The temperature difference, ΔT, becomes noticeably smaller and T_{out} becomes gradually closer to T_{fmax}. The thermal efficiency in this condition was calculated and the result is shown in Figure 2.33. It revealed that this operation can increase the thermal efficiency as well as improving the temperature difference, ΔT.

2.2.4.4 Discussion

The method of additional enthalpy combustion has been generally interpreted to be an effective way to enhance the thermal field in furnaces for increasing the heat transfer rate. The excessive increase of the maximum flame temperature, however, is a major problem causing overheating of materials in the furnaces. The result obtained here, as shown in Figure 2.33, revealed that high temperature additional

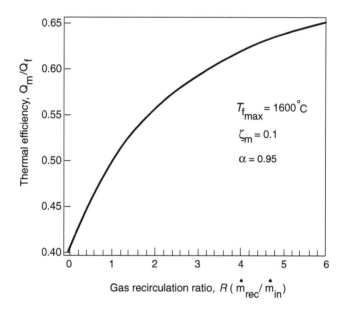

FIGURE 2.32 Combustion gas temperature in the model furnace affected by gas recirculation when the maximum temperature is kept constant.

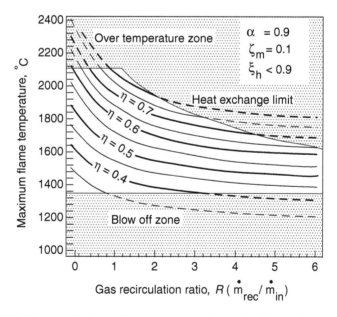

FIGURE 2.33 Thermal efficiency affected by gas and heat recirculation when the maximum temperature is kept constant.

enthalpy combustion can be effective under the condition of lowering the adiabatic flame temperature by gas recirculation. This improved heating method of combining high temperature additional enthalpy combustion with gas recirculation makes it possible to achieve a uniform temperature profile in furnaces while holding the maximum flame temperature constant.

A profile of thermal efficiency affected by gas recirculation ratio R and maximum flame temperature T_{fmax} was calculated to determine the optimized operating condition between thermal efficiency and NO_x emission. NO_x emission is in principle supposed to have a strong relationship with the peak flame temperature. A small amount of cold gas recirculation is actually one way to achieve low NO_x level even though it frequently causes unstable combustion, which results from a flame temperature decrease. It is assumed for this discussion that NO_x level is determined mainly by the maximum flame temperature.

The contour profile of the thermal efficiency obtained is shown in Figure 2.33 based on the assumption that the maximum heat exchange coefficient of inlet air and fuel, ξ_h, which was defined in Equation 2.2, can be less than 0.9. The lines of constant thermal efficiency in Figure 2.33 demonstrate that there is a trade-off point in terms of thermal efficiency maximization and NO_x emission minimization on the line of the heat exchanger limit line corresponding to $\xi_h = 0.9$. The lowest T_{fmax} is found on that heat exchanger limit line for a given thermal efficiency. For example, when a design thermal efficiency was fixed at 0.7, the lowest maximum flame temperature, 1730°C, was found on the boundary line of the heat exchanger limit with gas recirculation $R = 4.5$. If flexibility in operating temperatures is more important than minimizing NO_x emissions, then it is possible to maintain thermal efficiency by varying the gas recirculation ratio while changing the operating temperatures. Using the previous thermal efficiency as an example, 0.7 can be attained with the maximum flame temperature in the range from 1730 to 2150°C by varying the gas recirculation ratio. If the maximum flame temperature was fixed at 1800°C, it could be attained with any thermal efficiency in the range from 0.45 to 0.75 by controlling the gas recirculation ratio. Thus, maximum thermal efficiency can be obtained while minimizing the NO_x emission level or constant flame temperature can be maintained. The same profile was plotted with a three-dimensional plot in Figure 2.34 as a reference.

High temperature additional enthalpy combustion, combined with high temperature combustion gas recirculation taking place in the above-mentioned condition, is unique compared with normal combustion because the temperature increase has to be small due to the heat capacity increase. This unique combustion, recognized currently as high temperature air combustion, has completely different characteristics from a conventional operating method concerning flammability limits, combustion stability, combustion noise, and ignition process beyond the autoignition temperature.

This improved heating method using high temperature air combustion can provide several advantages such as control of the maximum flame temperature in responding to the properties of heated materials within the available zone and in principle very low NO_x levels.

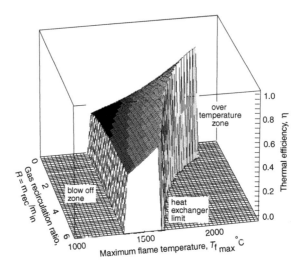

FIGURE 2.34 Thermal efficiency affected by gas recirculation and the maximum combustion gas temperature.

2.2.4.5 Summary

1. An improved heating method combining high temperature additional enthalpy combustion with gas recirculation was demonstrated, making it possible to have a uniform temperature profile in furnaces and to increase the average temperature.
2. A one-dimensional plug-flow type of furnace model was studied and the level of the heat loss from recirculating gas was proved to strongly affect the gas temperature in the furnace.
3. The operation increasing both gas recirculation ratio and heat recirculation can result in the reduction of both the maximum temperature and temperature gradients in the furnace. It can be expected this could, in principle, bring about NO_x reduction.
4. Thermal efficiency was improved by the operation of increasing gas recirculation and heat recirculation while maintaining a constant maximum flame temperature. It was found that an optimum point exists in terms of satisfying thermal efficiency improvement and NO_x abatement.

2.3 FUNDAMENTALS OF GASEOUS FUEL FLAMES

2.3.1 Extinction Limit and No_x in Laminar Diffusion Flame

It was commonly believed that a diffusion flame in a high temperature airflow generates a larger amount of NO_x than in room temperature air because of the higher flame temperature. For this reason, NO_x reduction has been pursued by lowering both the flame temperature and the oxygen concentration through exhaust gas recirculation. Recently, in the furnace industry, a combustion method using high temperature air has been used whereby air is preheated utilizing the sensible heat of high temperature exhaust gas.[5] The flames are stable in high temperature air, and the recirculation of the combustion gas lowers oxygen concentration, which is a principal agent in production of NO_x. Thus, it is possible to achieve both energy savings and NO_x reduction simultaneously. The high temperature air combustion brings about these advantages.[6-10]

Sobiesiak et al. constructed a new type of burner, in which complex mixing is done nonadiabatically, and realized low-pollution combustion with unusually low temperature flames using highly preheated air.[6] Some numerical simulations[11,12] and experimental research[13] have been reported on the flames in diesel engines where fuel is injected into high temperature and high pressure air. The fundamentals of the highly preheated air combustion have not been made clear. Therefore, we looked into the flame extinction/stabilization mechanism and the NO_x formation mechanism under the influences of the strain rate of the flow field of nonpremixed flames, using high temperature air.

2.3.1.1 Experimental Apparatus

Figure 2.35 shows the experimental setup. A rectangular combustion chamber with a sectional area of 3600 mm^2 (30 × 120 mm) was used for the experiments. A porous cylinder (diameter of 16 mm and a length of 30 mm) was installed at the center of the combustion chamber and the fuel was injected through it. The air was heated by 16 ceramic honeycomb elements (heat reservoirs), each 100 × 100 × 100 mm large, and they were heated to a prescribed temperature by premixed burners inserted upstream of the regenerator. The material of the heat reservoirs was aluminum titanate. The air was heated to a prescribed temperature through this heat exchanger with the heat reservoirs.

Downstream of the regenerator, comprising the heat reservoirs, the heated air was accelerated by a converging nozzle and then is supplied to the combustion chamber located vertically. The inner surfaces of the converging nozzle and a settling chamber were coated with ceramic fiber to avoid heat loss. There were ceramic balls and damping screens of fine mesh inside the settling chamber for eliminating large-scale turbulence. Therefore, the airflow entering the combustion chamber had a uniform velocity distribution and turbulence was negligible.

Figure 2.36 shows time histories of the temperature upstream of the temperature at the nozzle exit from the beginning to the saturated condition. The temperature downstream of the regenerator was set at 1490 K by regulating the fuel flow and equivalence ratio of the premixed burners. Whereas the temperature downstream of the regenerator quickly rose to the set value, the temperature at the nozzle exit rose slowly and saturated at 1323 K. When the airflow velocity at room temperature was 4.06 m/s, the retention time of the experiment was about 15 min. The temperature difference between the two curves is due to heat loss in the settling chamber and the nozzle.

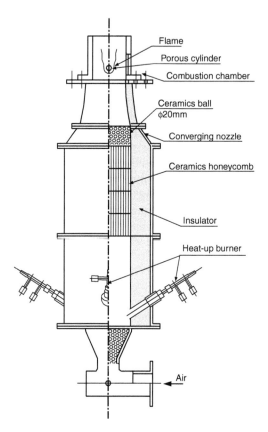

FIGURE 2.35 Experimental setup.

2.3.1.2 Velocity Field and Temperature Field

Supposing that a flow field is described by the potential theory, the strain rate in the stagnation region of the cylinder becomes $2V/R$, where V is the velocity of the uniform flow upstream of the cylinder and R is the radius of the porous cylinder. Figure 2.37 shows the velocity distribution and the temperature distribution over the nozzle exit when the combustion chamber is removed. It is noted that both distributions are almost uniform except for the boundary layers on the side walls.

The critical strain rate (velocity gradient) at flame extinction was defined as the strain rate when the flame extinguishes in the forward stagnation region and is suddenly converted into the wake flame. When the flame temperature was increased, the flame in the wake returned to the forward stagnation region, forming the envelope flame. Propane was used in these experiment.

The stagnation velocity gradient has a dimension of the so-called reaction frequency (time^{-1}) and, therefore, the critical stagnation velocity gradient can be used as a measure of overall reaction rate, which shows the flame intensity for each combination of reactants.

Combustion Phenomena of High Temperature Air Combustion

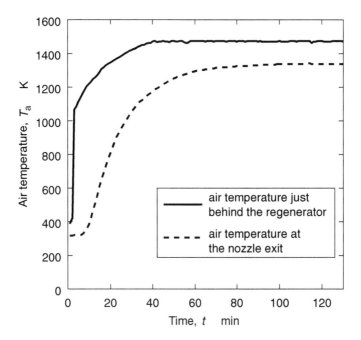

FIGURE 2.36 Time history of air temperatures just behind the regenerator and at the nozzle exit.

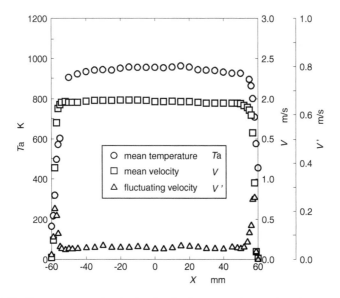

FIGURE 2.37 Temperature and velocity distributions over the nozzle exit.

2.3.1.3 Extinction and Re-ignition Temperatures of Laminar Diffusion Flame

Figure 2.38 shows the extinction limit of laminar flames expressed as a function of air temperature, T_a, where the parameter is the dilution amount of air by nitrogen. The larger the N_2 dilution amount, the lower the extinction strain rate becomes, which increases along with the air-preheating temperature. As the air-preheating temperature rises, the flame intensity increases. Whereas the critical strain rate at flame extinction, $(2V/R)_c$, shows mostly linear increase in the range below the spontaneous ignition temperature, it increases exponentially in a temperature range above the spontaneous ignition temperature. The velocity gradient at flame extinction becomes infinite at 1320 K and flames become stable above this temperature. This preheated air temperature is lower than the flame temperature at extinction by only 150 K. It is understood, accordingly, that the chemical reaction rate is faster than the diffusion rate between fuel and air.

Figure 2.39 shows the flame temperatures at extinction and re-ignition plotted against the flame strain rate. Flame strain rate increases and the flame temperature at flame extinction rises slightly as the airflow velocity increases. Transport of the reactants is accelerated as the flame strain rate increases and, consequently, the chemical reaction intensifies further as the reactants are consumed. This means that the flame temperature, T_f, rises to maintain the flames. The change from a wake flame to an envelope-type flame causes the flame temperature to rise to about 200 K above the extinction temperature. For this reason, the extinction/re-ignition phenomenon shows a hysteresis. See Figure 2.39.

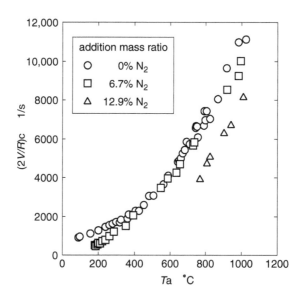

FIGURE 2.38 Extinction limits.

Combustion Phenomena of High Temperature Air Combustion

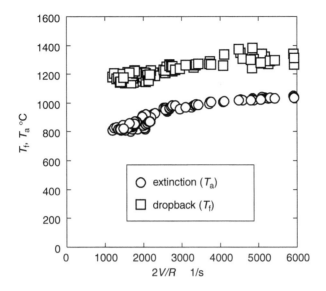

FIGURE 2.39 Extinction and drop back temperatures (hysteresis).

Figure 2.40 shows the relationship between NO_x concentration and flame temperature T_f at extinction when the flame strain rate is changed while the preheating temperature is kept constant. As the strain rate increases, the extinction flame temperature falls and its gradient becomes steeper. The lower the flame temperature

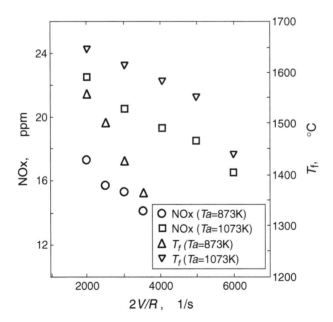

FIGURE 2.40 Dependence of flame temperature T_f and NO_x concentration on the strain rate $2V/R$.

becomes, the lower the NO_x concentration, and, hence, under a constant air-preheating temperature, incomplete combustion takes place and NO_x emission decreases as the flame strain rate rises. The flame temperature-dependency of NO_x suggests that the Zel'dovich mechanism is predominant.

Figure 2.41 shows changes of NO_x concentration and flame temperature T_f with the amount of the N_2 dilution (mass %). As more nitrogen is added, the flame temperature and the NO_x concentration decrease. NO_x concentration is decreased to 10 ppm when air is diluted with nitrogen by 30%. Because the air and nitrogen are heated considerably and the flame temperature is consequently high, flames can be maintained despite the air dilution with nitrogen; in other words, the flames are maintained even at a low oxygen concentration. Hence, this results in the reduction of NO_x concentration, presumably.

2.3.1.4 Distributions of Temperature and Concentrations of Species

Figure 2.42 shows the distributions of temperature and concentrations of NO, NO_2, NO_x (= NO + NO_2), O_2, CO and CO_2 along the stagnation streamline under the condition of the air preheat temperature of 1100 K and a flame strain rate of 3000 s^{-1}. Since diffusion flames are very stable under this condition, it was used as the reference condition in this study. Figure 2.43 shows also the distributions of flame temperature and the concentrations of the above-mentioned species near the flame extinction under the condition of the air preheat temperature of 1100 K and the flame strain rate of 6000 s^{-1}. It has to be noted that, here, the scale of abscissa is different from Figure 2.42 while the scales of ordinates are the same. This is because the distance between

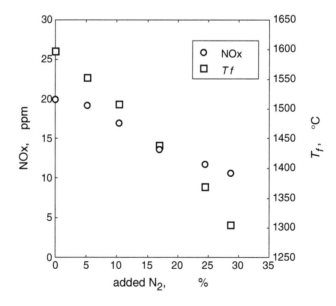

FIGURE 2.41 Effect of dilution of N_2 on the NO_x concentration and flame temperature.

Combustion Phenomena of High Temperature Air Combustion

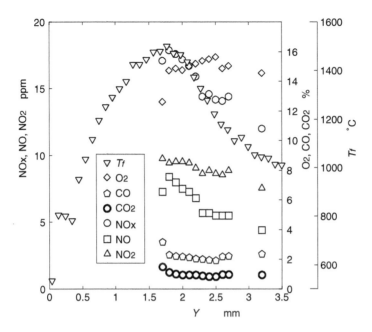

FIGURE 2.42 Distributions of flame temperature and species concentration ($T_a = 1100$ K, $2V/R = 3000$ s^{-1}).

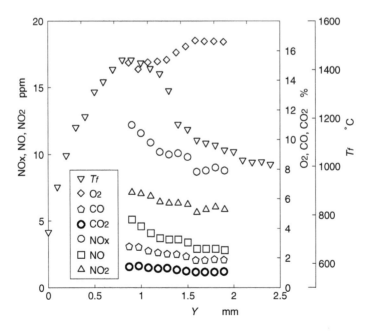

FIGURE 2.43 Distributions of flame temperature and species concentration ($T_a = 1100$ K, $2V/R = 6000$ s^{-1}).

the flame and the porous cylinder surface is narrower under the latter condition with the higher flame strain rate than under the reference condition.

As seen from the temperature distributions in the two cases, the thermal diffusion layer of the latter case is thinner than the reference case. The concentration measurements were made on the air side of the flame because the measurements in the fuel side of the flame were impossible because the diameter of the water-cooled probe was too large. Note that the flame temperature is lower in the latter case than in the reference case. NO_x is produced mainly in the flame zone in the counterflow diffusion flame. It is suspected that the NO_x formation would be much larger in practical burners because the residence time would be far longer. The higher the air-preheat temperature becomes, the larger the maximum NO_x concentration in the flames, because thermal NO_x produced through the Zel'dovich mechanism increases.

Comparing the two graphs, note that NO_x concentration in the flame zone is higher in the reference case than with the larger flame strain rate and that both concentrations of NO and NO_2 are lower in the latter case. NO_2 concentration is higher than that of NO in both cases. There is a possibility that NO was converted to NO_2 in the probe because, in the present experiments, sample gas was aspirated by the water-cooled probe made of stainless steel. The concentration of NO_x (= NO + NO_2), however, is not influenced by this conversion.

Concentrations of CO and CO_2 are far higher than the equilibrium values. This indicates that there is a high possibility of incomplete combustion taking place due to a large strain rate and a short residence time. A comparison of Figure 2.42 and Figure 2.43 shows that the concentrations in Figure 2.43 are higher than the corresponding values in Figure 2.42, due to the influence of the shorter residence time.

2.3.1.5 Effect of Flame Temperature on NO_x Formation

Figure 2.44 shows the concentration of NO_x (= NO + NO_2) expressed as a function of flame temperature in the reference case and with the larger flame strain rate. The flame temperature was measured with a thermocouple. The measurement procedure was as follows: first, temperature distribution was measured along a stagnation streamline; then NO_x concentration corresponding to the flame temperature T_f was measured. The results were plotted on the flame temperature vs. NO_x concentration plane for two different values of strain rate ($2V/R$) of 3000 and 6000 s^{-1} as the parameter. It is interesting that these two curves are similar to each other. An important point is that the NO_x concentration decreases as the flame strain rate becomes larger. This observation result is in conformity with the result of the numerical calculation made by Drake and Blint.[14] As stated above, the residence time becomes shorter as the flame strain rate increases, and as a result incomplete combustion takes place and the amount of NO_x formed in the flames and the diffusion layer becomes smaller. It is suspected, however, that NO_x formation is accelerated by the Zel'dovich mechanism when the combustion reaction proceeds to the equilibrium state in the wake.

NO_x concentration is almost constant in this temperature range from 1300 to 1625 K, which means that NO_x concentration does not depend on the flame temperature. This weak flame temperature dependency is inherent in the prompt NO

Combustion Phenomena of High Temperature Air Combustion

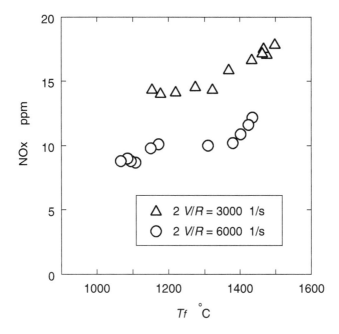

FIGURE 2.44 Relation between NO_x concentration and flame temperature.

mechanism; therefore, NO_x generated from the prompt NO mechanism is predominant in this low temperature range. NO_x concentration increases as the flame temperature increases in a temperature range higher than 1625 K, which is very close to the lowest temperature for maintaining flames where the surrounding is at room temperature. This indicates that the Zel'dovich thermal mechanism, which is strongly temperature dependent, becomes predominant in this temperature range.

2.3.1.6 Relationship between Flame Temperature and the Critical Velocity Gradient

As seen in Figure 2.38, above the spontaneous ignition temperature, the flame extinction limit increases exponentially with respect to the air-preheat temperature. This suggests that the overall chemical reaction rate can be expressed in the form of an exponential function of the flame temperature T_f. Flame extinction occurs when the chemical reaction rate becomes slower than the rate of reactant transport by diffusion. In the case of the porous cylindrical burner, the transport rate of reactants by diffusion is estimated by velocity gradient, $2V/R$. Accordingly, the axis of ordinate $(2V/R)_c$ is the critical velocity gradient and at the same time expresses the overall chemical reaction rate. To express the overall chemical reaction rate of propane in the form of the Arrhenius plot, it is necessary to know the flame temperature. In the present study, the flame temperature was assumed to be an adiabatic flame temperature, which is appropriate because the flame zone is so thin that it can be regarded as a flame sheet without thickness. The adiabatic flame temperature rises only by about 300 K if the preheated air temperature is raised from 300 to 1000 K. In the high temperature range, endothermic dissociation reactions become predominant. Consequently, the adiabatic flame temperature is not much affected by a rise in the preheated air temperature. For the Arrhenius plot, it is necessary to express the chemical reaction rate in the form of a function of flame temperature. As stated before, the critical velocity gradient can be used as a measure of the overall chemical reaction rate of propane.

Figure 2.45 is an Arrhenius plot of the data obtained in the present study. The axis of ordinate $\ln(2V/R)_c$ is a measure of the overall chemical reaction rate and the axis of abscissa is the reciprocal of adiabatic flame temperature. The highly linear gradient shows that the overall chemical reaction rate of propane can be expressed in the form of the Arrhenius plot. From this gradient, activation energy is calculated to be 380 kJ/kmol. It is understood from this that the oxidizing reaction of propane is strongly dependent on temperature and that NO_x production by the Zel'dovich mechanism is predominant.

2.3.1.7 Summary

In this study high-temperature air as combustion air is used for the purpose of investigations on extinction limit and NO_x formation in laminar diffusion flames. As a result, the mechanism of NO_x formation and NO_x reduction under the conditions of high temperature air combustion has been clarified.

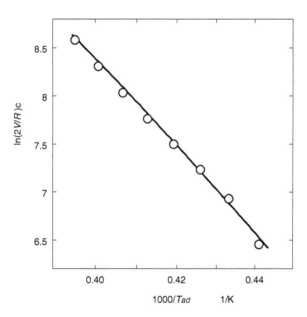

FIGURE 2.45 Arrhenius plot of data.

1. Flame extinction limit expressed by flame strain rate increases exponentially as the air preheat temperature becomes higher. But there is a limit to the preheated air temperature and, above the limit, flame extinction does not take place. This critical air preheat temperature is about 1320 K for propane.
2. Flame temperature becomes the adiabatic flame temperature when the air is preheated. This leads to the capability of the lean combustion even below the lower flammability limit. Thus, NO_x reduction is made possible using the high-temperature low-oxygen air as the combustion air.
3. Under the condition of constant air preheat temperature, both NO_x concentration and flame temperature decrease with an increase of the flame strain rate.
4. When the flame strain rate increases, the diffusion layer becomes thinner. Flame temperature decreases with the increase of the strain rate.
5. NO_x emissions become nearly constant in a flame temperature range from 1300 to 1625 K. This weak temperature dependency is inherent in the prompt NO mechanism. Accordingly, in the low temperature range the prompt NO becomes predominant.
6. The overall chemical reaction rate in the high temperature air combustion can be expressed as the Arrhenius type. The activation energy for propane in the high temperature air combustion is estimated as 380 kJ/kmol.

2.3.2 BURNING VELOCITY

Flames obtained by feeding fuel into air preheated to about 1000°C by recirculation of hot combustion gases are known not only to be efficient but also to show low NO emission.[15-17] Large fuel load, the amount of fuel burned per unit of volume and time, combined with simultaneous low emission of NO, is a desirable property of a flame in industrial devices. A high performance combustion process is required, especially in cases where the composition of the combustion gases is restricted to certain values by the properties of the material to be heated such that the permissible range of equivalence ratios is narrow.

Here, flat flames of methane with preheated and diluted air are treated theoretically to elucidate the mechanism by which preheated and diluted air leads to high efficiency and low NO emission flames. Since the calculations are done on flat flames, the fuel load is substituted with the fuel flux, the amount of fuel burned per unit time per unit area of the flame.

2.3.2.1 Simulation Model

In a flat flame, the fuel is not injected into a flow of hot air, but all the mixture is preheated and diluted. Dilution and preheating are obtained by assuming recirculation of products of complete combustion into the unburned fuel mixture, and then this is set to an initial high temperature at the burner.

The reaction scheme is that of Miller and Bowman,[18] with 226 elementary reactions occurring among 50 species. The governing equations for the flat flame are the following:

$$\frac{\partial \rho}{\partial t} + \frac{\partial}{\partial x}(\rho u) = 0 \qquad (2.16)$$

$$\rho c_p \left(\frac{\partial T}{\partial t} + u \frac{\partial T}{\partial x} \right) = \frac{\partial}{\partial x}\left(\lambda \frac{\partial T}{\partial x} \right) + \sum_i \phi_i h_i \qquad (2.17)$$

$$\rho \left(\frac{\partial \omega}{\partial t} + u \frac{\partial \omega}{\partial x} \right) = \frac{\partial}{\partial x}\left(\rho D \frac{\partial T}{\partial x} \right) + \phi_i \quad i = 1\ldots n \qquad (2.18)$$

$$p = \rho RT \sum_i \frac{\omega_i}{m_i} \qquad (2.19)$$

where x is the position, u the flow velocity, λ the thermal conductivity, w_i the mass fraction, ϕ_i the production rate, h_i the enthalpy, and m_i the molecular weight of the ith species. The above equations are solved with the CHEMKIN package.[19,20]

The dilution of the mixtures was simulated by assuming that combustion gases in the ratio composition $CO_2 : H_2O : N_2 = 0.0948 : 0.1896 : 0.7156$, the same as that of the gases after a complete combustion of a methane/air mixture of equivalence ratio = 1, are added to air. The dilution ratio, α, is expressed by the mole fraction of O_2 in diluted air. When not specified all flames investigated here have an equivalence ratio of unity.

2.3.2.2 Simulation Results and Discussion

2.3.2.2.1 Preheated but Not Diluted Premixed Flames

Figure 2.46 shows the variation of burning velocity with preheating temperature of flames with $\alpha = 0.2095$ (not diluted). It can be seen that the burning velocity increases abruptly for preheating temperatures T_0 higher than 1650 K. This is due to a change in the flame propagation mechanism that occurs around that temperature, reported previously.[21]

Together with the burning velocity,[22] the ignition delay time of a well-stirred mixture with the preheating temperature is also shown in the figure. The ignition delay time is defined here as the interval between $t = 0$ and the time at which dT/dt becomes maximum. The ignition delay times for $T_0 = 1600$ K and $T_0 = 1800$ K, for example, are 403 and 93.2 µs, respectively. In the flat flame, the time interval spent by the gases before reaching the flame front derived from the flow velocity and the distance from the burner to the flame front (defined as the position at which dT/dx is maximum) are 210 and 78 µs for $T_0 = 1600$ and 1800 K, respectively. For $T_0 = 1800$ K, the time interval in the flat flame is similar to the ignition delay time in the

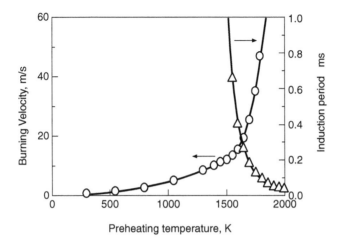

FIGURE 2.46 Burning velocity and induction period of preheated mixtures with $\alpha = 0.2095$.

well-stirred reactor. This is because at 1800 K reactions begin immediately after the mixture leaves the burner, while at 1600 K the position of the flat flame is more dependent on the diffusion of active species and the conduction of thermal energy to the unburned region. Especially the premature reactions at 1800 K, are fostered by considerable amounts of CH_3 and CHO formed at the moment the mixture leaves the burner. This means that the mixture can self-ignite and is the reason the burning velocity increases steeply from T_0 around 1650 K. Consequently the fuel flux also increases: the fuel flux for the flat flame at $T_0 = 1800$ K is about 17.5 times larger than that at $T_0 = 298$ K.

The imposition of preheating temperatures as high as 1800 K to mixtures is impracticable but theoretically it would be possible to obtain stable flat flames with burning velocities approaching 50 m/s, as shown in Figure 2.46. However, the NO levels also increase with T_0, as shown in Figure 2.47. Since NO forms continuously after the flame front in the adiabatic model, its concentration increases monotonically with distance. In a practical device, the temperature decreases after the flame front and the NO would have a maximum. In a practical device, the temperature decreases after the flame front and NO would have a maxima at a certain distance from the flame front. Considering this fact, the calculated values at a location of 10 mm from the flame front have been plotted as a reference NO concentration. The NO mole fraction exceeds 2300 ppm for $T_0 = 1600$ K.

2.3.2.2.2 Preheated and Diluted Premixed Flames

As seen above, preheating increases the burning velocity and consequently the fuel flux, but large amounts of NO are formed. An ideal combustion process would be obtained if the NO levels were lowered with the burning velocity kept large. This can be reached by controlling the chemical reactions, which are functions of temperature and the concentration of reactants. Here, to isolate the effects of temperature and concentration on the chemical reactions, the adiabatic flame temperature T_a was set to the fixed values of 1600, 1800, 2000, 2200, and 2400 K. Dilution ratio was

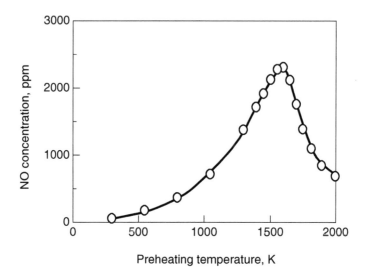

FIGURE 2.47 NO concentration in preheated flames with $\alpha = 0.2095$.

varied to obtain flames with those values for the adiabatic flame temperature and $T_0 > 298$ K. For example, for $T_a = 1600$ K, $\alpha \leq 0.13$; for $T_a = 1800$ K, $\alpha \leq 0.15$.

2.3.2.2.3 Fuel Flux

Figure 2.48 shows the variation of burning velocity with α and air preheat temperature for $T_a < 2200$ K. The burning velocity for $\alpha = 0.15$ is 1/9 of the value for $\alpha = 0.06$. From the figure, self-ignition can be thought to occur for conditions where $\alpha \leq 0.09$.

The variation of the fuel flux is similar to that of the burning velocity: from 5.42×10^{-4} mol/cm²·s for $\alpha = 0.06$ to 2.93×10^{-4} mol/cm²·s for $\alpha = 0.15$. The fuel flux is the product of the concentration of methane in the unburned mixture and the

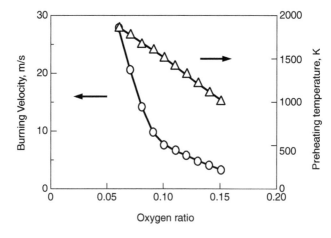

FIGURE 2.48 Burning velocity and preheating temperature of flames with $T_a = 2200$ K.

burning velocity. This product is equivalent to the reaction rate integrated through the reaction zone. According to the scheme employed here, all methane is converted first into CH_3 in the course of the oxidation process. Then the step $CH_4 \to CH_3$ is chosen because its rate gives the overall rate of the methane combustion in the investigated flames. In the Miller and Bowman scheme, the elementary reactions contributing to this step are

$$CH_4 + M \to CH_3 + H + M \qquad (R23)$$

$$CH_4 + O_2 \to CH_3 + HO_2 \qquad (R24)$$

$$CH_4 + H \to CH_3 + H_2 \qquad (R25)$$

$$CH_4 + OH \to CH_3 + H_2O \qquad (R26)$$

$$CH_4 + O \to CH_3 + OH \qquad (R27)$$

From these, R23 and R24 progress with small rates so that the other three define the total CH_4 consumption rate. The rate of each of the main reactions contributing to the $CH_4 \to CH_3$ step and the total CH_4 consumption rate for $\alpha = 0.06$ and 0.15 are plotted in Figure 2.49. The peak consumption rate of CH_4 at $\alpha = 0.06$ is 0.006 mol/cm^3·s, about 1/3 of the value at $\alpha = 0.15$. This is due to the dilution effect, despite a temperature at the peak consumption rate in the $\alpha = 0.06$ condition that is 340 K higher than that in the $\alpha = 0.15$ flame. Because the fuel flux at $\alpha = 0.06$ is about twice that at $\alpha = 0.15$, the reaction zone of the former flame has to be much thicker than that of the latter flame. For the purpose of comparison, the reaction zone thickness in flame under different conditions is defined here as the ratio between the difference $T_a - T_0$ and the maximum value of dT/dx. The variation of fuel flux and reaction zone thickness with α is shown in Figure 2.50; both are seen to change steeply for $\alpha < 0.09$.

The same was done for $T_a = 2000$ K flames, and the results are shown in Figure 2.51. The fuel flux shows a maximum, and the values are much smaller than those of Figure 2.50. This can be ascribed to the fact that no flames capable of self-ignition are obtained for $T_a = 2000$ K, so that preheating and diluting cannot increase the fuel flux.

2.3.2.2.4 NO Formation

The NO concentration at 10 mm after the flame front was found to be almost constant with α for $T_a \leq 2000$ K, but for $T_a = 2200$ and 2400 K, the NO level decreased as α decreased.

The rates of the main reactions producing NO and their sum are plotted in Figure 2.52. In contrast to the reactions belonging to the step $CH_4 \to CH_3$, the NO producing reactions are not inhibited by dilution of the mixture. This can be ascribed to the fact that as α decreases, the amount of N_2 in the mixture does not change and also the initial temperature increases. Many reactions related to NO formation comprise

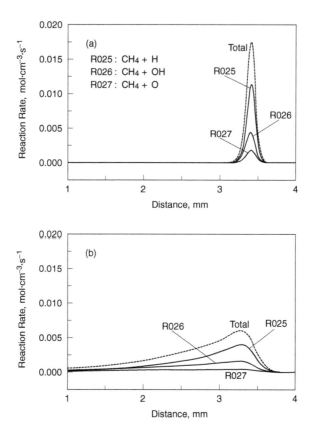

FIGURE 2.49 Rates of the CH_4 consumption in the two flames with $\alpha = 0.15$(a) and 0.06(b), and $T_a = 2200$ K.

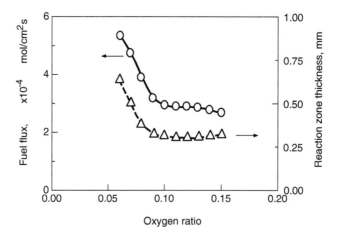

FIGURE 2.50 Fuel flux and reaction zone thickness of flames with $T_a = 2200$ K.

Combustion Phenomena of High Temperature Air Combustion

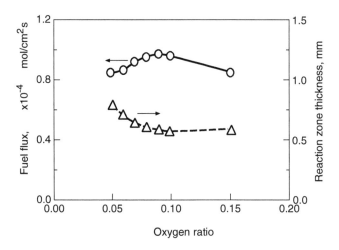

FIGURE 2.51 Fuel flux and reaction zone thickness of flames with $T_a = 2000$ K.

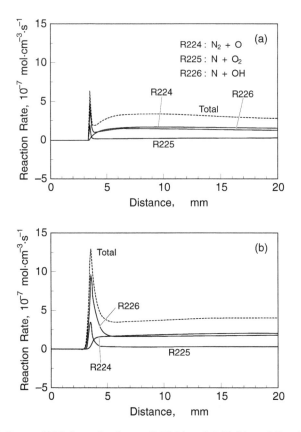

FIGURE 2.52 Rates of NO formation in $\alpha = 0.15$ (a) and 0.06 (b), and $T_a = 2200$ K flames.

the scheme employed here, but NO can be thought to be formed exclusively through the extended Zel'dovich mechanism:

$$N_2 + O \rightarrow NO + N \quad (R224)$$

$$O_2 + N \rightarrow NO + O \quad (R225)$$

$$N + OH \rightarrow NO + H \quad (R226)$$

because reactions other than these three occur with much smaller rates.

Although the NO formation rate increases as α decreases, the NO concentration at $\alpha = 0.06$ is 57.0 ppm, while at $\alpha = 0.15$ it reaches 114 ppm. The NO concentration is the product of the NO production rate by the reaction time. The reaction time is the ratio between the thickness of the NO production zone (because the concentration is that at 10 mm after the flame front, this thickness is 10 mm in all cases) and the flow velocity. Therefore, the low NO concentration in the $\alpha = 0.06$ flame is due to the large burning velocity at that condition.

The above discussion on NO formation is valid for other values of T_a. The same three reactions are the main NO-forming steps, and NO concentration is lower for smaller α. In Figure 2.53 the amount of NO emitted per unit of supplied fuel is plotted for various values of T_a and α. It can be seen that when an adiabatic flame temperature of 2200 K or higher is required, the NO emission is very sensitive to the dilution of the mixture with combustion gases, with optimum values of α being around 0.10 or lower. The same analysis was done for mixtures of equivalence ratio of 0.8 and 1.4. Fuel flux and NO emission varied with α and T_a in a similar way as the conditions of the equivalence ratio = 1.

2.3.2.3 Summary

The combustion of preheated and diluted mixtures of methane and air was simulated through a flat flame model with the following conclusions:

1. For cases where the adiabatic flame temperature is 2200 K or higher, the dilution of the mixture with combustion gases and the preheating to high initial temperatures were found to be effective in enlarging the fuel flux, the amount of fuel burned per unit area and time. However, to minimize the NO emission an optimization of the values of both the dilution ratio and preheating temperature is required, to have a large fuel flux with a low NO emission. In general, the reason preheated and diluted flames show high performance is that preheating increases the burning velocity by allowing self-ignition of the mixture. Dilution inhibits the combustion reactions but this effect is compensated by enlargement of the reaction zone thickness.
2. For cases where the adiabatic flame temperature is 2000 K or lower, the fuel flux and NO emission are virtually unaffected by preheating and dilution of the mixture. In these flames, the effect of preheating and dilution is just that of stabilizing the combustion of mixtures that could not burn stably under standard conditions.

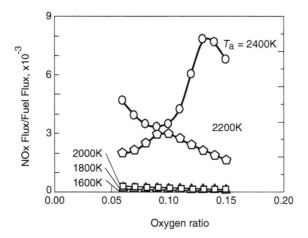

FIGURE 2.53 NO emission per unit of fuel as functions of α and T_a.

2.3.3 Mixing in Furnace

2.3.3.1 Jet Mixing

Mixing between fuel and air in the near-field of a burner or a flame stabilizer has been one of the key factors in designing burners of high performance. Coaxial diffusion flames, recirculating, and/or swirling flows are typical configurations which were used to conduct basic studies on the mixing processes occurring in burner flows, and we established large databases with many types of burners commercially available. We believed that once an ideal burner was designed, it would show the expected performance when installed in any type of furnace. When considering high temperature air combustion as the new combustion technology for industry, compatibility between high performance regenerative combustion and low nitric oxides emission is the final goal. Unfortunately, we have not established facts supported by experience or databases on combustion using highly preheated air. Observations in prototype furnaces showed that mixing processes between preheated air with flue gas or fuel in the furnace, not in the near-field of the burner, play a significant role in yielding preferable conditions for a desired combustion regime. Therefore we wished to conduct a basic experiment where we were able to obtain specific information on high temperature air combustion.

We obtained experimental results demonstrating that the emission of nitric oxides varies significantly by changing only the mixing process between fuel and air, while maintaining the same flow rate and temperature of air and fuel.[23]

Figure 2.54 shows the combustion chamber used in the experiment and the schematics of recirculating flow and the appearance of flame zone. The flow inside the chamber may seem relatively complicated, as we used an unusual flow configuration for the basic experiment. However, we should stand some distance away from the conventional burner experiments to observe the new phenomena of high temperature air combustion. The controlling factor of high temperature air combustion seems to be the mixing of preheated air with flue gas in the whole furnace; thus

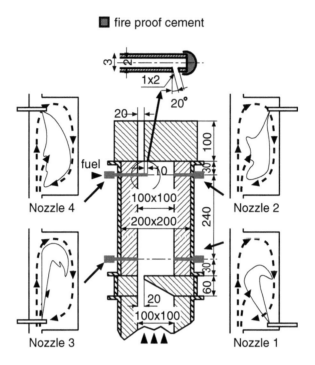

FIGURE 2.54 Combustion chamber used in the experiment.

the combustion rig was designed to demonstrate the different degrees of mixing modeling the flows in the furnace, not the flows in the near-field of the burner. The inlet of the combustion chamber of 300 mm length was once contracted and expanded again so that the flow forms a recirculating flow inside the chamber. The fuel used was natural gas and it was injected from one of four nozzles indicated in the figure. The tip of the fuel nozzle, 1.5 mm in diameter, was fixed 10 mm away from the heat insulator inside the wall for No. 1 and No. 2, and 30 mm for No. 3 and No. 4. The direction of fuel injection was 70°, upward for No. 1 and No. 3 and downward for No. 2 and No. 4, to the horizontal fuel tube. The fuel injection rate was kept constant at 0.06 ℓ_N/s, and the feed rate of air was varied between 1.5 and 4.0 ℓ_N/s at the standard condition, and the temperature between 1450 and 1100 K. The global excess air ratio (reciprocal of equivalence ratio) evaluated by the flow rates of fuel and air varied between 2.27 and 6.25.

The variation of emission index of nitric oxides, combustion air temperature and global excess air ratio is shown in Figure 2.55. A significant change in nitric oxide emission levels was observed by changing only the location of the fuel nozzle. For fuel nozzle 3 the tendency of emission index in terms of global excess air ratio was quite different from those for the cases of fuel nozzles 1 and 2. Further, for fuel nozzle 4 the emission level was extremely low for any operating conditions and seems independent of either global excess air ratio or combustion air temperature. Since no unburned hydrocarbons were detected at the exit of exhaust for all cases, the decreased nitric oxides emission should not be ascribed to imperfect combustion.

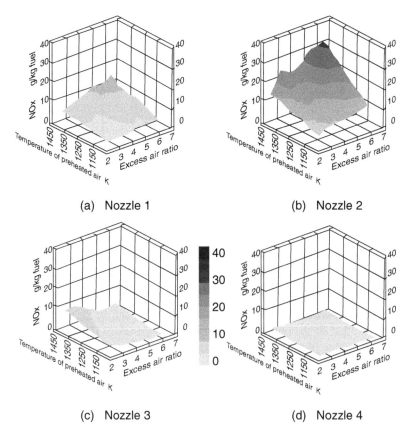

FIGURE 2.55 Influence of combustion air temperature and global excess air ratio on emission of nitric oxides.

It is thought that the flame structure or combustion regime was affected by the change of fuel nozzle location. Thus, the difference was in the mixing process of fuel in the combustor, because other experimental parameters, such as global excess air ratio and combustion air temperature, were kept the same.

A typical jet flame emitting intense radiation was observed for cases with high nitric oxides emission. The typical features of the flame for low nitric oxides were vague, with weak flames like a fog spreading everywhere, probably the "flameless oxidation" named by Wunning and Wunning.[24] Judging from the intensity of radiative emission, the maximum temperature level was relatively low because the flame never emitted bright emission. Therefore, we measured fluctuating temperatures with a fine thermocouple of 25 μm (Pt/Pt-Rh13%) coated with SiO_2, whose time constant of frequency response was electrically compensated.

The distributions of time-averaged temperature in the combustion chamber are shown in Figure 2.56. The maximum temperature appeared around the center of the recirculation zone for fuel injection nozzles 1 and 2. This fact agrees with the visual observation of flames mentioned above. The maximum temperatures were 1770 and

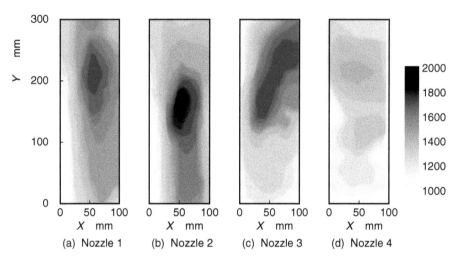

FIGURE 2.56 Distributions of time-averaged temperature (T_a = 1073 K).

1860 K for nozzles 1 and 2, respectively. Relatively high temperatures appeared in the downstream portion of the fuel jet for fuel nozzle 3, and its maximum was 1720 K. The lower half of the combustion chamber showed relatively low temperatures. In spite of the same inlet air temperature and velocity, the maximum value was only 1690 K for fuel injection nozzle 4. Intense mixing seems to yield a more flattened temperature distribution and lower maximum temperature. However, a much higher temperature would have been generated if fuel had been burned with pure air. Therefore, dilution of air with burned gases must have proceeded to some extent before combustion.

Figure 2.57a and b shows temperature fluctuations at the point of the highest time-averaged temperature. In spite of the same temperatures and flow rates of fuel and combustion air, both the time-averaged temperature distribution in the furnace and the root-mean-square of fluctuations are lower for the case of fuel nozzle 4 than for nozzle 2. The difference in emission index between the two cases is explained by cumulative frequency distribution of fluctuating temperatures shown in Figure 2.57c. The formation rate of nitric oxides becomes insignificant when the flame temperature is lower than 1800 K. The cumulative frequency corresponding to 1800 K reads almost 1.0 for nozzle 4, which means the temperature barely exceeds 1800 K. For nozzle 2, on the other hand, temperatures lower than 1800 K are 0.25, which leads to a higher emission level of nitric oxides.

As previously mentioned, intense mixing of combustion air with burned gases in the furnace, which is produced by high-momentum injection of combustion air, lowers the flame temperature significantly and yields distributed reactions as shown in the photograph on the cover of this book. This combustion regime in low concentration of oxygen can be sustained only by the supply of highly preheated air. If the temperature level of preheating is below the autoignition limit, the flame will be extinguished instantly. Therefore, in addition to the use of highly preheated air, the intense mixing of the air with sufficient burned gases before combustion is

Combustion Phenomena of High Temperature Air Combustion

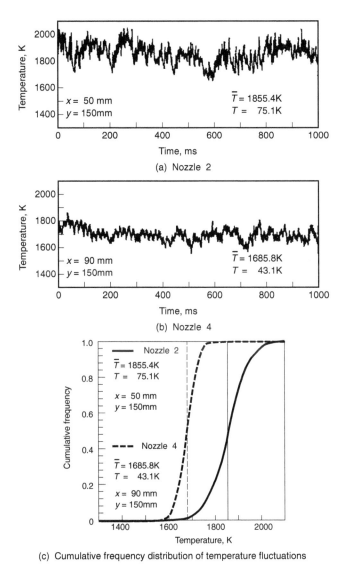

(a) Nozzle 2

(b) Nozzle 4

(c) Cumulative frequency distribution of temperature fluctuations

FIGURE 2.57 Temperature fluctuations at the maximum time-averaged temperature and their cumulative frequency distribution.

essential in advanced low NO_x technology, but is not applicable for low temperature air combustion.

2.3.3.2 Unmixedness

Local characteristics of a flame are dominated by the nature of turbulence in turbulent combustion, and so it is in furnaces because the flow regime in industrial furnaces is usually turbulent. Fluctuations in velocity, temperature, and concentration are influential factors to determine physical and chemical processes in a turbulent flame even in steady operation. Unmixedness is a degree of turbulent mixing, which tends to disappear with the dissipation of turbulence. Therefore, fluctuations in temperature indicate the existence of high and low temperature lumps of fluid passing through an observation point.

We tried to estimate flame uniformity and stability by observing the fluctuations in optical emissions from flames. Figure 2.58 shows an experimental rig and measuring devices consisting of a monochromator and a high-speed ultraviolet video camera. The emission radiated from a flame was first focused by the first lens and then collimated by the second lens prior to the light signal entering the monochromator. This makes images multiplex and avoids image distortion through the monochromator. The third lens to construct a monochromatic image of the flame, which is observed by the high-speed video camera, focuses the parallel beams emerging from the exit slit of the monochromator. The use of monochromator as a dispersion device offers significant advantages over the band-pass filter from the point of view of optical band-pass width and its easy variability.

The measured spontaneous molecular emission bands are C_2 (at 471 and 510 nm), CH (at 431 nm) and OH (at 306 nm). These chemical species are among the important radicals that provide an important role during combustion. The two C_2 bands were used to calculate the vibrational temperature as well as the distribution of temperatures of two dimensions within the flame based on a two-line method.

Figure 2.59 shows the calculated temperature profiles obtained with normal air at room temperature and highly preheated air having low oxygen concentration. The temperature profiles obtained spectroscopically indicate that the flame front exists only within the combustion reaction zone and is due to the emission radiated from C_2 molecules. A comparison between the two figures reveals that the temperature gradient becomes smaller with diluted air combustion.

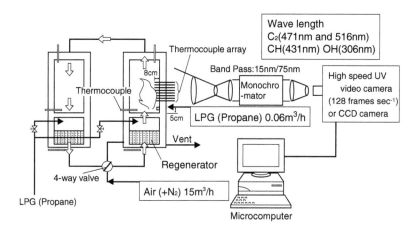

FIGURE 2.58 Regenerative combustion system with thermocouple array and high-speed spectro video camera.

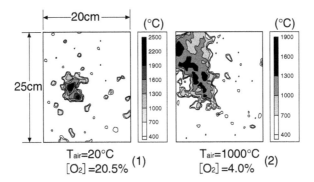

FIGURE 2.59 Profiles of C_2 vibrational temperature.

Figure 2.60 shows 2D distributions of the flame fluctuation for the three conditions. Note that high temperature air has a significant effect on decreasing the flame fluctuations in the main (middle) combustion region. The flame thermal uniformity is significantly improved with the use of low oxygen concentration and high temperature air. The use of high temperature and low oxygen concentration air results in broader spatial distributions of C_2 molecules and low peak temperatures and larger combustion volume. The lower flame temperatures are effective to decrease NO_x emission. The practical importance of the high temperature air combustion is, therefore, significant from the point of view of both thermal field uniformity and flame stability.

2.3.3.3 Well-Stirred Reactor

During the development of the advanced furnace utilizing high temperature air by regeneration, we concluded the nature of the combustion regime for high temperature air combustion. One opinion was that the regime resembles that of a well-stirred reactor (WSR), because large volumetric reactions, unlike a jet flame formed with a typical burner, dominated the furnace. In addition, the temperature distribution in the furnace became flatter and more uniform with the momentum increase of the

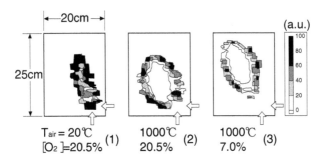

FIGURE 2.60 Profiles of flame fluctuation (4 Hz).

incoming combustion air. It is worth comparing these features with those in a stirred reactor from an engineering point of view.

The WSR is an idealized reactor that may be constructed using our concepts. It is a combustion vessel having inlet and outlet ducts, in which rapid combustion reactions proceed in the homogeneous mixture, in the molecular sense, of incoming reactants and hot burned products. The combustion efficiency of the reactor depends on its volume and the flowrate, that is, the residence time, and it can be related to chemical processes. Therefore, the temperature and composition in the reactor are theoretically uniform until its blowout limit. The actual characteristics of a real reactor, however, often vary with the nature of the flow inside and are not those of the well-stirred condition except in a small laboratory reactor. In that sense, the practical reactor should be called a jet-stirred reactor. Combustion states in a jet-stirred condition can be categorized with two parameters: i.e., a turbulent time scale, τ_m, and a chemical time scale, τ_{ch}. The ratio of those parameters is known as Damköhler number, $Da = \tau_m/\tau_{ch}$, by which we can classify ordinary burner combustion and combustion in a jet-stirred reactor. In general combustion processes the chemical time scale, τ_{ch}, can be overwhelmingly shorter than the turbulent time scale τ_m, which could be expressed as combustion with a large Damköhler number, that is, a flamelet regime. Accordingly, the idealized combustion, where rapid reactions take place in the homogeneous mixture even on the molecular level can be regarded as the low Damköhler number combustion sufficiently less than 1.0. It has been held that combustion in a jet-stirred reactor is characterized by low Damköhler number combustion, where uniformly distributed reactions, hence a uniform temperature distribution, prevail the flow in homogeneous mixture, in the scale of turbulence sense. There seem to be similar features observed in the combustion in a furnace operated with high temperature air. However, in a practical furnace, rapid mixing for producing uniform mixtures, which is caused by high momentum jet flows, shows a gap between idealized and actual combustion. The latter experiments actually revealed that decomposition and dissociation of fuel were proceeding simultaneously with combustion reactions at around the outer edge of fuel jet issued into the furnace, yielding nonuniform weak reaction zones. The high temperature air combustion seems to realize a smaller Damköhler number by decreasing the chemical time scale, τ_{ch}, with the use of low oxygen concentration air, although a turbulent time scale, τ_m, is mostly the same as usual combustion. Further, the flow inside the furnace would not be the same in a jet-stirred reactor because the furnace is far too large for the incoming jet momentum to stir uniformly. However, even though high temperature air combustion is not the same as in a jet-stirred reactor, knowledge about the similarity and difference between the two would still be useful for further improvement of high temperature air combustion.

2.3.4 Pollutant Formation

2.3.4.1 Nitric Oxides

Oxides of nitrogen, referred to as NO_x, are emitted from combustion processes, mainly as one of the pollutant species in burned products. The two major oxides of nitrogen emitted from combustion systems are nitric oxide, NO, and nitric dioxide, NO_2. NO_2 is usually lower in concentration than NO in burned products, but it is one of the most toxic gases and can be a serious health hazard. Although relatively large concentrations of NO_2 can be formed in the flame zone, most of them are converted to NO in the subsequent postflame region. Therefore NO_2 is considered to be a transient species existing at flame conditions. It has been reported that rapid mixing of hot burned gas and cold air surrounding the turbulent flame can result in a rapid quenching of the NO_2, resulting in relatively large NO_2 concentrations in the cooled downstream regions of the flame. With HiTAC in furnaces, however, local and temporal high temperatures, often seen in open flames, do not appear. A relatively uniform temperature occurs throughout the flow in the furnace, and the rapid quenching in the mixing layer with cold air does not occur. Accordingly, NO can be considered as the major species of NO_x in the exhaust gas produced by HiTAC systems.

As the major species of oxides of nitrogen, NO is considered to be formed in one of the following three ways:

1. N_2 reacts with an O-atom to form thermal NO in high temperature circumstances, such as flames.
2. Prompt NO is formed considerably fast in flame fronts through a mechanism other than the thermal NO.
3. Fuel NO is formed at comparatively low temperatures from the nitrogen released from the nitrogen-containing compounds in the fuel.

Thermal NO is the principal source of NO emission from combustion systems. The principal reactions governing the formation of thermal NO from molecular nitrogen during the combustion of lean and near stoichiometric fuel air mixtures are given by the Zel'dovich or extended Zel'dovich mechanism.

$$N_2 + O \Leftrightarrow NO + N \qquad (R1)$$

$$O_2 + N \Leftrightarrow NO + O \qquad (R2)$$

$$N + OH \Leftrightarrow NO + H \qquad (R3)$$

Reaction R3 becomes important only in near stoichiometric and rich flames where high temperature is held long enough to produce significant amounts of NO. The thermal NO formation rate is generally slower than that of combustion reactions and most of the NO is formed in the postflame region. Therefore, the NO formation process is often decoupled from the combustion processes and the NO formation rate is conveniently calculated using the Zel'dovich mechanism assuming equilibrium of the combustion reactions. Because the Zel'dovich mechanism shows strong nonlinear dependency on temperature, calculation of NO based on the time-averaged temperature is not appropriate to predict the emission level of turbulent combustion where instantaneous temperature usually fluctuates depending on the intensity of turbulence. The probability density function of temperature fluctuations, or at least its intensity, should be taken into account in the calculation of NO.

Heavy oil and coal generally contain significant amounts of organic nitrogen compounds, and the fuel nitrogen can be converted into fuel NO even at relatively low temperatures around 1000°C. The amount of fuel nitrogen converted to NO is considered high for lean combustion and low in rich combustion. Although the nitrogen content in fuels varies significantly, no nitrogen content is mostly assured when gaseous hydrocarbon fuels are used. Prompt NO is first recognized by the discrepancy between actual NO concentration in flames and the concentration predicted by the Zel'dovich mechanism even at temperatures as low as approximately 1300°C. The term prompt NO was initially used because of the rapid formation of NO in the flame, but it means NO formed in flames by mechanisms other than the Zel'dovich mechanism in a more general sense.

Prompt NO is now considered to be formed via an HCN group, that is, N atoms are formed mainly through Reactions R4 through R9. In turn, N atoms are consumed by Reactions R2 and R3, and converted to HCN through Reactions R10 and R11.[25]

$$CH + N_2 \Leftrightarrow HCN + N \tag{R4}$$

$$C + N_2 \Leftrightarrow CN + N \tag{R5}$$

$$CH_2 + N_2 \Leftrightarrow HCN + NH \tag{R6}$$

$$CH_2 + N_2 \Leftrightarrow H_2CN + N \tag{R7}$$

$$CO + N_2 \Leftrightarrow NCO + N \tag{R8}$$

$$H + N_2 \Leftrightarrow NH + N \tag{R9}$$

$$N + CH_3 \Leftrightarrow H_2CN + H \tag{R10}$$

$$H_2CN + M \Leftrightarrow HCN + H + M \tag{R11}$$

Therefore, the presence of a C atom plays an important role in the scheme. Consequently, prompt NO must be taken into account only in hydrocarbon flames. Since thermal NO predicted by the Zel'dovich mechanism exhibits strong dependency on temperature, the discrepancy between the predicted thermal NO and the actual emission level has been ascribed to the presence of prompt NO, which becomes large particularly in lean combustion or low temperature flames. Recently, it has been found that the Zel'dovich mechanism predicts too much NO also in the higher temperature range compared with the value obtained by the full kinetic scheme including the chemistry of prompt NO. The coupling of thermal and prompt mechanisms should always be important in correct prediction of NO emission.[25] It was proved that both thermal and prompt mechanisms would produce a large concentration of NO with the increase of flame temperature. Adding to these facts, we know from long-term experience that temperature is the most important factor governing the formation of NO in most practical combustion devices, and that significant reductions in NO emissions are achieved by reducing both local and overall temperature levels.

In Section 2.3.3 we showed that flame temperature in HiTAC varies depending on the mixing process in the furnace, even if it had the same air excess ratios. Further, the temperature level is not always high compared with that in ordinary combustion in spite of the use of preheated combustion air. The primary reason NO emission was reduced significantly is that the reduced flame temperature resulting from the dilution of combustion air with the burned product yielded a comparatively distributed heat release in the low oxygen concentration atmosphere. These characteristics are studied numerically on a laminar diffusion flame between fuel and air or air diluted with combustion products in counterflow configuration using a detailed database of chemistry called CHEMKIN.[26,18]

Figure 2.61 shows the variation of temperature distribution and NO concentration in flames for different preheating of combustion air. The peak temperature increases with the preheat temperature of the air and results in high concentration of NO. According to the numerical predictions done by Ju and Niioka,[25] NO is mainly formed through the thermal mechanism on the high temperature air-side of a flame and through the prompt mechanism in the narrow flame zone, and the emission index increases dramatically with preheated air temperature.[25]

The influence of oxygen concentration in nitrogen-diluted air on NO formation is demonstrated in Figure 2.62. The peak temperature in the flame and corresponding NO concentration show drastic decrease with the oxygen concentration in spite of the constant preheating temperature of diluted air at 1400 K. In the case of an oxygen concentration of 10%, the maximum flame temperature falls to almost the same level as for pure air of ambient temperature. NO concentration far below that of the non-preheated flame is predicted for an oxygen concentration of 5%. We can see that the concentration of oxygen is also a dominant factor in NO formation as far as diffusion combustion occurring in furnaces is concerned.

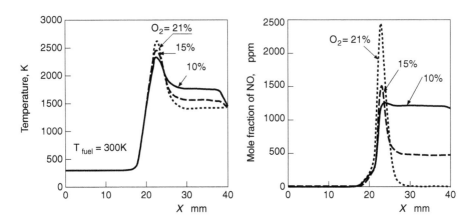

FIGURE 2.61 Effect of initial NO concentration in diluted air.

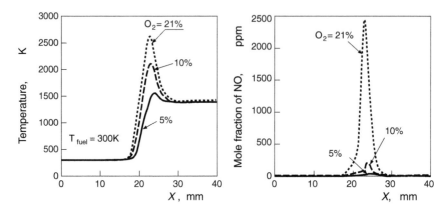

FIGURE 2.62 Distributions of temperature and NO in flat flames for various oxygen concentrations in diluted air.

Once considerable NO is formed by combustion with high temperature pure air, the resultant combustion products may contain high concentrations of NO. Therefore, combustion air diluted with burned gas may sometimes contain relatively high concentrations of NO prior to combustion. It is actually predicted that the decomposition rate of NO overcomes its formation rate for some cases. That is, when the concentration of NO in the diluted air exceeds a certain level, noticeable reduction of NO may occur in the flame as shown in Figure 2.63. In spite of the possibility of reduction of NO by hydrocarbons to a certain level, we cannot always expect further reduction of NO when its concentration becomes relatively low.

2.3.5 Pollutant Formation and Emission

Flames obtained by injection of fuel into a jet of preheated air are known to burn with high efficiency and show low NO emission.[23,27-29] In a flame obtained by

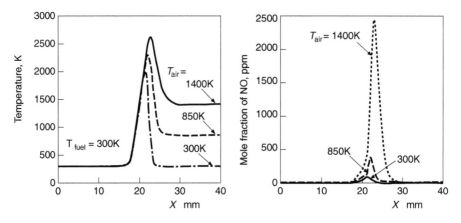

FIGURE 2.63 Distributions of temperature and NO in flat flames for various air preheat temperatures.

injection of propane into hot air, polycyclic aromatic hydrocarbons (PAH) species are formed before ignition of the mixture,[30] and there is a possibility that these species play a role in lowering the NO levels by reacting with the nitrogen oxides. Here the propane ignition in the hot air jet is simulated through a well-stirred reactor model, and the mechanism of PAH formation is analyzed. Methane, as the simplest hydrocarbon fuel, could also be considered, but propane flames are expected to form PAH species more easily than methane flames. This is because in the former flames the concentration of C_3 species, important PAH precursors, is higher.

Richter et al.[31] suggested a chemical reaction scheme for predicting PAH and fullerene formation in rich flames of aromatic hydrocarbon fuels. The thermochemical and transport parameters for each species are also provided in this valuable work. They have tested the scheme with experimental data from a benzene/oxygen/argon flat flame and have concluded that their scheme reproduced the experimental results reasonably well. Here the same scheme (slightly modified by inclusion of propane oxidation reactions) is closely applied to a well-stirred reactor situation in which rich mixtures of propane and air are preheated to a high initial temperature and allowed to react and ignite. The well-stirred reactor is an idealized simulation target, but it is very useful because its modeling does not require parameters such as diffusion coefficients, thermal conductivities, and viscosities of each species. Accurate dependencies on temperature are very difficult to obtain, especially for species as large as PAH molecules and radicals. Thus, the chemical reaction scheme and the thermochemical properties of the species included are the only data required to simulate the system. Although the one-dimensional flat flame model is more realistic, it has a greater propensity to lead to errors caused by inaccuracies in the transport parameters. The well-stirred flame model allows study of the flame chemistry, isolated from the physical phenomena involved in combustion processes.

2.3.5.1 Calculation Method

The calculations were performed with the SANDIA CHEMKIN[19] package. The chemical reaction scheme was that suggested by Richter et al.[31] with the addition of reactions of propane combustion taken from GRI-Mech.[32] The resultant combined scheme comprises 907 reactions occurring among 251 species.

2.3.5.2 Results and Discussion

Ignition of ϕ = 5 Mixture

First, a mixture of equivalence ratio ϕ equal to 5 was investigated. This is a very rich condition, rarely seen in practical devices. However, because we are interested in simulating the combustion reactions in a flow of propane injected into hot air, which forms a diffusion flame in which the equivalence ratio varies from zero to infinity depending on the position in the chamber and time, the ϕ = 5 condition chosen here is reasonable.

In addition, PAH species are known to be formed in the postflame region of a rich flame, where the local equivalence ratio would be even greater, because the concentration of oxygen there is expected to be very low.

The variation of temperature and concentration of propane, oxygen, and other main species with time for the case in which the initial temperature was set to $T_0 = 1000$ K is shown in Figure 2.64. In spite of the large value of ϕ, propane decomposes completely into H_2, CO, H_2O, C_2H_4 and C_2H_2.

The concentration of C_2H_4 was found to increase early, from $t = 0$ s, while the production of the other species became apparent much later. C_2H_2 is seen to be formed after ignition and its presence in the post-ignition period is very important for the PAH formation reactions, as will be discussed below. The concentrations of H_2, CO, H_2O and CO_2, the main combustion products, remain almost constant after 1.8 s.

The temperature shows a peak at around $t = 1.8$ s, revealing that exothermic reactions occur first and endothermic reactions occur subsequently. It also shows that combustion reactions are not sufficiently intense to provoke a "real" ignition of the mixture, as the term is usually understood. However, the temporal profiles of species concentrations and reaction rates do support the use of the term *ignition* here. The time at which the temperature peaks is then referred to as the ignition time, for convenience. The most important exothermic reactions at early stages are R564 and R617. The main reactions in the scheme are listed in Table 2.3, with their rate coefficient parameters.

One property of the flame under investigation is the absence of H, OH, and O from the scene. The most abundant of these three active species, H atoms, have a maximum of just 62 ppm; OH and O are, at most, 3 and 1 ppm, respectively. This confirms that combustion reactions do not occur intensely, as suggested by Figure 2.64.

The variation of concentration of PAH-related species with time is shown in Figures 2.65 through 2.68. A detailed analysis of the reaction rates reveals that propane is first decomposed into C_3H_7 mainly through reactions R837 and R838. A fraction of the formed C_3H_7 then breaks into C_2H_5 through reaction R847, and from C_2H_5, reactions belonging to the well-known C_2 route occur, forming C_2H_4, C_2H_3, and C_2H_2. There are no reactions in the scheme linking C_3H_7 to C_3H_6, so that the latter species is formed not through H abstraction from the former, but through reactions involving C_2 species, such as $C_2H_4 + CH_2$ (R513).

PAH-related reactions are numerous, each having its own characteristic temporal spectra, so that it is difficult to group them into temporal stages. To analyze them methodically, it is convenient to group the reactions according to the period in which they show their largest rates. The chemical reactions are described here as if they occurred in three different stages. From these, the time interval around the ignition time is very well characterized and this stage can be easily isolated from the others. The reaction rates in this first stage are large in comparison with those at later stages. The first stage lasts for about 30 ms. The main starting species for the PAH formation reactions are C_2H_2 and C_2H_3. These species, through R457 and −R482(−: reverse reaction), give rise to C_4 species such as CH_2CHCCH_2 and H_2CCCCH_3. CH_2CHCCH_2 is converted into C_4H_4 through R451 and R459, and a considerable fraction of the latter species forms H_2CCCCH through R466. H_2CCCCH, however, is oxidized by molecular oxygen into $CH_2CO + HCCO$ (R479). C_2H_2 combines also with CH_3 to give C_3H_4 (allene) through M R533.

TABLE 2.3
Main PAH (Polycyclic Aromatic Hydrocarbons) Formation Reactions. $k = AT^n \exp(-E/RT)$ in units of mol, cm, s, cal/mol, K

No.	Reaction	A	n	E
5.	$C_{10}H_7 \rightarrow$ INDENE* + CO	740E+11	0.0	43850
7.	INDENE + OH $\rightarrow$ INDENE* + H_2O	3.43E+09	1.2	–447
10.	$C_7H_7 + C_2H_2 \rightarrow$ INDENE + H	3.20E+11	0.0	7000
12.	$C_{10}H_7OH + H \rightarrow C_{10}H_7OH + H \rightarrow C_{10}H_7O + H_2$	1.15E+14	0.0	12400
13.	$C_{10}H_7OH + H \rightarrow C_{10}H_8 + OH$	2.23E+13	0.0	7929
19.	$C_{10}H_7S + C_2H_2 \rightarrow$ A2R5 + H	3.98E+13	0.0	10100
67.	$C_{10}H_7P + C_2H_2 \rightarrow$ A2YNEP + H	3.98E+13	0.0	10100
68.	A2YNEP + H $\rightarrow$ A2YNEP*S + H_2	2.50E+14	0.0	16000
70.	A2YNEP*S + $C_2H_2 \rightarrow$ A3*S1	3.98E+13	0.0	10100
82.	A3*S1 + $C_2H_2 \rightarrow$ ACEPHA + H	3.98E+13	0.0	10100
84.	$C_{10}H_7S + C_6H_6 \rightarrow$ FLTHN + H + H_2	4.00E+11	0.0	4000
95.	ACEPHA $\rightarrow$ FLTHN	8.51E+12	0.0	62860
101.	A3 + H $\rightarrow$ A3*S2 + H_2	2.50E+14	0.0	16000
102.	A3*S2 + $C_2H_2 \rightarrow$ PYRENE + H	3.98E+13	0.0	10100
106.	CPCDPYR + H $\rightarrow$ CPCDPYR*S + H_2	2.50E+14	0.0	16000
109.	CPCDPYR*S + $C_2H_2 \rightarrow$ COR + H	3.98E+13	0.0	10100
155.	FLTHN + H $\rightarrow$ FLTHN- + H_2	2.50E+14	0.0	16000
157.	BGHIF + H $\rightarrow$ BGHIF- + H_2	3.98E+13	0.0	10100
236.	BGHIF – + $C_2H_2 \rightarrow$ COR + H	3.98E+13	0.0	10100
287.	OH + $C_6H_6 \rightarrow C_6H_5OH + H$	8.94E+19	–2.0	13310
322.	$C_6H_5 + C_6H_6 \rightarrow C_{12}H_{10} + H$	1.00E+12	0.0	7000
323.	$C_{12}H_{10} + H \rightarrow C_{12}H_9 + H_2$	2.50E+14	0.0	16000
325.	$C_{12}H_9 + C_2H_2 \rightarrow$ A3 + H	3.98E+13	0.0	10100
336.	$C_{10}H_8 + H \rightarrow C_{10}H_7S + H_2$	2.50E+14	0.0	16000
337.	$C_{10}H_8 + H \rightarrow C_{10}H_7P + H_2$	2.50E+14	0.0	16000
340.	$C_8H_6 + H \rightarrow$ A1YNE* + H_2	2.50E+14	0.0	16000
343.	A1YNE* + $C_2H_2 \rightarrow C_{10}H_7S$	3.98E+13	0.0	10100
358.	$C_6H_5 + C_2H_2 \rightarrow C_8H_6 + H$	3.98E+13	0.0	10100
363.	$C_7H_7 + H \rightarrow C_7H_8$	4.81E+20	–2.1	1986
367.	$C_7H_8 + H \rightarrow C_7H_7 + H_2$	1.20E+14	0.0	8235
368.	$C_7H_8 + H \rightarrow C_6H_6 + CH_3$	1.20E+13	0.0	148
378.	$C_6H_5 + CH_3 \rightarrow C_7H_7 + H$	5.00E+13	0.0	0
380.	$C_5H_5 + C_2H_2 \rightarrow C_7H_7$	1.73E+17	–1.9	10231
382.	$C_4H_4 + C_2H_2 \rightarrow C_6H_6$	4.47E+11	0.0	30090
383.	$C_6H_6 + H \rightarrow C_6H_5 + H_2$	2.50E+14	0.0	16000
388.	$C_3H_4 + H_2CCCH \rightarrow C_6H_6 + H$	2.20E+11	0.0	2000
430.	$C_5H_5 \rightarrow C_5H_5(L)$	4.09E+47	–10.4	54874
431.	$H_2CCCH + C_2H_2 \rightarrow C_5H_5(L)$	5.62E+32	–7.3	6758
451.	$CH_2CHCCH_2 + O_2 \rightarrow C_4H_4 + HO2$	1.20E+11	0.0	0
457.	$2C_2H_3 \rightarrow CH_2CHCCH_2 + H$	4.00E+13	0.0	0
459.	$CH_2CHCCH_2 + H \rightarrow C_4H_4 + H_2$	3.00E+07	2.0	1000
466.	$C_4H_4 + H \rightarrow H_2CCCCH + H_2$	3.00E+07	2.0	5000
479.	$H_2CCCCH + O_2 \rightarrow CH_2CO + HCCO$	1.00E+12	0.0	0

TABLE 2.3 (CONTINUED)
Main PAH (Polycyclic Aromatic Hydrocarbons) Formation Reactions. $k = AT^n \exp(-E/RT)$ in units of mol, cm, s, cal/mol, K

No.	Reaction	A	n	E
482.	$H_2CCCCH + H_2 \rightarrow C_2H_2 + C_2H_3$	5.01E+10	0.0	20000
495.	$C_3H_6 + CH_3 \rightarrow C_3H_5 + CH_4$	2.21E+00	3.5	5675
513.	$CH_2 + C_2H_4 \rightarrow C_3H_6$	9.03E+13	0.0	0
514.	$C_3H_4 + H \rightarrow C_3H_5$	1.20E+11	0.7	3007
520.	$C_3H_5 + C_2H_3 \rightarrow C_3H_4 + C_2H_4$	2.41E+12	0.0	0
521.	$C_3H_5 + C_2H_5 \rightarrow C_3H_4 + C_2H_6$	9.64E+11	0.0	−131
523.	$C_2H_4 + HCH \rightarrow C_3H_5 + H$	3.19E+12	0.0	5285.4
524.	$C_3H_4 \rightarrow C_3H_4P$	1.01E+28	−4.6	63183
529.	$C_3H_4 + M \rightarrow H_2CCCH + H + M$	1.00E+17	0.0	70000
530.	$C_3H_4P + M \rightarrow H_2CCCH + H + M$	1.00E+17	0.0	70000
531.	$C_3H_4 + CH_3 \rightarrow H_2CCCH + CH_4$	2.00E+12	0.0	7700
532.	$C_3H_4P + CH_3 \rightarrow H_2CCCH + CH_4$	2.00E+12	0.0	7700
533.	$C_3H_4 + H \rightarrow C_2H_2 + CH_3$	2.00E+13	0.0	2400
544.	$C_3H_4P + O \rightarrow HCO + C_2H_3$	7.50E+12	0.0	2102
551.	$C_2H_2 + HCCO \rightarrow H_2CCCH + CO$	1.10E+11	0.0	3000
564.	$2CH_3 + M \rightarrow C_2H_6 + M$	3.18E+41	−7.0	2762
617.	$C_2H_3 + O_2 \rightarrow CH_2O + HCO$	4.00E+12	0.0	−250
837.	$H + C_3H_8 \rightarrow C_3H_7 + H_2$	1.32E+06	2.5	6756
838.	$OH + C_3H_8 \rightarrow C_3H_7 + H_2O$	3.16E+07	1.8	934
847.	$HO_2 + C_3H_7 \rightarrow OH + C_2H_5 + CH_2O$	2.41E+13	0.0	0

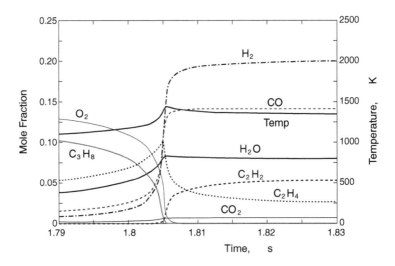

FIGURE 2.64 Temporal profiles of temperature and main stable species around "ignition" at $\phi = 5$.

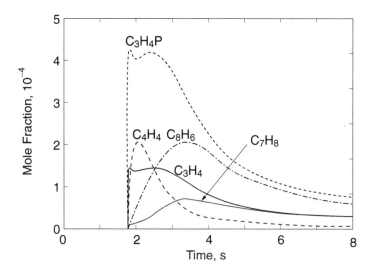

FIGURE 2.65 Temporal profiles of some important species in the first 8 s of PAH formation mechanism at $\phi = 5$.

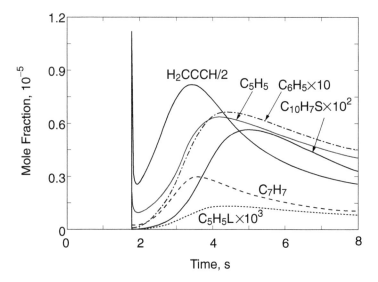

FIGURE 2.66 Temporal profiles of some important radicals in the first 8 s of PAH formation mechanism at $\phi = 5$.

Part of C_3H_4 is converted to C_3H_4P (propyne) through R524. Furthermore, C_2H_2 combines with HCCO through R551 to give H_2CCCH (propargyl radical), which goes partially into C_3H_4 P (–R530). C_3H_4 and C_3H_4P play a central role in the formation of the first aromatic ring. The importance of C_3 species as PAH precursors has been already demonstrated in previous investigations.[33-36] Here, too, C_3 species

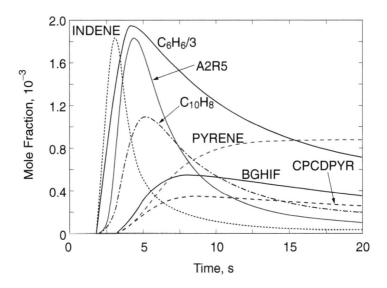

FIGURE 2.67 Temporal profiles of the most abundant species at $\phi = 5$ up to 20 s.

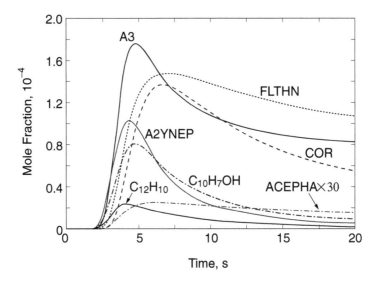

FIGURE 2.68 Temporal profiles of species containing two, three, and more aromatic rings at $\phi = 5$.

are seen to be precursors of PAH, rather than C_4 species, which are oxidized in a large amount. Thus, only a small fraction of the total carbon atoms form PAH at this condition, because a large amount of them oxidized through $C_4H_4 \rightarrow H_2CCCCH \rightarrow CH_2CO + HCCO$. Only a small fraction of C_4H_4 will later contribute to the formation of benzene through R382. Along with C_3H_4 and C_3H_4P, C_3H_5 is formed from C_2H_4 through R523 and from C_3H_6 through R495. Reactions of C_3H_5 (R520, R521, R514) are additional sources of C_3H_4. Figure 2.69 displays the ways in which the first aromatic species form. After the first stage, the reactions proceed at slower rates through much longer time intervals. Reaction R530, which in the first stage proceeded in the reverse direction $H_2CCCH \rightarrow C_3H_4P$, now proceeds in the forward direction, producing H_2CCCH. In addition, R532 also forms H_2CCCH, from C_3H_4; the rate of this reaction in the second stage is about 1/3 of that at the first stage, but it lasts now for more than 3 s. H_2CCCH combines then with either C_2H_2 (R431) to form C_5H_5 (L), or C_3H_4 (R388) to form C_6H_6 (benzene). The linear C_5H_5L becomes cyclic (C_5H_5, cyclopentadienyl radical) through M R430. C_5H_5 reacts with C_2H_2 to form C_7H_7 (radical of toluene) through R380. Further addition of C_2H_2 to the latter species forms INDENE (R10). Indene is formed up to 1.5 s after ignition, but later R10 proceeds in the reverse direction, and indene decomposes back into $C_7H_7 + C_2H_2$.

The time at which R10 changes sign coincides with that at which the above cited reactions R532, R431, R430, and R380 peak. This time interval of about 2 s after the first stage can be referred to as the second stage, where indene also peaks, as can be seen in Figure 2.67. Indene can form indenyl radical (INDENE*) through

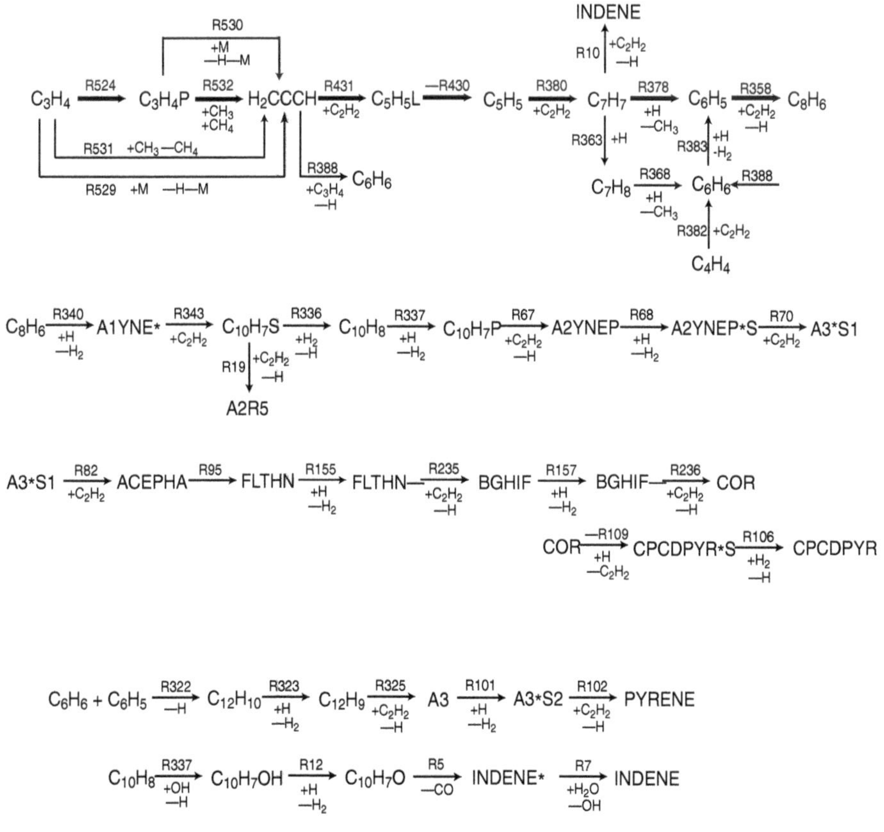

FIGURE 2.69 Main reaction routes in the second and third stages of $\phi = 5$.

H abstraction, but because it does not participate in reactions leading to larger PAH molecules, its concentration is largely defined by R10 only, having a maxima close to 2%. Yet, in the second stage, two routes of decomposition of C_7H_7 are observed: one into C_6H_5 (phenyl radical) through −R378 and the other into C_6H_6 after formation of C_7H_8 (toluene) (−R367, R368).

From the third stage, in which the reactions peak about 1 s later than the reactions of the second stage, larger aromatic species form almost exclusively through HACA (hydrogen abstraction, C_2H_2 addition) reactions. From C_6H_6, H abstraction (R383) and C_2H_2 addition (R358) give C_8H_6 (phenyl acetylene). From C_8H_6, H abstraction (R340) and C_2H_2 addition (R343) form $C_{10}H_7S$ (1-naphthyl radical). H addition through reaction of H_2 with $C_{10}H_7S$ (R336) gives $C_{10}H_8$ (naphthalene). C_2H_2 addition into 1-naphthyl (R19) forms A2R5 (acenaphthalene).

From naphthalene, H abstraction (R337) gives $C_{10}H_7P$ (2-naphthyl radical) and C_2H_2 addition (R67) to the latter species leads to A2YNEP ($C_{12}H_8$, 2-naphthylacetylene). Naphthalene can also be oxidized via reaction with OH radicals (−R13) with formation of $C_{10}H_7OH$, whose concentration is considerable, as can be seen in Figure 2.68. From 2- naphthylacetylene, H abstraction (R68) and C_2H_2 addition (R70) lead to A3*S1 (1-, 8-, 9-, and 10-phenanthryl radical). Further C_2 addition (R82) forms ACEPHA (acephenanthrene). Fluoranthene (FLTHN) is then formed mainly from an intramolecular rearrangement reaction (R95) from ACEPHA and partly through combination of $C_{10}H_7S$ and C_6H_6 (R84).

From fluoranthene, H abstraction (R155) and C_2 addition (R235) give BGHIF (benzo[ghi]fluoranthene). Further H abstraction (R157) and C_2 addition (R236) lead to COR (corannulene). COR can decompose into CPCDPYR (cyclopenta[cd]pyrene) through reaction with H (−R109, R106).

In parallel with the above reactions, in the third stage phenyl and benzene combine themselves to form $CZ_{12}H_{10}$ (bi-phenyl) through R322. H abstraction (R323) and C_2 addition (R325) form A3(phenanthrene). Further H abstraction (R101) and C_2 addition (R102) give PYRENE. Reactions of the third stage proceed with rates about a tenth of those proceeding at the second stage but they last for more than 10 s and lead to the broad temporal species profiles seen in Figures 2.67 and 2.68.

Larger species also form, but their production rates are small and not analyzed here. Benzene and indene have concentration profiles peaking early and decreasing later. Richter's calculation results on a flat flame failed to reproduce the indene decrease at the postflame region. There are no reactions in the scheme linking indene to larger molecules, but even with this lack of reactions consuming indene, this species is found here to decrease through the reverse of the same reaction by which it was produced at earlier stages. The benzene decrease is due to reactions such as R322 and R84 and oxidation reactions like R287.

Ignition of $\phi = 2$ Mixture

Next, the analysis of the ignition of a mixture of equivalence ratio equal to 2 is described. The temporal profiles of temperature and main stable species are plotted in Figure 2.70. Here, the temperature is seen to increase monotonically with time up to the adiabatic flame temperature, which exceeds 2180 K. According to

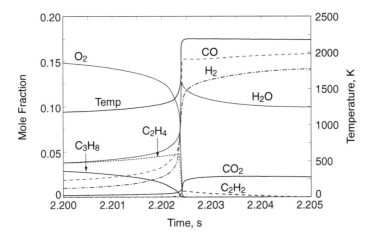

FIGURE 2.70 Temporal profiles of temperature and main stable species around ignition at $\phi = 2$.

CHEMKIN, the ignition time is 2.2023 s. In comparison to the $\phi = 5$ case, the concentrations of CO_2, CO and H_2O are high after ignition, indicating a more intense combustion. The concentrations of active species such as H atoms and OH radicals are also much larger than at $\phi = 5$: their maximum values are close to 1 and 0.2%, respectively. C_2H_4 continues to be produced from the beginning of the process, but its concentration reaches only 5%, lower than the 9% reached at $\phi = 5$. The peak of C_2H_2 is also lower, 5%. The reaction rates at this condition are much larger; for example, R513, by which C_3H_6 is produced from $C_2H_4 + CH_2$, is about 100 times faster than in the $\phi = 5$ case. And the reaction linking C_3H_4 to H_2CCCH is more than ten times larger. However, no PAH species form. Even benzene has a peak concentration of just 1.4×10^{-8}. C_3H_4 and C_3H_4P were found to decompose through reactions R533 and R544 just after ignition; H_2CCCH keeps a significant concentration at later stages, but the rate of the step $H_2CCCH + C_2H_2 \rightarrow C_5H_5(L)$ (R431) is about 100 times smaller than at $\phi = 5$. The rate coefficient of this reaction is expressed as $5.62 \times 10^{32} T^{-7.3} \exp(-6758\text{cal/mol})/RT)$, as shown in Table 2.7. Because at $\phi = 2$ the temperature is much higher than at $\phi = 5$, the rate coefficient of R431 becomes very small in the post-ignition period, due to the $T^{-7.3}$ factor. The inhibition of PAH formation at small values of equivalence ratio is then attained not through oxidation reactions, but through a thermal effect that seems artificial. This is a point of the scheme that can be improved in future work. Experiments should be done to check if the predictions given by the scheme are correct; if the PAH formation in a mixture of constant equivalence ratio were actually much more intense at temperatures as low as 1500 K in comparison to temperatures as high as 2000 K, the scheme, then, although in an artificial way, could be considered as able to predict correctly the PAH formation in rich hydrocarbon flames. In that case, the combustion of preheated mixtures would show low soot emission, in addition to low NO emission.

SUMMARY

The PAH formation mechanism in preheated mixtures of propane and air with equivalence ratios of 5 and 2 was investigated through a well-stirred model. The employed chemical reaction scheme comprised 902 reactions occurring among 251 species, and the results can be summarized as follows:

1. C_4 species were not seen to be precursors of PAH because, although formed at large rates, they are oxidized through reaction with molecular oxygen. C_3 species like allene (C_3H_4), propyne (C_3H_4P), and propargyl radicals (H_2CCCH) are the main precursors of the first aromatic ring.
2. Naphthalene and other larger species containing more than one aromatic ring are seen to be formed through the HACA (H abstraction C_2 addition) mechanism. An exception is fluoranthene, which forms mainly through a rearrangement reaction from acephenanthrene.
3. When the equivalence ratio was set to 2, the PAH species did not form. At this condition, the temperature raised to a value higher than 2180 K. Such a high temperature caused decomposition of C_3H_4 and C_3H_4P and a significant decrease in the rate coefficient of the step $H_2CCCH \rightarrow C_5H_5L$, which is in the main route for the PAH formation. The expression for the rate coefficient k of this step has a $T^{-7.3}$ factor that causes k to decrease as temperature increases. This seems to be an artificial way to inhibit PAH formation when combustion reactions proceed at large rates. The step $H_2CCCH \rightarrow C_5H_5L$ is a potential target for future work and a better understanding of this reaction can lead to improvements in Richter's scheme. At the same time, if this inhibition of PAH formation at high temperatures were supported by experimental evidences, then the scheme could be considered able to predict correctly the PAH formation in hydrocarbon flames, with low soot emission — an additional property of the combustion of preheated mixtures.

2.3.6 RADIATION

There are two kinds of combustion flames: luminous flames and nonluminous flames. While the former includes suspended solid particles and gray approximation can be applied to its radiation effect, the latter has to be analyzed under a nongray condition (different consideration of wavelength bands involved in radiation and those not involved in it; see Figures 2.71 and 2.72), because its radiation is mainly gas radiation from CO_2 and H_2O. Because radiation has to be handled in actual practice, naturally, as a three-dimensional phenomenon, the analysis of nonluminous flame radiation is, however, not easy and it requires some kind of simplification. Further, the overlapping of radiation wavelength bands has to be taken into account regarding a gas body containing CO_2 and H_2O, and hence it is insufficient for obtaining the total radiation effect by simply adding up individual heat transfer effects of CO_2 and H_2O, each calculated separately.

In other words, to analyze the radiation effect of gas or nonluminous flames of a gas body containing CO_2 and H_2O, it is necessary to calculate the absorption coefficient for each of the radiation wavelength bands, considering the overlapping of the radiation from CO_2 and that from H_2O (see Figure 2.73). Radiation heat transfer of gas or flames alone is unrealistic because radiation effects of solid surfaces are inevitably involved in the heat transfer inside a furnace with highly preheated air combustion. Overlapping with the radiation wavelength bands of furnace walls as shown in Figure 2.74, therefore, has also to be taken into account in the analysis and, accordingly, analysis procedures will be different depending on whether gray or nongray approximation is applied to the furnace walls.

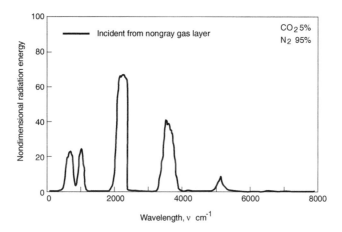

FIGURE 2.71 Wavelength characteristics of CO_2.

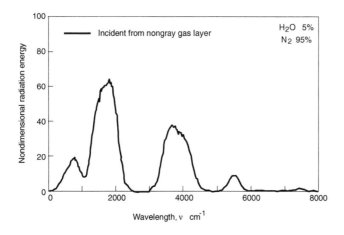

FIGURE 2.72 Wavelength characteristics of H_2O.

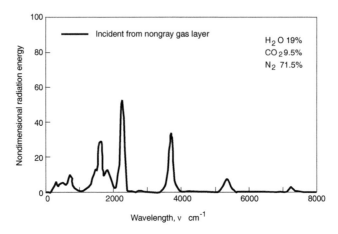

FIGURE 2.73 Wavelength characteristics of CO_2/H_2O mixed gas.

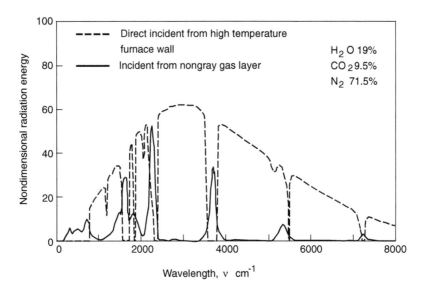

FIGURE 2.74 Radiation characteristics of nongray gas and solid surfaces.

Although the nongray radiation effects were studied using conventional analytical techniques, many of the study results related to cases where temperature and partial pressure of the gas were uniform and few looked into cases where they were unevenly distributed. But in actual cases, temperature and the partial pressure are never uniform but change inevitably near furnace walls or the objects to be heated, and for this reason there are many problems in analyses that neglect this fact.

Further, because a study of radiation heat transfer effects cannot be applied for actual purposes unless it is done in a three-dimensional condition, conventional analytical techniques are not adequate. For this reason, with a three-dimensional radiation, analysis of nongray gas has to be in the form of a numerical solution.

When studying the radiation effect of nongray gas, the value of monochromatic absorption coefficient has to be calculated. Introducing the Elsasser model and using wave number á 1/m, the coefficient is expressed as below:

$$k_{v_i} = \rho\left(\frac{e}{D}\right)\frac{\sinh(2\pi C/D)}{\cosh(2\pi C/D) - \cos(2\pi v*/D)} \quad (2.20)$$

where ρ kg/m³ is density of the gas, e is nondimensional line intensity, D m is line interval, C m is half line width, and $v* = |v - v_0|$, where v_0 1/m is typical wave number of an absorption band (see Table 2.4). The following equations are derived when the notion of exponential broadband model of Edwards et al. is introduced to Equation 2.20:

$$\left(\frac{e}{D}\right) = \left(\frac{A}{W}\right)e^{v*/W} \quad (2.21)$$

$$\left(\frac{\pi C}{D}\right) = BP_e \quad (2.22)$$

where A is nondimensional band intensity, W is a band width parameter, B is a line broadening parameter and P_e is a pressure parameter. In cases where the cosine term of Equation 2.23 can be neglected, then the following will be true:

$$k_{v_i} = \rho\left(\frac{A}{W}\right)e^{-v*/W}\tanh(2BP_e) \quad (2.23)$$

TABLE 2.4
Typical Wavelengths of Absorption Bands

CO_2 μm	H_2O μm
15.0	Rotational
10.4	6.3
9.4	2.7
4.3	1.87
2.7	1.38
2.0	

Accordingly, the monochromatic absorption coefficient of nongray gas k_v 1/m can be calculated as a summation of the absorption coefficient k_{vi} 1/m of every absorption band as follows:

$$k_v = \sum_{i=1}^{n} k_{v_i} \qquad (2.24)$$

where n is the total number of the absorption bands. Because the monochromatic absorption coefficient of nongray gas is a function of wave number and temperature and pressure of the gas as is clear from Equation 2.20 or 2.23, it is necessary to understand the difference from the case where the coefficient is regarded constant as the mean absorption coefficient in gray approximation.

In an actual numerical analysis, nongray gas can be treated in the same manner as gray gas by dividing the entire range of radiation wave numbers into many bands of monochromatic radiation wave numbers, and the behavior of the radiation energy particles can be calculated regarding the wave number bands involved in the radiation and those not involved in it. When it is necessary to apply the nongray conditions to solid surfaces, behavior of the radiation energy particles can be calculated in the same manner as gray bodies by dividing, similarly, the entire radiation wave number range into many bands of monochromatic radiation wave numbers.

It would, of course, be necessary to take suitable measures such as use of a common division of the radiation wave number range for both the nongray gas and the nongray solid surfaces. The greater the number of bands the radiation wave number range is divided into, the more accurately the nongray characteristics will be calculated.

Regarding the radiation characteristics of nongray gas and gray gas, whereas the emissivity value of gray gas is 1 when its mean effective thickness is infinite, that for nongray gas is less than 1 even when the mean effective thickness is infinite. As a consequence of a supposition that the absorption coefficient of gray gas does not change depending on its temperature, the coefficient is kept constant because the change of the absorption effect by density fluctuation is compensated for by temperature. But considering that the monochromatic absorption coefficient of nongray gas is strongly related with density, as is clear from Equation 2.20 or 2.24, it can be concluded that the assumption that the absorption coefficient is constant regardless of density change is contradictory to reality.

The biggest problem is that CO_2 and H_2O, which actually emit the radiation, are nongray gases, and that some part of the radiation energy emitted from solid surfaces to which gray approximation is applicable passes through these gas without being absorbed in some wavelength bands. This phenomenon is the same when the mean effective thickness is larger. During the radiation analysis, due attention should be paid to it as a fundamentally different point from the cases of the gray gas approximation.

In some practical cases where simplified analyses are acceptable, the gray gas approximation may be applied, but it may not be easy to calculate furnace temperature distribution together with the amount of radiation heat transfer. Although it is necessary to typify the objects of analysis to some extent in accordance with the practical requirements, it is not desirable to apply conditions such as linear radiation approximation or gray approximation simply for the reasons of the analysis technique employed.

Regarding then applications of the nongray or gray characteristics to solid surfaces in a furnace, whereas gray conditions may be generally applicable to bright metal surfaces despite their low emissivity, nongray conditions should be applied to the analysis of objects such as molten glass. Apparent emissivity of objects with rough surfaces such as furnace walls and oxidized metal surfaces tends to be considerably greater than the value of the component material itself due to reflection effects of the surface roughness. Thus, the nongray characteristics of the material are attenuated, and the gray approximation becomes applicable.

As described above, gray approximation is applicable to solid surfaces in the case of industrial furnaces, especially metal heating furnaces, but glass melting furnaces and the like require application of nongray properties to the objects to be heated; otherwise, practically useful results cannot be obtained.

When high temperature air combustion is used for an industrial furnace, the flames will be nonluminous flames especially with gaseous fuels, and a homogeneous high temperature field will be formed inside the furnace in most cases. Although the final temperature of the heated objects is not much different from the general furnace temperature in the case of metal heating furnaces, the furnace temperature is low near the heated surfaces because the temperature of the heated surfaces is low in the case of boilers or petrochemical furnaces. Because the temperature of the heated objects is not very high, either, in the case of metal annealing furnaces or at the entry side of metal heating furnaces, the furnace temperature falls to some extent there.

For the purpose of radiation analysis envisaging acceleration of heat transfer in industrial furnaces, it is essential therefore to introduce analysis techniques capable of dealing with furnace temperature distribution with any combustion processes including the highly preheated air combustion. This point should not be overlooked in developing the computer software. Influences of distributions of temperature and partial pressure on nongray gas radiation are shown in Tables 2.5 and 2.6 in comparison with those in the case of even temperature and partial pressure. These examples make clear that the difference cannot be overlooked, that is, the heat flux of either CO_2 or H_2O fluctuates, depending on the distribution pattern of temperature and partial pressure, as widely as from 0.7 to 2.8 times the heat flux under even distributions. Consequently, there are many problems in an abridged calculation of radiation effects using mean temperature and mean partial pressure.

In these tables, many cases of paired boxes are shown. When high temperature region in furnace lies and burned gas stays far from the wall, thermal radiation is smaller, but when combustion takes place near the wall, radiative energy is larger because of the thin optical thickness. An amount of the radiative energy from the combusting gas reaches wall to wall. In short, the uniform distribution of temperature and burned gas like high temperature air combustion may not be more advantageous, but in these figures, the area of the wall is not considered.

In general, the area of high temperature wall contributes more part of total radiative heat transfer. Uniform gas distribution makes the available wall area wide.

TABLE 2.5
Heat flux of H_2O in Consideration of Wedge Shape Distribution (in the case of 1-m mean effective thickness)

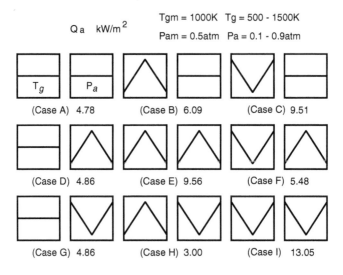

TABLE 2.6
Heat flux of CO_2 in Consideration of Wedge Shape Distribution (in the case of 1-m mean effective thickness)

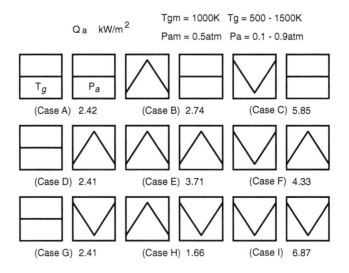

With regard to collision flames, the effect of convection heat transfer in the furnace is not negligible. The total heat transfer effect in such a case can be obtained by combining the radiation effect with the amount of convection heat transfer calculated separately. It is, of course, important to look at it as a combined radiation/convection effect, but it is sufficient for the purpose to add the convection effect as an adjustment to the predominant radiation effect.

It is usually the case that nonluminous flames are formed in the highly preheated air combustion with gaseous fuels, but radiation heat transfer analysis needs to take formation of luminous flames into consideration when solid or liquid fuels are used. In this case, existence of solid particles in the flames is regarded as the cause of the formation of luminous flames, and radiation characteristics of the solid particles will have to be studied together with the radiation of gas. Handling of this factor, however, is different depending on whether the diameter of the particles is considerably larger than the radiation wavelength or nearly the same. In the former case, the radiation is scattered by reflection effect at the particle surfaces, whereas in the latter case the scattering is caused by interference. However, when the particle diameter becomes smaller yet, the scattering effect becomes nearly identical to that of gas radiation.

2.4 FUNDAMENTALS OF LIQUID FUEL FLAMES

2.4.1 LIQUID FUEL FLAME CHARACTERISTICS AND STABILITY

An experiment on a spray combustion was carried out in highly preheated, diluted air with a view to comparing the results of LPG gas as the fuel. The spray pressure to be jetted out from the nozzles, the oxygen content, and the measuring temperature were changed to investigate the spray combustion. Further, photographs of the flame forms were taken in detail by a digital camera within the combustible range to study what differences in the combustible range existed according to both the highly preheated, diluted air temperature and the oxygen content. Both the changing state of the flame form and the changing of its color were also investigated.

2.4.1.1 Experimental Apparatus

The spraying experiment was carried out using an injection nozzle that was arranged to go straight to the high temperature air generator, manufactured by Nippon Furnace Kogyo NFK (see Foreword). The injection nozzle was designed compactly about 10 mm in diameter, to minimize disturbance of the flame flow. For the vessel, a pressurized oil storage vessel with an accumulator was prepared, instead of the usual vessel with a spill-type nozzle, to permit control of both the injection pressure and the injection volume.

2.4.1.1.1 Spraying Device
Schematic diagram of the whole device. The accumulator was filled with kerosene and was used instead of a fuel tank. The following injection method was adopted. A T pipe on the upper portion of the accumulator was set using a copper pipe 6 mm in diameter. The required pressure was supplied using a high pressure air tank. In the case of this injecting method, the time required for applying pressure was not

needed, and the necessary pressure could be obtained steadily by controlling the opening of the regulator. The piping diagram is shown in Figure 2.75. The spray nozzle was connected to the furnace by a pressure hose, and it was secured on a slider that was set up to a traverser, to prevent it slipping out of the bayonet angle during testing and to enhance the reproducibility.

2.4.1.1.2 Combustion Device

A high temperature gas generator (the trade name is Airenthalpy Intensifier, AI, manufactured by NFK) was used as the combustion device that can produce heated air at a high temperature in a short time and that can also dilute simply the inside of the furnace. Schematic diagrams of the combustion device are shown in Figure 2.1 and the configuration of the heat reservoir and regenerator used in the present experiment in Figure 2.76. The heat reservoir is made of ceramic (aluminum titanate and cordierite) and is in a honeycomb state. Meshes of $300/in^2$ are set up in this heat reservoir, and two of the heat reservoirs are arranged in each furnace, respectively.

2.4.1.1.3 Spray Nozzle

The nozzle used in the experiment was of a hollow-cone spray-flow type. It is recognized that this type of nozzle can be used even in a furnace with a narrow range to produce a stable flame at a lower outflow rate of the spray. The hollow-cone spray flow is a flow of the fuel as follows. When liquid fuel is jetted out conically at an angle to the liquid axis, a liquid film in a conical shape is atomized into small particles, resulting in a hollow-cone spray. Since the pressure in the vicinity of the central axis of such a spray is lower than the surrounding pressure, the ambient air is forced in and small liquid droplets are transferred to the central part by this airflow. As a result, a classification of droplet diameters occurs, that is, the comparatively smaller liquid droplets concentrate at the central part and the larger droplets at spray boundary.

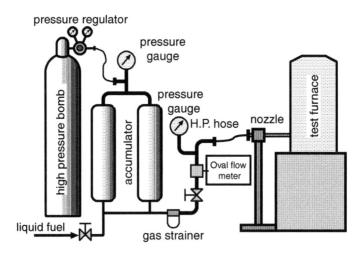

FIGURE 2.75 Schematic diagram of the whole installation for liquid fuel injection.

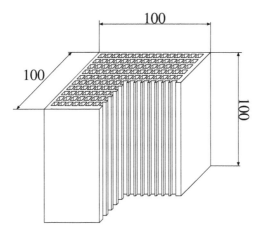

FIGURE 2.76 Configuration of heat reservoir.

Further, if the ambient pressure becomes higher and the density of ambient gas is also increased, the divergent angle of spray becomes narrow, resulting in the state of spray becoming nearer to an axial spray having a high flow rate of liquid droplets even at the central portion of flow. This is because, when the gas density becomes higher, exchange of the respective momentum of both the liquid droplets and the gas is quickened, and therefore the pressure lowering rate at the central portion of the spray flow increases. The divergent angle (spraying angle) of the spray is decided by the two factors of the liquid axis direction at the outlet of nozzle and the velocity component in the rotating direction. The divergent spray cone angle of 60° was adopted in the present experiment.

In the figure, a hole for inserting the spray nozzle is arranged at the side of the furnace. The reason for arranging this hole in such a way is to compare the flame of liquid fuel in a cross-flowing state with results previously obtained for gaseous fuel. The fuel spray flow is injected through this hole at a right angle to the diluted airflow already preheated at a high temperature. However, in the case of the experimental furnace, because the respective blower outputs of absorption and exhaust are sometimes out of balance to produce a possible negative pressure, a great quantity of air outside of the furnace might be sucked in through the hole during measurement. This would result in greater influence upon the oxygen content in the furnace. Therefore, the diameter of the hole for inserting the nozzle was made close to the diameter of the nozzle. The gap was designed to be covered by a heat insulating material by filling it in at the same time as the nozzle was inserted.

2.4.1.2 Experimental Method

2.4.1.2.1 Air Preheating

A heat exchanger of a high-speed switching type with a generative system was used to heat the air and to generate the high temperature preheated air. The air preheating method is as follows. First, the air is drawn in by a blower. The drawn-in air passes through the directional control valve on the left side facing the furnace to be measured (as shown in Figure 2.1). It then passes through the heat reservoir in the furnace to be measured, which by this time has become cooled. The air that has passed through the heat reservoir in the furnace to be measured is then heated by both the pilot burner and the heat-up burner.

The heated air enters the furnace on the right side as shown in the test section of combustion chamber shown in Figure 2.1. Then, the heated air passes through the cool heat reservoir on the right side as shown in the figure. At this time, the valves of both the pilot burner and the heat-up burner are closed.

When the heated air passes through the heat reservoir, which is cool, a heat exchange is carried out between the heated air and the heat reservoir, heating up the cooled heat reservoir. After exchanging the heat, the air passes through the directional control valve and then is discharged by the blower as waste gas. The above-mentioned process is defined as half cycle, and this is carried out continuously for about 20 s.

After completing a half cycle, the directional control valve is switched to the direction of the right side of the furnace. The supplied air passes through the heat reservoir on the right side, which has already exchanged the heat to be heated, and is further heated by both the pilot burner and the heat-up burner. The drawn-in air is heated by exchanging the heat with the heat reservoir and by both the pilot burner and the heat-up burner.

The heated air enters the furnace to be measured, and passes through the heat reservoir to exchange the heat. The air having been heat-exchanged passes through the directional control valve to be discharged. After completing this process, the directional control valve is switched. This half cycle is carried out repeatedly to heat the air.

Experimental conditions for carrying out tests are shown in Table 2.7. The expression of the oxygen content is as follows:

$$\text{Oxygen} = 0.21 \frac{(\text{Air})}{(\text{Air} + \text{Nitrogen})}$$

TABLE 2.7
Experimental Conditions

Items	Specifications
Fuel	JIS No. 1 Kerosene
Spray nozzle	Hollow-cone type
	Spraying angle: 60 degrees
	Rated capacity: 0.3 gal/h
Spraying pressure	0.5 MPa
Oxygen content	3, 5, 8, 10, 12, 15%
In-furnace airflow rate	15 m_N^3
Preheating temperature	250 – 1100°C
Gas	Fuel: LPG,
	Dilute: Nitrogen

Here, Oxygen is the volumetric content of oxygen, Air is the volumetric flow rate of air, and Nitrogen is the volumetric flow rate of nitrogen. In the present experiment, the volume flow rates of both the diluting nitrogen and the air were added together, to set the in-furnace airflow rate as 15 m³/h at normal condition.

2.4.1.2.2 Spray Pressure

It has been already confirmed that the finer the fuel droplets, the stabler the combustion. Further, in the case of the pressurized spraying method, the spraying pressure should be increased, because the fuel droplets are made fine when the spraying pressure is high enough. However, even in the case of the experiment where the injecting device with the accumulator was applied, the fuel flow rate was increased as the spraying pressure was increased further. This resulted in easier generation of incomplete combustion in a smaller test furnace to be measured. Therefore, tests were carried out under the conditions where the ideally fine fuel droplets were used and the spraying pressure was maintained around 0.5 MPa, where the characteristics of the spray flame in diluted air could be easily observed. Under these experimental conditions, the spraying pressure and the fuel flow rate were proportional. The fuel flow rate is linearly varied with the spraying pressure.

2.4.1.2.3 Spraying Method

The following types of spraying methods can be used for finely atomizing the liquid fuel: the oil hydraulic spraying type, which injects the liquid fuel at a high speed; a twin fluid spraying type, which sprays the liquid fuel by gas jet diffusion; a rotating type, in which the liquid fuel is sprayed by a centrifugal force; a colliding nozzle type, in which the liquid fuel is collided to make fine particles; an oscillating type, in which the liquid fuel is disrupted by such oscillation ultrasonic waves; and a static electricity type, in which the liquid fuel is atomized by high voltage static electricity.

The twin fluid spraying method makes measuring complicated because it takes in air, steam, and other contaminants. Both the centrifugal force spraying method and the ultrasonic wave spraying one are difficult to apply in an atmosphere of high temperature air. Therefore, the commonly used oil hydraulic spraying method was adopted to carry out the experiment. This spraying method has the following advantages: it can be applied most easily from the viewpoint of devices in a laboratory environment and nozzle clogging caused by the carbonizing of the fuel through an oxygen-free thermal decomposition in the high temperature air rarely occurs. In any case, the spraying pressure exerts a wide influence upon the diameter of liquid particles to be obtained.

With the purposes of easily controlling the spraying pressure and of obtaining a stable pressure, a method of accumulating the pressure to an accumulator was adopted in the present experiment. The accumulator was used instead of a pressure tank, filled with kerosene, and was pressurized by a high pressure air tank arranged on top of the accumulator. By adopting this pressurizing method, the time required for accumulating pressure using a device such as a pressurizing pump can be eliminated, and the necessary stable pressure can be obtained by simply controlling the opening of the regulator.

The nozzle is inserted in the hole on the right side facing the furnace to be measured. It may take a little longer, because the capping of the hole arranged on the furnace is tightly closed because it would be dangerous if the hole opened during the course of raising the temperature. Then the pressure of the spraying device is raised to the level of experimental conditions, and immediately after sliding the nozzle to secure it in the spraying position in the furnace, the valve is opened to start measuring without delay. Great care should be taken in this procedure.

2.4.1.2.4 Measurement of Flame

For investigating the characteristics of the flame, a digital camera was fixed in front of the furnace on a tripod to take photographs of the flame. To eliminate any background luminescence, it was necessary that the physical relationship between the furnace and the camera should be fixed during testing. However, this practice was not carried out in the present experiment.

The photographs of the flame were taken at intervals of 50°C in the combustible temperature range from 1100 to 250°C. However, since some time was required for inserting the nozzle and diluting the air, the temperature fell slightly during these operations. Therefore, the in-furnace temperature was raised to 1200°C to carry out such measuring practices in time, and then the measuring was done. The falling rate of the in-furnace temperature changed with every experiment due to such factors as the warming-up state of the furnace. Therefore, in the present experiment, the warming-up operation of the furnace to be measured was fully carried out before measuring.

Further, the in-furnace wall surface was designed to be covered by a thick heat-insulating material, so that the wall temperature was the same as the diluted air temperature. Therefore, it was assured that there would be no influence caused by the dispersion of the wall temperature, when measuring the flame in the process of its self-cooling.

2.4.1.3 Experimental Results

2.4.1.3.1 Temperature of Blowout

It could be observed that the flame in the case of the spray combustion in the high temperature preheated diluted air differed clearly from the flame in the air at ordinary temperature. Investigation was carried out on how the range, where the diffused flame could be steadily combusted, was affected by both the temperature of the high temperature preheated diluted air and its oxygen content. The results of the experiment are shown in Figure 2.77, in which the broken line shows the results obtained when LPG with none of the evaporating process was burned. Under the conditions of the experiment, i.e., in high temperature preheated and diluted air, the flame formed a lifted flame, and it was curtailed as the in-furnace temperature went down, resulting in blowing out. Further, the blowout of the flame occurred at a higher temperature, as the air dilution rate was greater.

In the case of using kerosene and under the conditions that the oxygen content was 15%, the flame blowout temperature could not be measured in the temperature range below 250°C. However, it is considered that the curves shown in Figure 2.77

Combustion Phenomena of High Temperature Air Combustion

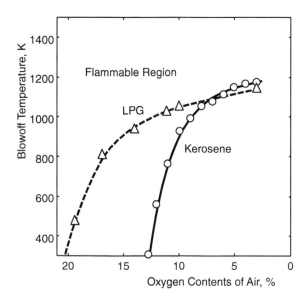

FIGURE 2.77 Relationship between the oxygen content and the flame blowoff temperature.

would vary depending on a flame to be formed and the structure of the combustion device. If the temperature of autoignition, which is a specific characteristic of the fuel itself, can be decided according to the air dilution rate, it could be considered that these curves are to be dependent on only the fuel itself.

Photographs of the states of spray flame combustion in the high temperature preheated diluted air, when the oxygen content is changed, were taken with the digital camera. The photographs are shown in Figures 2.78 through 2.81. It was observed that the more red the color is, the higher the temperature, and the bluer the color is, the lower the temperature. Based on this fact, it is easy to observe the difference in temperature within the flame and the state of the flame in the violet color that is peculiar to combustion in the high temperature, preheated, diluted air.

2.4.1.3.2 Flame Form and Flame Color

The flame turned to form clearly a lifted flame at a distance from the nozzle, and a red flame generated on top of a violet flame. An orange flame was generated at first, and then a violet flame was generated from the bottom portion. As the temperature went down, the violet flame gradually expanded to the top, and the flame turned all violet, then blew out. The states of the flame at 1050, 1000, and 950°C, when the oxygen content was 8%, are observed in Figure 2.79. Further, it has been made clear that, under the conditions where the in-furnace temperature was more than 1000°C and the oxygen content less than 5%, the external appearance of the flame changed greatly. Figure 2.80 shows changes of the flame form according to the oxygen content at 1000°C of in-furnace temperature.

As shown in the figure, turbulence scarcely appears in the flame, and the flame is nonluminous. However, if the oxygen content is higher, the flame form changes with time. The causes for this phenomenon might be due to turbulence of the high

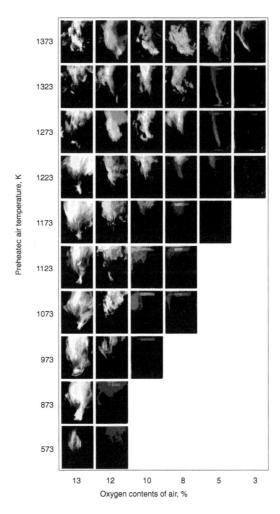

FIGURE 2.78 Photographs of the spray flame in the high temperature preheated and diluted air with the oxygen content in air (T_g = 300 to 1100°C).

temperature airflow, a slight change of the spray in a short time, or due to turbulence caused by an uneven generation of the combustion at some local regions. However, these possible causes have not been investigated in the present experiment.

2.4.1.4 Discussions

2.4.1.4.1 Blowout of Flame

A flame burning at a small distance from the exit of a nozzle, that is, a lifted flame, was formed and blew out as the in-furnace temperature decreased. It can be considered that the reason for the flame blowout is that, as the in-furnace temperature decreases, the fuel droplets form a steam layer, which prevents the fuel droplets from gaining enough energy to cause autoignition.

Combustion Phenomena of High Temperature Air Combustion 115

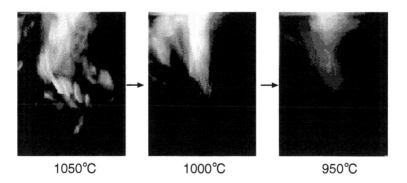

FIGURE 2.79 Changes of the flame form.

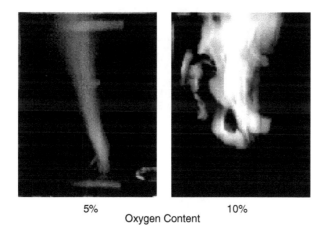

FIGURE 2.80 Changes of the flame form by oxygen content (left: 5%, right: 10%).

Furthermore, the blowout of the flame occurred at a higher temperature, with higher air dilution rate. However, the curves shown in Figure 2.77 as a result of the experiment may change depending on the flame to be formed and the structure of the combustion device. If the autoignition temperature as a specific characteristic of the fuel can be determined according to the air dilution rate, it can be considered that these curves are dependent only on the fuel. Even in this case, the type of measurement device should be designated.

2.4.1.4.2 Changes in Flame Form and Flame Color

The flame was like a lifted flame at some distance from the nozzle, and a red flame was generated on top of a violet flame. At first, an orange flame was generated, and then a violet flame was generated from the bottom part. As the temperature fell, the violet flame expanded gradually to the upper side, and the flame changed to all violet. This violet flame was clearly stable compared with the other.

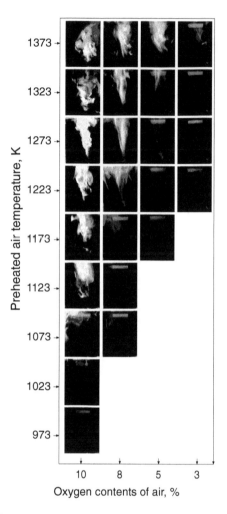

FIGURE 2.81 Spray flame at 0.3 MPa of the spray pressure in the high temperature preheated diluted air.

The blue-green color originates from a $C_2 \cdot CH$ radical emission. The color of the flame with the strong C_2 radical emission is a rich green, and the color of the flame with the strongCH radical emission is a rich blue. Further, it has been recognized that, as the air ratio decreases, the C_2 radical emission increases and the CH radical emission decreases. The blue-green flame obtained in the high temperature preheated diluted air has a wide distribution of the C_2 radical emission. This had been already confirmed by experiments carried out before.

From the results obtained in observing changes of the flame state, shown in Figure 2.79, it has been confirmed that around the central portion of the flame the temperature is higher, and there is little difference in temperature between these flames. Further, it has been made clear that the external appearance of the flame changes greatly under the conditions at a temperature higher than 1000°C and at an oxygen content lower than 5%.

2.4.1.4.3 Spray Combustion in the High Temperature Preheated Diluted Air

It has been confirmed that the combustion stabilizes when the fuel oil droplets become finer. Therefore, to make the fuel oil droplets fine, the spraying pressure should be increased. However, the fuel flow rate is already very high for this small scale experimental apparatus due to limitations in the size of the atomizer. Thus spraying pressure could not be increased, and proper atomization of the droplets was not possible in this case. To decrease droplet size, the fuel flow rate in the in-furnace flowing field should be increased more than usual, or the rated capacity of the nozzle should be lowered if a sufficiently small atomizer can be manufactured. Further a spraying installation that is resistant against spraying pressure should be required. Figure 2.81 shows the forms of flame by each oxygen content at 0.3 MPa of spray pressure in high temperature preheated and diluted air.

2.4.1.5 Summary

Observation of the flames reveals that the spray combustion in high temperature preheated, diluted air has its own shape and its own combustion mechanism. These are both quite different from the normal spray combustion. The characteristics of the spray combustion are as follows:

1. Changes of NO_x value. NO_x emissions reduce in the same manner as gaseous fuel when the highly preheated air is lower than 15% in oxygen content.
2. Changes of flame form and flame color. The flame forms a clearly lifted flame at some distance from the nozzle, and a violet flame is generated from the bottom portion after formation of an orange flame. Then, a bluish-green flame appears on the top of a violet flame. As the temperature goes down, the violet flame gradually expands to the upper side, and the flame changes to all violet in color. Compared with a normal flame, the blue-green flame, which is peculiar to the case of combustion in high temperature preheated, diluted air, is more stable like a laminar flame. In the case of the violet color flame that is obtained in high temperature preheated, diluted air, the distribution of the C_2 radical emission is spread widely.
3. Blowout of flame. The flame is reduced and then blown out, as the high temperature preheated and diluted air temperature in the furnace lowers. Furthermore, the greater the dilution rate, the higher the temperature at which the blowout of flame occurs.

2.4.2 EMISSIONS IN LIQUID FUEL FLAME

2.4.2.1 Emissions on Liquid Fuel Combustion

This section introduces the results of the high temperature air combustion tests conducted at NKK Keihin with various types of fuels and combustion methods. The

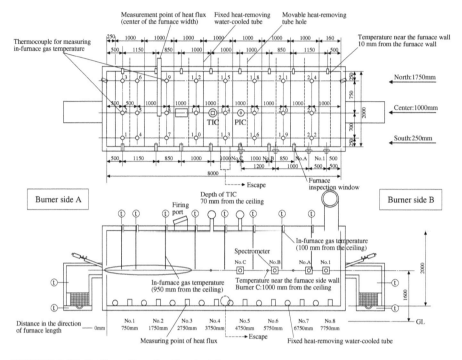

FIGURE 2.82 Outline of combustion test furnace.

types of fuels included heavy oil A, town gas, and COG, and the types of combustion methods included the dispersed combustion type burner method and the concentrated combustion type burner method. Figure 2.82 shows the test furnace. Figure 2.83 shows the burners. Figure 2.84 shows the conditions of flame combustion. Figure 2.85 shows the results of the NO_x density as measured in the test. Figures 2.86 and 2.87 show the in-furnace temperature distribution. Figure 2.88 shows the in-furnace heat flux distribution.

1. Outline of combustion test furnace (see Figure 2.82)
2. Outline of test burners (see Figure 2.83)
3. Flame pictures with different types of fuels (see Figure 2.84)
4. NO_x emission characteristics (see Figure 2.85)
5. In-furnace temperature distribution (see Figures 2.86 and 2.87)
6. In-furnace heat-flux distribution (see Figure 2.88)

2.5 FUNDAMENTALS OF SOLID FUEL FLAMES

2.5.1 SOLID FUEL FLAME CHARACTERISTICS

The problem of global environmental pollution has become more serious and, therefore, in the field of solid fuel combustion the development of advanced combustion technologies with environmental safeguards to burn more efficiently while restricting

Combustion Phenomena of High Temperature Air Combustion

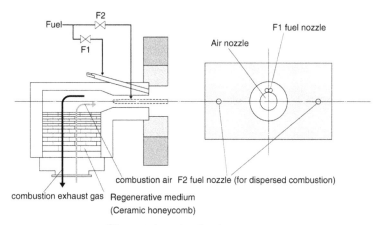

Dispersed combustion type

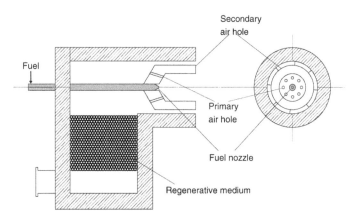

Concentrated combustion type

FIGURE 2.83 Outline of test burners.

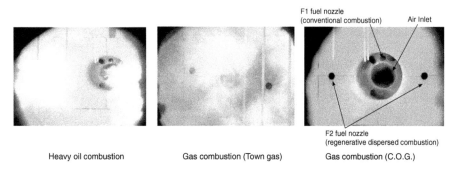

Heavy oil combustion   Gas combustion (Town gas)   Gas combustion (C.O.G.)

FIGURE 2.84 Conditions of flame combustion on types of fuels.

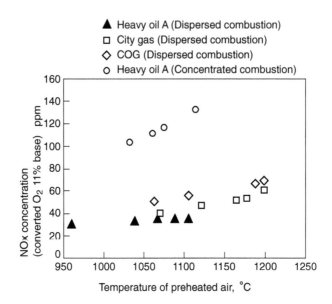

FIGURE 2.85 Comparison of preheated air temperatures and NO_x emission by fuels and combustion methods (in-furnace O_2 = 2.2 ~ 2.8%, in-furnace temperature (TIC) = 1270 ~ 1320°C). (TIC represents Temperature Indicator Controller.)

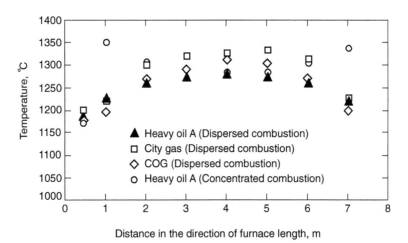

FIGURE 2.86 A 10-min time averaged in-furnace temperature distribution on the center line of burners. (In-furnace temperature (TIC) = 1255 ~ 1265°C, heat input = 1.28 MW (1.1 Gcal/h)). (TIC represents Temperature Indicator Controller.)

emissions of pollutants is a necessity. But even if discussing only solid fuels, we have many kinds such as biomass, coal, solid wastes, or rocket fuels and can find various combustion types according to the types of solid fuels. In this section, we review the combustion technologies for coal as a typical example of solid fuel and explore its fuel properties that are necessary for estimating combustion behavior.

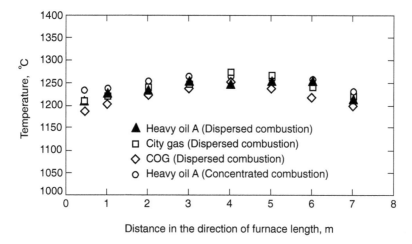

FIGURE 2.87 Horizontally in-furnace temperature distribution on 750 mm from the center line of burners. (In-furnace temperature (TIC) = 1255 ~ 1265°C, heat input = 1.28 MW (1.1 Gcal/h)). (TIC represents Temperature Indicator Controller.)

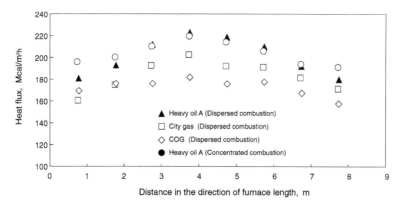

FIGURE 2.88 Comparison of in-furnace heat flux distributions by types of fuel and combustion methods. (In-furnace temperature (TIC) = 1255 ~ 1265°C, heat input = 1.28 MW (1.1 Gcal/h)). (TIC represents Temperature Indicator Controller.)

We will also discuss the main issues regarding combustion phenomena both inside and surrounding solid particles and the final stage of combustion. Finally, we will briefly discuss the reaction phenomena of pulverized coal under high temperatures air combustion conditions.

2.5.2 COMBUSTION PROCESS OF COAL

The combustion phenomena of coal are shown in Figure 2.89[37] in the case of pulverized coal. Combustion phenomena proceed in the order of the following stages. First, pulverized coal is preheated. Then, its volatile matter is emitted and ignited.

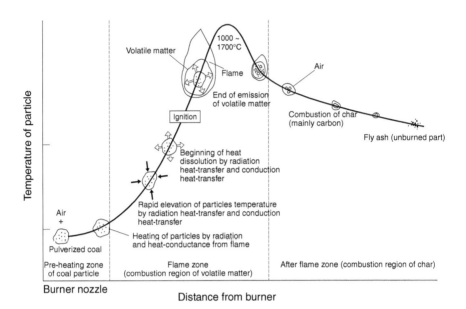

FIGURE 2.89 Combustion phenomena of pulverized coal particles.

Next, both volatile matter and fixed carbon burn after reaching the maximum temperature of particles. Finally, the coal turns to ash through complete combustion. To understand these phenomena clearly and to investigate them more precisely, we take each stage of combustion separately such as the initial properties of coal, the ignition phenomena, the emission of volatile matter, the combustion of char, and the burnout phenomena. The fundamentals of each phenomenon will be explained in this order.

2.5.2.1 Properties of Coal

The combustion phenomena of coal are clearly affected by its fuel properties. If the characteristics of the combustion of coal can be estimated without combustion tests, the total cost of tests of combustion will be reduced as combustion tests on fundamental parts will be unnecessary. The important methods for the purpose are to estimate quantitatively the properties of coals in detail. The results of the industrial chemical analysis based on the JIS standards can be an index to indicate the combustion characteristics of coal. But at present more detailed properties analysis is required, because many kinds of coal are imported to Japan in large quantities from all over the world.

Figure 2.90[38] is an example of a series of property analyses prediction index of combustion characteristics. In addition to the industrial chemical analysis, various data such as chemical composition of metals in ashes, maceral compositions (fine structure), base compositions containing oxygen, calorific value, carbon structure, mineral compositions, molecular weight and chemical compositions of volatile matter, specific surface area, densities, porosity, functional base analysis, and extraction characteristics are listed in this figure. This information is useful for considering

Combustion Phenomena of High Temperature Air Combustion

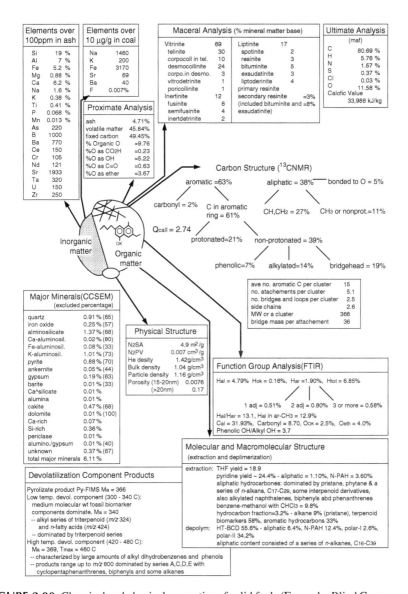

FIGURE 2.90 Chemical and physical properties of solid fuels (Example: Blind Canyon coal).

theoretically the results of reaction experiments. The newest analysis instruments are necessary for obtaining some of these data.

2.5.2.2 Combustion Phenomena around Particles

The most fundamental combustion of solid fuels is combustion of carbon. A model of its combustion phenomena is shown in Figure 2.91.[39] In case of a low surface temperature, the surface reaction (I) $C(S) + O_2 \rightarrow CO_2$ occurs predominantly as shown

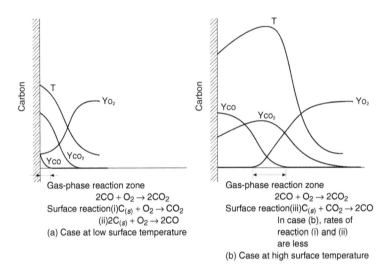

FIGURE 2.91 Combustion model of carbon.

in Figure 2.91a, because the combustion rate of carbon is low and O_2 diffuses to the surface of carbon particles. Moreover, reaction (II) $2C(S) + O_2 \rightarrow 2CO$ occurs simultaneously, through which CO is oxidized to CO_2 immediately near the surface. In the case of high surface temperatures, the surface reaction (III) $C(S) + CO_2 \rightarrow 2CO$ becomes predominant and the generated CO is oxidized to CO_2 in the atmosphere in the same way. Accordingly the temperature of the gas phase, some distance from the surface, becomes higher than that of the particle surface. The former reaction (a) is called the Single Film Model, the latter reaction (b) the Double Film Model.

Ignition of coal is accompanied by the emission process of volatile matter. Combustion of pulverized coal takes place at comparatively high temperatures, so that its emission process is accomplished as quickly as under 100 ms, but the reaction is an important process to obtain stable and low NO_x combustion.

In general, the emission process of volatile matter depends strongly on the initial and succeeding changes in the properties of coal during combustion. Therefore, an exact modeling of coal combustion has not yet been realized. But the simplest model is the following Equation 2.25, assuming that the emission process is a first-order reaction.

$$\frac{dV}{dt} = K(V^* - V), \quad K = A_0 \exp\left(-\frac{E}{RT}\right) \qquad (2.25)$$

where V is the volume of volatile matter emitted previously; V^* is the volume of volatile matter emitted during time, $t \rightarrow \infty$; and E and A_0 are the activation energy and the frequency factor of the process, respectively.

Figure 2.92[40] shows the relation between the rate constant obtained by combustion tests on various kinds of coals and combustion conditions and reciprocal of absolute temperature. These are examples on Arrhenius plots of the emission process of volatile

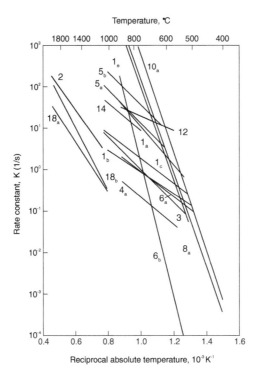

FIGURE 2.92 Rate constant of emission process of volatile matter by the first-order reaction model. (The numbers on the curves indicate types of coal and suffixes indicate combustion condition.)

matter. This figure shows that the rate constant differs according to the types of coal and the combustion conditions. The above-mentioned scattering of data is due to the fact that the rate constants are not normalized by the specific surface area of the particle and are not corrected by actual particle temperatures because the rate of the emission process of volatile matter is controlled by the heat transfer.

Further, the multistage parallel model is proposed by Ubhayakar et al.,[41] Pitt,[42] Kobayashi et al.,[43] etc. The FLASHCHAIN model proposed by Niksa, et al.[44] is mainly used for modeling of the emission process of volatile matter. What is important in this case is that the volume of volatile matter is obtained by industrial analysis only at the temperature of 1173 K. The temperature of actual combustion would often be different from analysis temperature and therefore it is necessary to confirm temperature dependence of the amount of evaporation in advance. In this case, Q factor (VM*/VM) is often taken into account for the analysis (VM* is volatile matter at 1173 K).

Currently, there are few reports on the research of clarifying experimental distribution of gas concentration around solid particles in the region of volatile matter combustion because of the difficulty of its measurement. Therefore, we estimate the general combustion characteristics in a furnace by sampling combustion gas. But the results of gas analysis obtained in the particles dispersion field such as combustion of pulverized coal is generally the average value of all mixed gas around the particles.

Hence investigations[45-47] that deal with mathematical analysis of flame structure using the modeling of single particles are reported.

Figure 2.93[46] shows an example of the calculation of distribution of temperature and gas concentration around particles by the mathematical analysis of the single particle model. This figure is the result of analysis assuming that volatile matter is evolved uniformly from around the particle. This figure shows that the flame temperature becomes higher than that of the particle surface because volatile matter burns separately from the surface. Under this condition, combustion of volatile matter is predominant and little fixed carbon burns.

The analytical calculation indicates that after emission of volatile matter, fixed carbon (char) begins to burn at the particle surface. However, it is reported[48,49] that the emission of volatile matter occasionally occurs in a jet-like state, in which combustion of fixed carbon occurs simultaneously because oxygen can diffuse onto the particle surface.

2.5.2.3 Combustion Phenomena inside a Particle

The cases necessary to consider the combustion phenomena inside particles are the following:

- Char combustion after the process of emission of volatile matter in case of pulverized coal
- Combustion of lump coal that is used in stoker boilers and coal stoves and the case to analyze reaction phenomena inside desulfurization reagent in furnace desulfurization

Various technological models on the reaction phenomena occurring inside particles are proposed because the structure inside particles cannot be identified exactly. The simplest models are: (i) homogeneous reaction model; (ii) nonreacted nucleation model; and (iii) diffusion model which is the case between (i) and (ii). Case (i) corresponds to the model where the diffusion rate of gas in a particle is sufficiently fast compared to the reaction rate, which is comparatively slow, and that the rate-determining factor is reaction in particles, which corresponds to the low apparent density of the particle. However, case (ii) corresponds to the model where the particle is dense, that the rate-determining factor is diffusion in particles, that the gas-solid reaction occurs at the point of reaching the reaction gas, and finally the reacted shell and nonreacted core are formed. Case (iii) corresponds to the model where the reaction proceeds at a rate comparable to that of diffusion, in which case as the reacted portion of particles becomes less, the more inner the reaction of the particle.

2.5.2.4 Final Stage of Combustion

Various models are proposed to estimate the burnout tendency or the complete combustion time of coal particles. In this section, the research summarized by Essenhigh[50] will be introduced. Essenhigh proposes the complete combustion time in the following equation:

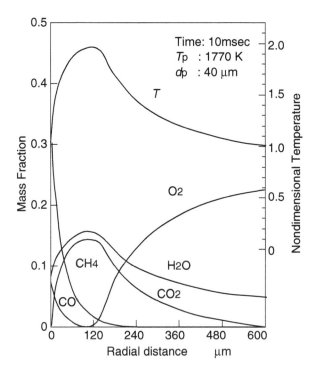

FIGURE 2.93 Flame structure around pulverized coal particles by the single particle model.

$$t^* = \frac{\rho_0}{2C_{A,b}k_s} d_{p,0} + \frac{\rho_0}{8C_{A,b}D_m} d_{p,0}^2 \qquad (2.26)$$

where ρ_0 is the net density of particle. Equation 2.26 coincides with the second-order rule of Nusselt,[51] as shown in the following equation:

$$t^* = \frac{\rho_0}{8C_{A,b}D_m} d_{p,0}^2 \qquad (2.27)$$

Figure 2.94[50] shows the relation between the complete combustion time of coal and the diameter of particle. Combustion temperature is 1773 K. The relation coincides very well with Equation 2.27 in the case of 100-μm particles, which shows clearly the diffusion defined process.

2.5.2.5 Combustion Behavior of Coal at Synthetic Air Condition of High Temperature

To clarify the reaction behavior of coal under the condition of high-temperature air, it is necessary to design a combustion furnace for estimating quantitatively the emission rate of volatile matter. In this section, an example of the reaction behavior

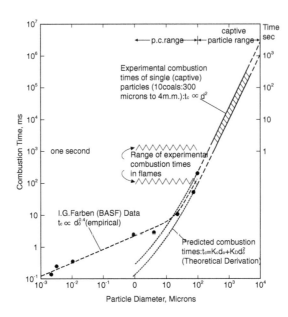

FIGURE 2.94 Relation between complete combustion time and diameter of solid particle.

of coal under the condition where it is actually reacted under synthetic high temperature air is introduced.

This test plant is a horizontal type reaction furnace using pulverized coal under high temperature and high oxygen content, as shown in Figure 2.95.[52] With this furnace, the emission behavior of volatile matter can be simulated. Under such variable factors as the types of coal, the temperature of combustion air, and the oxygen content, the oxygen ratio, tests are conducted by means of sampling and

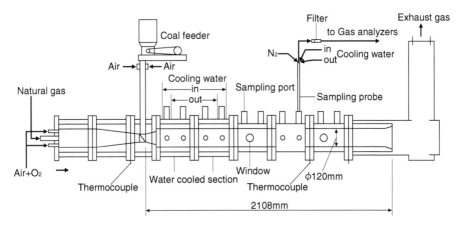

FIGURE 2.95 Outline of horizontal reaction furnace of pulverized coal of high temperature and high oxygen content.

Combustion Phenomena of High Temperature Air Combustion

analysis of reacting particles and gases and by optical measurement of instantaneous particle temperature.

Figure 2.96[52] shows the changes of generated gas concentration and the average temperature of particles and the reaction ratio of volatile matter (VM) and fixed carbon (FC) in the direction of the center of furnace under the test conditions: the kind of coal (WT coal); the air temperature (1300 K); the oxygen ratio (0.8); and the oxygen content (21%). This figure shows that pulverized coal, when injected into the furnace, abruptly emits volatile matter and burns, and according to the elevating particles temperature, O_2 is consumed and CO_2, CO, and H_2 are generated.

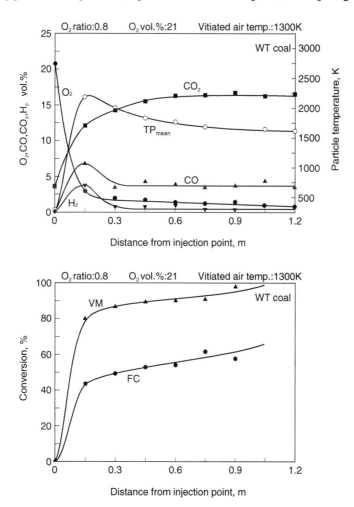

FIGURE 2.96 Change of reaction ratio of concentration of generated gas, average particles temperature, volatile matter (VM), and fixed carbon (FC) as a function of distance from the injection point along the center of the furnace.

As shown in this figure, as abrupt emission of volatile matter occurs under such high temperature reaction as these test conditions, we assumed the following rate constant model (2.28) for quantitative evaluation of emission rate of volatile matter from the ratio of remaining volatile matter in reacting particles obtained by the first sampling port.

$$\frac{x}{t} = k \qquad (2.28)$$

where x is the reaction ratio of volatile matter, t is the time, and k is the reaction rate constant.

$$k = k_0 \exp\left(\frac{-E}{RT_p}\right) \qquad (2.29)$$

where k_0 is the frequency factor, E is the activation energy, R is the gas constant, and T_p is the average temperature of particles.

When pulverized coal is reacted under various test conditions, the relation between the obtained rate constant of emission of volatile matter and reciprocal of particle temperature, which is the so-called Arrhenius plot, can be obtained, as shown in Figure 2.97.[52] Comparing the three kinds of coal in this figure, little difference can be observed. This means the emission rate of volatile matter at the reaction field of high particle temperature over 2000 K was not affected by the type of coal.

2.5.2.6 Summary

In the future, the interest in the combustion of solid fuel will remain centered on the technology related closely to energy and environmental issues. Therefore, it will be increasingly necessary to develop highly advanced efficient processes and to research the basic technologies that will incorporate environmental solutions. At the same time, the topping cycle complex electric power generator that combines combustion and gas generation is now under development and consequently it is important to accumulate the basic knowledge not only in the combustion field but also in the gas-generation and heat-dissolving processes.

2.5.3 EMISSIONS IN SOLID FUEL FLAMES

A pilot trial on high temperature air combustion[53,54] of coal, herein called HTAC99, was carried out by the IFRF in July 1999. This section reports the results of this trial on coal high temperature air combustion comparing them with the results obtained in the previous trials with oil and natural gas. In the section, the high temperature air combustion experiments carried out with natural gas and those with light and heavy fuel oil will be referred to as HTAC97[55-57] and HTAC98,[58] respectively.

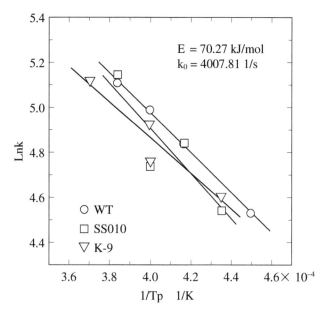

FIGURE 2.97 Arrhenius plot of reaction rate constant of volatile matter.

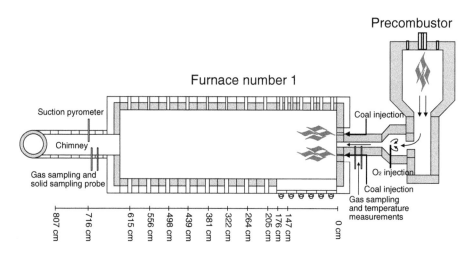

FIGURE 2.98 Experimental setup for the HTAC99 experiments.

2.5.3.1 The Furnace Setup

The HTAC99 trial was carried out in the IFRF furnace No. 1, which consists of 11 water-cooled segments.[56] For each segment the wall temperatures are monitored using thermocouples located on the top and the side of the furnace. The furnace heat extraction is monitored measuring the volumetric flow rate and the temperature rise of the cooling water circulating in each segment. The furnace has a 1-foot square cross-section of 2×2 m for an internal length of 6.25 m. The furnace is refractory lined.

Measurements are carried out using the ports located on both sides of each segment. A pressure transducer is mounted on segment five to monitor the furnace pressure. To avoid air ingress, the furnace is kept in overpressure.

The regenerative air is simulated using a precombustor. The precombustor is operated using natural gas (NG) as fuel. Oxygen is added to the precombustor flue gas to keep the oxygen level to 21% vol wet. After the oxygen addition, the comburent is injected into the furnace. The temperature and composition of the comburent are monitored using a suction pyrometer and a gas-sampling probe, respectively.

An endoscope with a video camera is mounted on the roof of the furnace at 30 cm from the front wall. The image is displayed on a monitor. Selected parts are recorded on video to study the flame behavior.

A schematic of the burner block is shown in Figure 2.99. It consists of a 124-mm air channel located at the center of the burner block. The comburent is injected with an axial velocity of ú 65 m/s, at a temperature of ~ 1350°C. Two coal guns can be located at three different distances from the centerline:

- At 280 mm (the same position of the NG guns and oil injectors in the HTAC97 and HTAC98 trials)
- At 175 mm
- At 385 mm

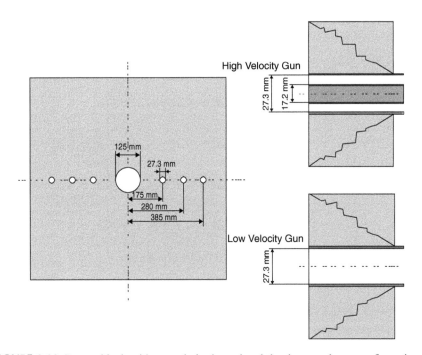

FIGURE 2.99 Burner block with central air channel and the three coal gun configuration.

During the trial two different coal guns were tested:

1. Low-velocity coal gun
2. High-velocity coal gun

The geometric data of the two guns are shown in Figure 2.99.

2.5.3.2 Fuel Properties (Natural Gas/Coal)

Natural gas — The NG used in the precombustor is supplied by CORUS. The NG composition and properties are listed in Table 2.8.

Coal — Guasare coal has been selected for the experiments. This coal is classified as a "high volatile bituminous A coal" by ASTM. Its properties are reported in Table 2.9. The coal was milled to give a particle size distribution of 80% < 90 μm. This coal has been chosen for the experiments because it has a high volatile and low ash content.

Coal plant — The pulverized coal is prepared in the IFRF coal plant. Figure 2.100 shows a sketch of the coal-feeding system used during the HTAC99 trial. The pulverized coal is fed through a compressed air line into the furnace. The distributor feeds the coal mass flow in the two coal guns. The pressure is monitored in the coal lines to assure the same coal mass flow in the two guns.

2.5.3.3 Experimental Program

The experiments consisted of detailed in-flame measurements and input/output measurements. Detailed in-flame measurements were taken for the baseline flame with the coal gun positioned at 280 mm from the centerline. In-flame measurements include mean and rms axial velocities, gas temperature, CO_2, O_2, CO, NO_x, CnH_m, solids, solid concentration, total radiance, and total radiative fluxes at the furnace wall.

TABLE 2.8
Natural Gas Composition and Properties

	Vol. %
CH_4	87.82
C_2H_6	4.59
C_3H_8	1.59
C_4H_{10}	0.52
C_5H_{12}	0.13
O_2 + Ar	0.02
CO_2	1.65
N_2	3.68
C/H ratio (by mass)	3.20
O_2 requirement (kg$_2$O/kgfuel)	3.57
Density kg/m_N^3 at 15°C	0.822

TABLE 2.9
Guasare Coal Properties

	Ultimate Analysis Dry Basis					
	C	H	N	O	S	Ash
wt %	78.41	5.22	1.49	10.9	0.82	3.3

	Proximate Analysis Dry Basis			
	% Moisture (105°C)	Volatiles %	Fixed Carbon %	LCV
wt %	2.9	37.1	56.7	31.74 MJ/kg

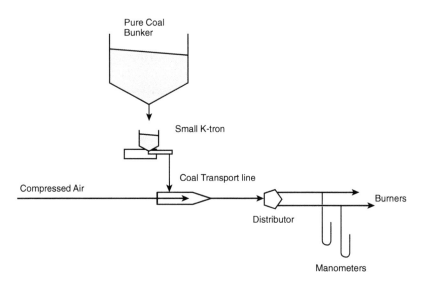

FIGURE 2.100 Coal feeding system for the HTAC99 experiment.

Velocity measurements (mean velocity and rms) were performed using the IFRF watercooled laser doppler velocimetry (LDV) probe. In-flame temperatures were measured using a suction pyrometer equipped with a type B thermocouple (Pt 6%Rh/Pt 30%Rh). Local in-flame gas compositions were measured using a gas-sampling probe. Measurements of total radiance and total radiative heat fluxes were performed using standard IFRF narrow angle radiometer and ellipsoidal radiometer probes, respectively. Solid sampling was performed using the IFRF static solid sampling probe. A detailed description of the probes for detailed in-flame measurements can be found in References 56, 58, and 59. In the input/output measurements the furnace flue gas temperature, CO_2, O_2, CO, NO_x, C_nH_m, burnout were measured for 15 different flames. The parameters that were examined are:

- Position of the coal guns
- Amount of the coal transport air
- Transport air velocity
- NO_x level in the comburent from the precombustor

2.5.3.4 In-Flame Measurements

Detailed in-flame measurements were performed for a baseline flame, in which the coal guns were positioned in the same configuration used for the oil and NG injectors. In the baseline configuration the low velocity coal gun was used with 130 kg/h transport air resulting in an air–coal ratio of 1.96 kg_{air}/kg_{coal} and an injection velocity of 26 m/s. The air–coal ratio and the coal velocity injections have values typically used in pulverized coal combustion applications. The furnace wall temperatures were around 1250°C (see Figure 2.101).

The natural gas and light fuel oil flames examined in the HTAC97 and HTAC98 trials, respectively, were similar in appearance. Both types of flames were difficult to distinguish from background radiation with no clear visible flame.[56,58] In contrast to the NG and LFO flame, the HFO test revealed two luminous jets approximately 2 m in length. During the HTAC99 trial a visible flame was clearly distinguishable from the background radiation. Figure 2.102 shows the flame for the baseline configuration. It is similar to the HFO flame.

All the in-flame measurements were performed in the plane of the coal guns and their results are tabulated in the appendices. In the following sections the main results will be presented. Each graph consists of two parts: contour lines are plotted on the left-hand side of the figure, whereas profiles are plotted on the right-hand side. To facilitate visual understanding, an outline of the furnace and burner geometry is drawn overlaying the measurements.

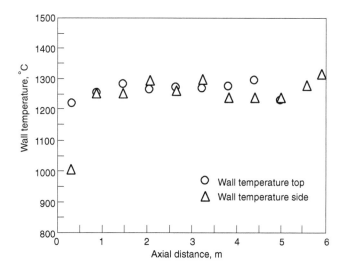

FIGURE 2.101 Wall temperature (top and side).

FIGURE 2.102 Baseline coal flame.

2.5.3.4.1 Heat and Mass Balance

Detailed heat and mass balances are shown in Figure 2.103 for HTAC97, HTAC98, and HTAC99 experiments.

The composition of the comburent going into the furnace is similar for the three cases. During the HTAC97 preheated air was used in the precombustor resulting in higher NO_x concentration in the comburent. The fuel thermal input in the HTAC99 is 580 kW as in the previous trials. The coal burner is operated at 20% excess air, giving a stoichiometric ratio of $\lambda = 1.2$. The furnace efficiency is calculated as:

$$\eta = \frac{H_{ext}}{H_{in}} \times 100$$

2.5.3.4.2 Gas Composition

The gas composition profiles show smooth gradients as was already found in the previous HTAC trials for natural gas and oil. Almost the whole furnace volume is filled with combustion products with the flue gas composition. Combustion seems to take place in large furnace volume. To compare the previous trials with the HTAC99 in the following figures the results of the HFO flame are reported together with the gas composition profile in the coal flame.

Oxygen — Figure 2.104 shows plots of the O_2 measurements in the coal flame on the left side and in the HFO flame on the right side. The furnace is filled with combustion products at 3% O_2. Practically everywhere in the furnace this O_2 level is measured. Only at the comburent inlet high O_2 concentrations are measured (up to 23% vol. dry).

Carbon monoxide — Figure 2.105 shows the CO concentration in the furnace for the coal flame and HFO flame. It reported the 5000 ppm contour line that may be considered the flame boundary. On the fourth traverse (133 cm from the front wall) high CO concentrations were measured; in the fifth traverse (205 cm) CO up to 3000 ppm was measured. From these measurements, a flame length of almost 2 m can be estimated. This flame length matches with the "visible flame" length; a clear flame was present until the fourth segment (205 cm). Carbon monoxide

Combustion Phenomena of High Temperature Air Combustion

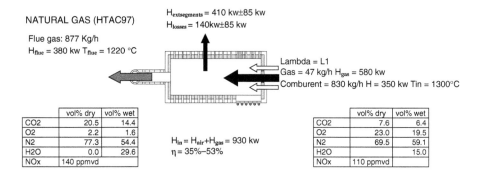

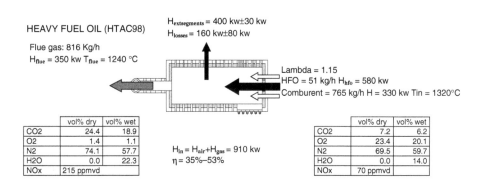

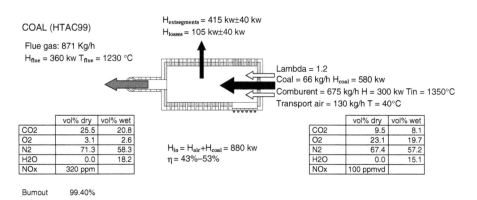

FIGURE 2.103 Heat and mass balance of the furnace.

emissions were below 50 ppm at the chimney. The HFO flame shows a similar flame length. However, on the first traverse a second peak in the CO concentration was measured (on the oil jet boundary at the wall side). The NG and LFO flames had a clearly different CO profile. High CO concentrations were measured farther

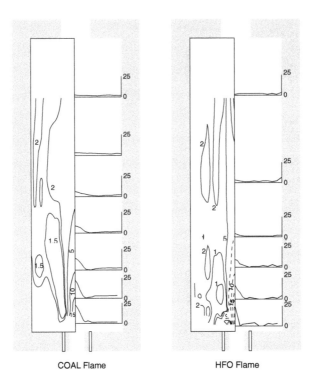

FIGURE 2.104 Oxygen concentration coal flame (Vol%, dry).

downstream in the furnace and both NG and LFO flames seemed to have a much larger flame volume. The highest input velocity of the fuel jets may explain this difference. The NG jets had an input velocity of 100 m/s. Because of the higher momentum of the jets, the entrainment of recirculation products before mixing with the combustion air is higher for the NG and LFO flames.

NO$_x$ — The NO$_x$ profile shows a clear peak on the coal jets (Figure 2.106). Farther downstream the profile is flatter and closer to the outlet value.

Unburned hydrocarbons — No high C$_n$H$_m$ concentration was measured in the coal flame (Figure 2.107). The peak concentration of 1520 ppm was measured in the second traverse. No C$_n$H$_m$ was found at the chimney. In the HFO flame the C$_n$H$_m$ concentrations have a similar profile but much higher values (as can be seen in Figure 2.107, where the C$_n$H$_m$ for HFO are presented in vol% dry).

2.5.3.4.3 Temperature Measurements

Figure 2.108 shows the temperature contour and profile in the furnace for the coal flame (on the left side) and for the HFO flame (on the right side). The temperature field in the coal flame is uniform almost in the entire furnace and is similar to the NG and oil flames. The whole furnace is filled with combustion products at temperatures in the range of 1350 to 1500°C.

Temperature gradients were measured only in the coal jets at the first two traverses. The coal jets are immersed into hot combustion products and this produces

Combustion Phenomena of High Temperature Air Combustion

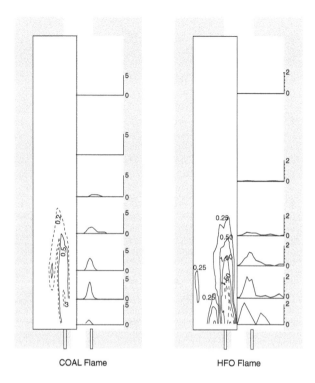

FIGURE 2.105 Carbon monoxide concentration (Vol%, dry).

a very fast heat up and devolatilization of the coal particles. The visible ignition occurs at 10 to 15 cm from the coal injection. The strongest gradient in the temperature profile was found in that region at the first (15 cm) and second (43 cm) measured traverses.

At the first traverse the "lowest temperature" (1250°C) was measured and at the second traverse the peak temperature was 1497°C. The peak flame temperature is 150°C higher than the preheated temperature of the comburent. In this region the most intense combustion takes place.

2.5.3.4.4 Velocity Measurements

Figure 2.109 shows the velocity contour and profiles in the furnace. At the first three measured traverses both the comburent and coal jets are clearly visible. The strong central air jet enters the furnace with a high velocity of ~65 m/s.

The central jet is mainly producing the flow pattern in the furnace. The jet is expanding and it builds a large recirculation zone. The coal jets enter the furnace in the recirculation zone of the central air jet with a velocity of 26 m/s. They are no longer distinguishable from the central jet after the third traverse.

The measured peak velocity in the central air is lower than the expected velocity. The peak measured velocity is on traverse 2 (47 m/s); on the first traverse the highest measured velocity is 42 m/s. In the first and second traverses, a velocity very close to the inlet one (65 m/s) was expected. According to the free jet theory in the second

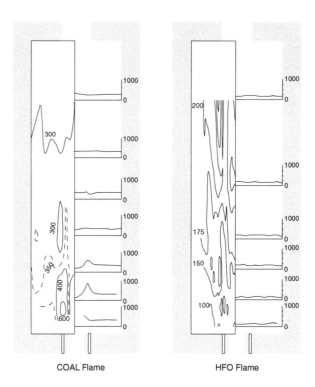

FIGURE 2.106 NO_x concentration measured in the experiment at HTAC99 (ppmvd).

and third traverses the jet is in the potential core region (x/d = 1.2 and 3.4, respectively), where velocity remains equal to the initial velocity. The measured low velocities may be explained considering that during the measurements the central air jet was not seeded. In the first two traverses, only a small amount of coal particles is entrained by the air and accelerated from the recirculation velocity up to a high inlet air velocity. That resulted in a poor signal in the LDA measurements in the center of the air jet at the first two measured traverses and in a lower measured velocity.

These considerations are confirmed by the calculation of the entrained mass flow in the air jet. Figure 2.110 shows the entrained mass flow in the central jet calculated as the integral of the measured velocity within the jet boundary.

The calculation was performed only in the first four traverses because farther downstream it was not possible to define the central air jet boundary. In the first traverses the central jet boundary was considered the zero tangent condition in the velocity profile. Such an integration is always associated with inaccuracies due to the integration procedure, the assumption of the jet axial-symmetry and the errors associated with temperature and gas composition measurements.[56] In the calculation the gas composition is assumed to be similar to the flue gas, which has a density of 0.2 g/m^3. Finally, the mass flow is corrected for the temperature effect of density.

Figure 2.109 shows that in the first two traverses the calculated mass flow is less than the input mass flow. At the third and fourth traverses, the calculated entrainment

Combustion Phenomena of High Temperature Air Combustion 141

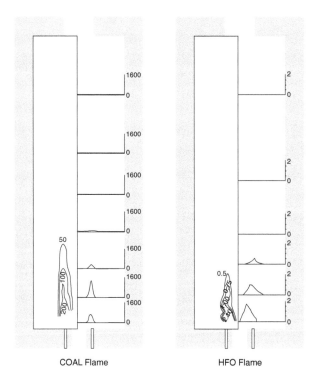

FIGURE 2.107 Hydrocarbon concentration (ppmvd for coal flame at HTAC99 and vol% dry for HFO flame at HTAC98).

is consistent with the free jet theory. At that location the velocity measurements are more accurate. That confirms that the velocity measurements at the first two traverses in the center of the air jet were not fully representative of the gas phase velocity.

2.5.3.4.5 Burnout

Solid samplings were taken at the chimney and on the coal jet center line. The burnout level was calculated using the ash tracer technique assuming that the mineral matter is conserved during the combustion. The burnout is defined as follows:

$$\text{Burnout} = \frac{1 - \dfrac{\text{Ash}_{initial}}{\text{Ash}_{sample}}}{1 - \text{Ash}_{initial}}$$

Figure 2.111 shows the burnout profile on the center line of the coal jet. The visible ignition was at ~10 cm from the front wall. No measurements were performed before the ignition; the first measured port was at 73 cm with a burnout level of ~65%. High burnout level was found at the chimney (99.4%).

This burnout value corresponds to a carbon-in-ash content of 25%. This carbon-in-ash content is too high for commercial use of the fly ash (less than 5%

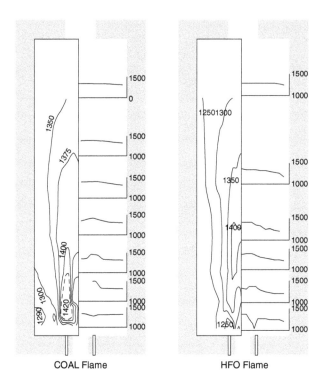

FIGURE 2.108 Temperature coal flame (degree centigrade).

carbon-in-ash is required). However, considering the very low ash content of the coal, carbon-in-ash problems were expected and it seems to be very difficult to reach lower carbon-in-ash levels.

2.5.3.4.6 Solid Concentration

In Figure 2.112 the solid concentration at axial distance of 205 cm from the front wall is reported. In view of the time necessary to carry out each measurement (~1 h/point), only one traverse and the flue gas solid concentration were measured.

Solid concentration measurements were performed with the standard IFRF solid sampling probe. This technique requires measurements of both the mass of solid collected in the filter (msolid,amb) and the dry volumetric flow rate ($V_{gas,amb,dry}$):

$$C_{solid} = \frac{m_{solid, flame}}{V_{gas, flame}} = \frac{m_{solid, amb}}{V_{gas, amb, dry}} \frac{T_{amb}}{T_{flame}} \left(1 - X_{H_2O}\right)$$

where:

$m_{solid,flame}$ mass of the sampled solid in the flame
$V_{gas,amb,dry}$ dry volume of gas sampled in the flame at ambient temperature
T_{amb} ambient temperature during the measurements
T_{flame} flame temperature at the measured point

Combustion Phenomena of High Temperature Air Combustion

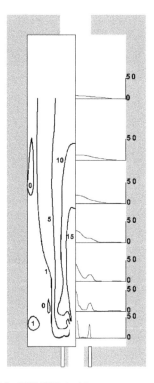

FIGURE 2.109 Velocity field in HTAC99 (m/s).

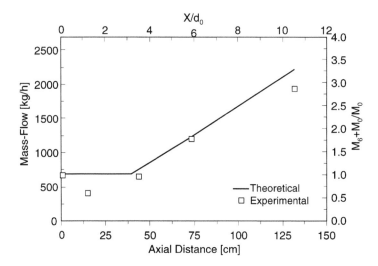

FIGURE 2.110 Central air jet entrainment.

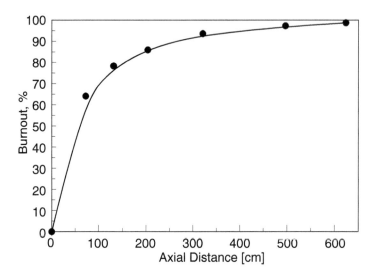

FIGURE 2.111 Burnout profile at the coal jet center line.

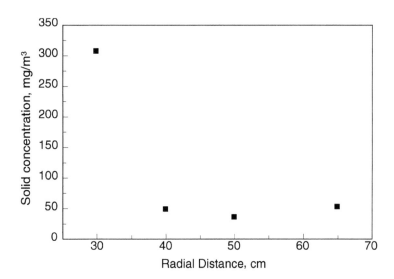

FIGURE 2.112 Solid concentration at axial distance 205 cm from the front wall.

For every measured point:
- T_{flame} the measured in-flame temperature (T_{flame} ~1600 K)
- X_{H_2O} calculated from the measured gas composition (X_{H_2O} ~0.17)
- T_{amb} 25°C

The solid concentration measurements show a peak of 305 mg/m³ at the coal jet center line. These measurements were used to perform the analysis of the total radiance measurements, as will be further discussed. The solid concentration in the flue gas must be calculated in mg/m_N^3 to enable comparison with legislation; the formula to calculate the solid concentration in mg/m_N^3 is as follows:

$$C_{soot}\left[mg/m_N^3\right] = C_{meas}\left[mg/m^3\right] \frac{T_{flame}}{273.15}$$

The flue gas solid concentration was found around 1327 mg/m_N^3. In the HFO flame the solid particle in the flue gas was estimated to be about 2400 mg/m_N^3. The lower value in the coal flame may be explained considering the low ash content of the coal and insufficient atomization in the HTAC98 trial.

2.5.3.4.7 Total Radiative Heat Flux

Figure 2.113 shows the total radiative heat flux for all four fuels, NG, LFO, HFO, and coal. The heat flux profiles are very flat for all four fuels examined.

The heat flux values for the coal flame vary between 350 and 390 kW/m² and the profile is very similar to that of the oil flame. The lower heat fluxes for light fuel oil had already been discussed in the HTAC98 report.

The measured heat fluxes are high compared with the value in the normal combustion system (~250 kW/m² for coal combustion and 150 kW/m² for NG combustion). Enhancement in the heat fluxes characteristic of the high temperature air combustion was observed in the coal flame as well but not in the same magnitude as in the NG flame. The higher heat fluxes measured in the coal flames compared to the NG flame are due to the presence of solid/soot particles. However, the radiative heat fluxes measured in the coal flame were not much higher than in the NG flame (as it can be seen in Figure 2.112). In both flames the furnace temperature conditions were very similar, and it may be inferred that the high values are largely due to wall radiation.

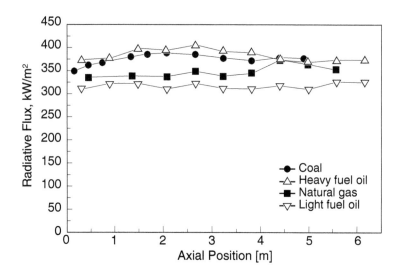

FIGURE 2.113 Total radiative heat fluxes for NG, LFO, and HFO flames.

2.5.3.4.8 Total Radiance

Total radiance measurements were performed by traversing the flame with a cold target sighted by a narrow angle radiometer probe,[60] which remained in a fixed position at the opposite wall of the furnace (see Figure 2.114).

To minimize reflections at the target surface, this was blackened using Zynolite paint with a total normal emissivity of 0.94. The paint was shielded from the hot combustion gases using an N_2 purge. During the measurements, the cold target temperature was kept between 25 and 45°C. Total radiance measurements have been carried out 205 cm downstream of the burner outlet, and have been compared with model predictions (Figure 2.115).

Total radiance predictions have been performed using the Exponential Wide Band Model[61] and the Chan and Tien scaling method.[62] This model is formulated for nonhomogeneous and nongray gas/soot mixtures.[63] The total radiance leaving a nonscattering media of length L is given by the following equation:

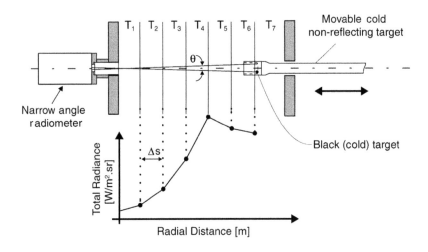

FIGURE 2.114 Schematic of the measurement method.

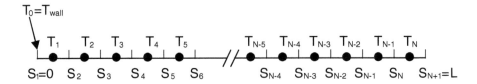

FIGURE 2.115 Discretization of the gas layer for solving Equation 2.30.

$$I(L) = \int_{v=0}^{\infty} I_v(0)\tau_v(0,L)dv + \left\{ \int_{v=0}^{L} \frac{\partial \tau_v(s',L)}{\partial s'} I_v^0(s')ds' \right\} dv \quad (2.30)$$

where $I(L)$ is total medium intensity in W/(m² sr), $I_v(0)$ is the spectral intensity incident upon the medium, I_v^0 is the blackbody spectral intensity in W/(m²·sr·cm⁻¹), and v is the wave number in cm⁻¹. $\tau_v(s',L)$ is the spectral emissivity of a column of gas of length $(L - s')$, and is defined as:

$$\tau_v(s',L) = \exp\left(-\int_{s'}^{L} k_s^v(s'')ds''\right) \exp\left(-\int_{s'}^{L} k_a^v(s'')ds''\right) = \tau_s^v \tau_g^v$$

where k_s^v and k_a^v are the spectral absorption coefficients of soot and gas, respectively. Similarly, τ_s^v and τ_g^v are the spectral soot and gas emissivity.

As schematized in Figure 2.114, the solution of Equation 2.30 was obtained upon discretizing the gas layer into N homogeneous cells, over which temperature, total pressure, and gas composition were assumed constant. The values of temperature and species concentration that are required for the calculations are those obtained from the in-flame measurements. These were taken at the same axial distance as the total radiance measurements.

Under the following assumptions:

1. The radiation incoming from the wall is well approximated by a blackbody emitter at the temperature T_w so that: $I_v(S = 0) \cong I_v^0(T_w)$
2. The variations in the blackbody function are small over the spectral intervals.
3. The mechanisms of gas emission–absorption take place in M non-overlapping bands.

Equation 2.30 becomes:

$$I(L) = I_{soot}(L) + \sum_{j=1}^{M} \sum_{k=0}^{N-1} \left[I_{v_{e',j}}^0(T_{k+1}) - I_{v_{e',j}}^0(T_k) \right] A_{nhj}(S_{k+1}, S_{N+1}) \tau_{S_{k+1}}^{v_{e',j}} \quad (2.31)$$

$$I_{soot} = \frac{15}{\pi^4} \frac{\sigma T_0^4}{\pi} \Psi^3\left(1 + \frac{C_0 \beta_N T_0}{C_2}\right)$$

$$+ \frac{15}{\pi^4} \sum_{k=1}^{N} \frac{\sigma T_k^4}{\pi} \left[\Psi^3\left(1 + \frac{C_0 \alpha_{k+1} T_k}{C_2}\right) - \Psi^3\left(1 + \frac{C_0 \alpha_k T_k}{C_2}\right) \right] \quad (2.32)$$

where

v_c	the frequency at the center of the band in cm^{-1}
σ, C_2	the Stefan-Boltzmann constant and the second Planck constant, respectively
A_{nh}	the non-homogeneous total band absorptance, A_{nh}, is calculated using the Chan and Tien scaling method
$\tau_{sk+1}^{v_c'j}$	the soot emissivity on the jth gas band over the path length $L - S_{k+1}$. This term accounted for the effect of overlapping bands between the gas mixture and the soot. The soot emissivity is calculated using the soot spectral absorption coefficient, approximated by the relation:

$$k_s^v = \frac{C_0 f_v v}{10^4} \qquad (2.33)$$

where f_v is the soot volumetric fraction in [m³/m³], v is the wave length in cm^{-1}, and C_0 is a constant

ψ^3	the pentagamma function
α_k and β_k	the integrals of the soot volumetric fraction over the path length $L - S_k$ and $S_{k+1} - 0$, respectively
I_{soot}	the total radiance from soot only

First, total radiance predictions have been performed using Equation 2.31 and considering gas phase, only. Comparison between measured and predicted radiance profiles shows that the presence of particulate significantly enhances radiation.

Second, total radiance calculations have been performed using the measured particulate concentration at the axial distance of 205 cm. No solid concentration measurement has been performed between $R_D = -25$ cm and $R_D = +25$ cm. The particulate concentration in this area has been supposed equal to the recirculation zone particulate concentration.

For the sake of simplicity, it was considered that the particles radiate as soot particles. The density of particulate was taken equal to 1800 kg/m³. The constant C_0 typically lies between 4 and 10 for soot. The value 10 has been chosen arbitrarily for the predictions.

Good agreement between measurements and calculations has been obtained (see Figure 2.116). The experimental values and prediction curve follow a similar trend. However, the model does not capture the large variation of total radiance occurring in the flame region between $R_D = +0.0$ cm and $R_D = +20$ cm. The lack of solid concentration measurements between these two radial distances does not permit a more accurate analysis.

2.5.3.5 Input/Output Measurements

During the input/output measurements the effects of different parameters have been investigated. The furnace conditions have been kept at the same level for all the

Combustion Phenomena of High Temperature Air Combustion

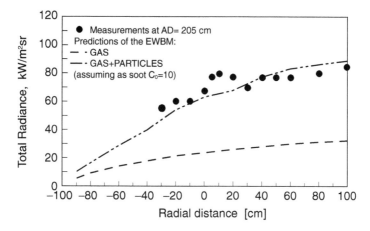

FIGURE 2.116 Comparison between total radiance predictions and total radiance measurements.

TABLE 2.10
Baseline Output Conditions for the Parameters Study

Temperature, °C	1100
Oxygen, vol. % dry	2.9
NO_x, ppm	216
Burnout, %	98

measured flames, the flue gas temperature was ~1100°C and wall temperature ~1150°C. Table 2.10 reports the outlet quantities for the baseline flame under these furnace conditions.

Table 2.11 reports the input conditions in the different configurations. The parameters that were investigated are:

1. Gun position
2. Transport air velocity
3. Transport air mass flow
4. Precombustor NO_x level

The position of the coal guns (configuration) has been studied for the low- and high-velocity coal guns (flames 1, 2, 3 and 8, 9, 10 in Table 2.11). The transport air mass flow has been studied in the baseline configuration for high and low coal gun velocity (flames 2, 4, 5 and 5, 6, 7 in Table 2.11). The NO_x level at the inlet has been investigated in the baseline configuration with high-velocity coal gun (flames 11, 12, 13, 14, 15 in Table 2.11).

TABLE 2.11
Input/Output Conditions Measurements

Flame No.	Configuration	Coal Gun	Transport Air Mass Flow, kg/h	Transport Air Velocity, m/s	Precombustor NO_x Level, ppm
1	o●o○o●o	Low velocity	130	26	100
2	o●o○o●o	Low velocity	130	26	100
3	●oo○oo●	Low velocity	130	26	100
4	o●o○o●o	Low velocity	215	45	100
5	o●o○o●o	Low velocity	80	15	100
6	o●o○o●o	High velocity	130	43	100
7	o●o○o●o	High velocity	215	71	100
8	o●o○o●o	High velocity	80	28	100
9	oo●○●oo	High velocity	130	43	100
10	●oo○oo●	High velocity	130	43	100
11	o●o○o●o	High velocity	130	43	100
12	o●o○o●o	High velocity	130	43	146
13	o●o○o●o	High velocity	130	43	155
14	o●o○o●o	High velocity	130	43	130
15	o●o○o●o	High velocity	130	43	140

2.5.3.5.1 Coal Gun Position

Three different gun positions have been tested (for both low- and high-velocity coal guns) as has already been described. Figure 2.117 shows the flue gas NO_x level for the three coal gun positions.

NO_x emissions decrease from configuration 1 (coal gun position close to the central air jet) to configurations 2 and 3 (coal gun farther from the air jet). The same trend was found for low-velocity gun and high-velocity gun. This effect is due to the oxygen availability in the primary zone of the flame. Figure 2.118 shows pictures of the flame in configurations 1 and 3 for the low-velocity guns.

In configuration 1 the coal is mixing faster with the central air jet, and combustion is taking place at the boundary of the central air jet in a region with O_2 vol% dry higher than 3% (flue gas level). The flame is practically attached. Configuration 3 shows a different flame shape. The coal is injected far from the central air, it entrains a large amount of combustion products with 3% O_2 before the ignition. The flame was lifted; the ignition stand off distance of the coal jet was ~50 cm from the burner. The combustion mainly takes place in a lower O_2 % environment (fuel-rich zone) resulting in a lower NO_x flame. Figure 2.119 shows a sketch of the effects of the separation between central air and coal injections on the mixing between fuel and comburent.

The burnout level was high in all three coal gun configurations. Figure 2.120 shows the burnout % for high- and low-velocity coal guns in the three configurations.

The burnout measurements show for both high- and low-velocity coal guns a similar trend slightly decreasing from configuration 1 to configuration 2 and then increasing in configuration 3. Configuration 2 has an O_2 profile on the center line

Combustion Phenomena of High Temperature Air Combustion 151

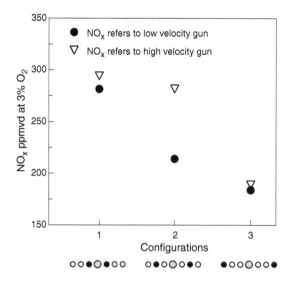

FIGURE 2.117 NO_x emissions vs. the coal gun position for low- and high-velocity coal guns.

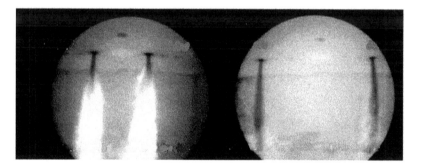

FIGURE 2.118 Flame picture for configurations 1 and 3.

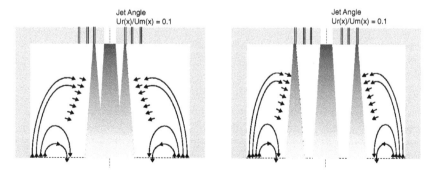

FIGURE 2.119 Mixing in the three coal gun positions.

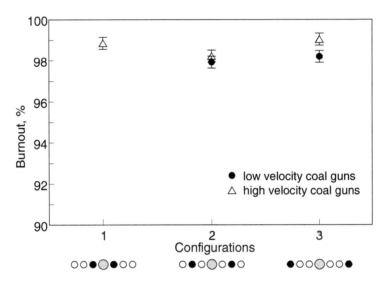

FIGURE 2.120 Burnout in the three different configurations for high- and low-velocity coal guns.

passing for a zero oxygen concentration (see Figure 2.104); no in-flame measurements were performed for the other two configurations. It may be inferred that in configuration 1 the fast mixing with the central air jet and in configuration 3 the delayed ignition (high entrainment before the combustion starts) probably reduce (or avoid) this low oxygen region in the coal jet. The intermediate configuration 2 has the more relevant drop in the oxygen concentration on the centerline of the coal jet resulting in a lower outlet burnout level. In configuration 3 the flame is more detached compared with configuration 2 and the char burnout reaction starts later. However, the char reaction starts with high O_2 availability, and before the post flame zone reaches a higher burnout level than in configuration 2.

2.5.3.5.2 Coal Transport Air Mass Flow

Three coal transport air mass flow had been tested for high- and low-velocity coal guns keeping the coal feeding rate at 66 kg/h and the coal gun position in the baseline configuration (coal gun at 280 mm from the central air jet):

1. 80 kg/h (1.21 kg of air/kg of coal)
2. 130 kg/h (1.97 kg of air/kg of coal)
3. 215 kg/h (3.26 kg of air/kg of coal)

The flames with low transport air mass flow(80 kg/h) were not stable. The low momentum coal jet was affected by the strong central air jet and entrained in the recirculation zone of the central air jet. Figure 2.121 shows the photographs for the low transport air mass flow and baseline transport air mass flow (130 kg/h) with high-velocity coal guns.

Combustion Phenomena of High Temperature Air Combustion 153

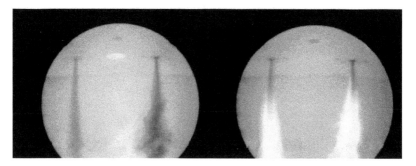

FIGURE 2.121 Low coal transport air mass flow (left) and baseline transport air mass flow (right) with high-velocity coal guns.

The lower O_2 availability (lower transport air) and the low momentum of the coal jet determine that this is not a fully stable condition. The resulting burnout level was low for both the low- and high-velocity coal guns (see Figure 2.122).

The two flames with high transport air mass flow were similar (see Figure 2.123). Both flames showed an ignition lifted 35 to 40 cm from the front wall; the ignition stand-off distance was slightly longer for the high-velocity coal gun. Burnout level in the flue gas for both the flames was higher than for the baseline case (see Figure 2.122).

At all three tested transport air mass flows, the burnout shows an increase from the low-velocity coal gun to the high-velocity coal gun. The coal jet momentum effect on the flue gas burnout level is summarized in Figure 2.124. In this figure the

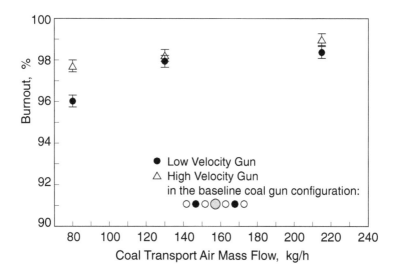

FIGURE 2.122 Burnout level for high- and low-velocity coal guns at different coal transport air mass flow.

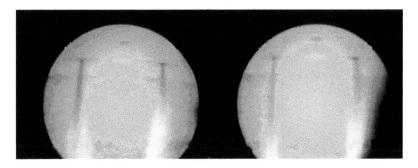

FIGURE 2.123 High coal transport air mass flow for low- (left) and high- (right) velocity coal gun.

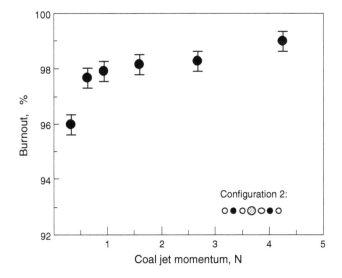

FIGURE 2.124 Burnout vs. the coal jet momentum.

coal jet momentum is reported on the x-axis; it should be pointed out that in flames considered here the momentum is changed in two different ways:

1. Changing the coal gun type (low- and high-velocity coal gun)
2. Changing the transport air mass flow (from 80 up to 215 kg/h)

The two flames at lower coal jet momentum are the two unstable flames already discussed (low transport air mass flow). For the other four flames, an increase in the burnout level is measured, increasing the momentum. This effect is due to the higher oxygen availability in the primary flame zone. The high momentum flames are more lifted, at the ignition front more oxygen is available, and the burnout reaction starts faster than in the low momentum flame.

The effect of the different ignition position is clear on the NO_x emissions as well. In the high momentum flames (transport air 215 kg/h), the coal jet entrains more oxygen before ignition takes places and more oxygen is available from the transport air (higher mass flow).[64] Thus, the local stoichiometry in the flame zone is higher compared to attached (or less lifted) flames; the oxygen availability at the ignition is higher compared to the baseline condition (130 kg/h transport air) resulting in higher NO_x emissions (see Figure 2.124).

The higher local stoichiometry in the primary flame zone may be due both to the higher transport air mass flow and to the higher momentum. The second parameter is thought to be more important as can be seen from Figure 2.125. At the same transport air mass flow the NO_x emissions for the high-velocity gun flames are higher. High momentum flames are more lifted and entrain more oxygen before the ignition.

2.5.3.5.3 Precombustor NO_x Level

The effect of the different NO_x levels in the comburent was studied in the baseline configuration (coal gun position at 280 mm from the central air and transport air 130 kg/h) for the high-velocity coal gun. The comburent flow was kept at the same temperature changing only the NO_x concentration. Figure 2.126 shows the NO_x emissions vs. the inlet NO_x concentration in the comburent.

As expected, an increase in the flue gas NO_x emissions with an increase in the inlet NO_x concentration was found. However, this correlation is not linear. It seems that the flue gas NO_x emissions tend to level off at an NO_x input of 100 ppm, tending to an asymptotic value slightly lower than 250 ppm for zero NO_x concentration in the comburent. Furthermore, it may be concluded that small changes on the NO_x input around 100 ppm (the level during the input/output measurements) do not change the NO_x level significantly in the flue gas.

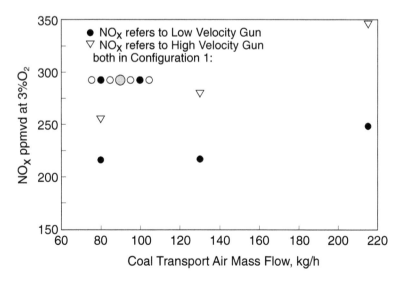

FIGURE 2.125 NO_x level for high- and low-velocity coal guns at different coal transport air mass flow.

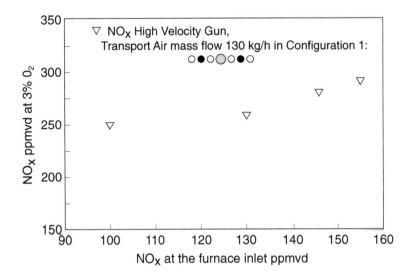

FIGURE 2.126 Flue gas NO_x emissions vs. inlet NO_x concentration.

2.5.3.6 Summary

During the HTAC99 trial, the high temperature air combustion of coal was examined. A 580-kW burner was used in IFRF furnace No. 1 in a geometric configuration similar to the previous HTAC trials (the same distance between central air and fuel injection). Detailed in- flame measurements were performed in one flame (baseline flame) and the effects of several parameters on the flue gas emissions were studied.

Analysis of the collected data in the three HTAC trials shows that high temperature air combustion works for NG, LFO, HFO, and coal. The LFO and NG flames were very similar, and no visible flame was observed. With firing coal and HFO a visible flame was observed under all the conditions.

All the examined fuels showed similar in-flame gas composition and temperature fields. The gas composition and temperature tend to be uniform all over the furnace. The combustion is slow and takes place after the fuel jets have been diluted by the entrainment of hot combustion products. The in-flame peak temperature is largely reduced compared with standard combustion techniques with such a highly preheated air temperature.

The peak in-flame temperature was measured around 1500°C for NG and coal (slightly lower for the HFO and LFO). That value is surprisingly low considering the high preheated level of the comburent between 1300°C (NG) and 1350°C (coal). The temperature rise between air inlet and peak in-flame temperature is as low as 150°C; in standard combustion technique this value is usually larger than 1000°C. The in-flame temperatures are between 1300 and 1500°C. At this temperature level the thermal NO_x mechanism is nearly suppressed.

All along the furnace the measured heat fluxes were flat, high, and in similar ranges for all the fuels that were tested. The heat flux was slightly lower for LFO. In particular for NG combustion these heat flux values were surprisingly high.

The particulate concentration in the flue gas for the HFO and so for the coal flame is probably too high to operate the technique with a regenerative heat exchanger. As was already concluded in the HTAC98 report, for the HFO the atomization should be improved to reduce the particulate emissions.

It can be concluded the coal high temperature air combustion is a promising technique with low emissions and high heat fluxes. The NO_x emissions were low compared with standard combustion techniques under similar furnace conditions. NO_x emissions were measured in the range 250 to 350 ppm; in standard combustion techniques a much higher value (on the order of 600 to 700 ppm) may be expected.

Further studies are needed on "air preheating" to apply high temperature air combustion to coals. The ash deposition in the regenerative part should be investigated or a "different" source of hot air should be tested (as has already been studied for firing unclean fuels with high preheated temperature[65]).

In high temperature air coal combustion, different conditions have been tested to determine the influence of the following parameters on the NO_x and burnout in the flue gas:

1. Coal jet velocity
2. Transport air mass flow
3. Gun position

Burnout problems were found only with a very low coal jet momentum and low transport air mass flow. Under these conditions the flame was not stable. In all the other conditions no large differences in the burnout level were noticed.

As expected, an essential condition to achieve low NO_x emissions was oxygen availability in the primary combustion zone. The fuel injection too close to the central air (fast mixing with the combustion air) and the high coal jet momentum (more lifted flame) both determine an oxygen-rich primary combustion zone that leads to high NO_x emissions. The lowest NO_x emissions were found for the coal gun positioned farther from the central air both for low- and high-velocity coal injections.

These results are related to the specific coal used in the trial. The coal characteristics are thought to be important for the technique especially the ignition behavior of the coal. The coal type effects on the above-discussed results should be further investigated.

2.5.4 Combustion Rate of Solid Carbon

With the clarification of the combustion mechanism of gaseous fuels and liquid fuels to a certain extent, systematic studies on energy saving and low-pollution gas emission in connection with high temperature air combustion have been conducted. However, when it comes to solid fuels such as coal, which is used by certain industries, there have been few basic studies or research on the effects of high temperature combustion air, partly because the properties of coal differ substantially depending on the regions where the coal is produced.

This section discusses features of solid carbon (graphite) combustion, placing emphasis on the combustion of dry-distilled coal, which is the most typical fuel in coal combustion systems, and clarifies the influence of high temperature air combustion techniques on the combustion mechanism. The following sections present the results of experiment and analysis.

2.5.4.1 Combustion Field and Solid Carbon Specimens

The high temperature air generator, manufactured by NFK (see Foreword) and shown in Figure 2.127, was used for the combustion test in this study. This alternate changeover combustion type equipment can continuously generate high temperature airflow on the order of 1280 K for a duration of about 3 min by accumulating heat in a ceramic honeycomb (regenerative media, ③ in the figure) and subsequently exchanging the heat with the airflow. The high temperature airflow thus generated is discharged into the atmosphere through a nozzle (② in the figure, having a diameter of 12 mm) at a uniform flow rate (max 50 m/s) and a uniform temperature.

Two dimensional stagnation flow formed with the uniform airflow mentioned above containing a solid carbon (graphite) was used as the combustion field (Figure 2.128). The flow field in the front stagnation region is uniquely defined by the use of the velocity gradient $a(=4V/d)$, which is represented with the constant velocity

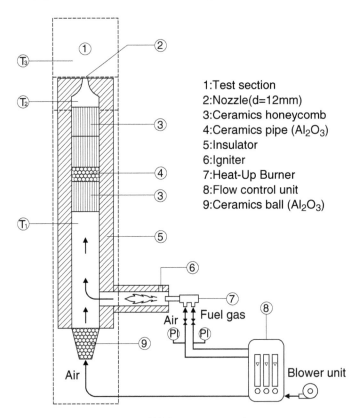

FIGURE 2.127 General appearance of high temperature air generator.

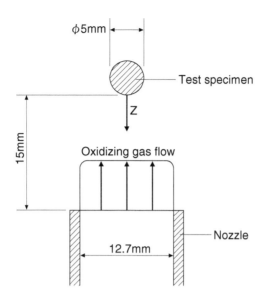

FIGURE 2.128 Combustion field in stagnation flow.

V and the test specimen diameter d. Further, it should be noted that the temperature and concentration distributions in this region are both one dimensional.

The test used a ground bar of artificial graphite (5 mm in diameter, 40 mm in length, with a concentration of 1.25×10^3 kg/m^3, and 44% porosity) as the solid carbon test piece. The reactivity of the material has been analyzed by a series of studies.[66,67] Electric heating using an AC current (16 V, max. 1625 A) was used in the test and it enabled us to obtain combustion rates under the condition of a uniform surface temperature regardless of the endothermic reaction or exothermic reaction that is incidental to chemical reactions. Surface temperature was measured with a two-color thermometer.

2.5.4.2 Experimental Results

When obtaining combustion rates from surface regression speeds, the combustion of solid carbon (graphite) is greatly influenced by such factors as velocity gradient, surface temperature, oxidizing agent temperature, and oxygen concentration. This section first presents the results of the test utilizing room temperature airflow and then discusses the results of the test using high temperature airflow.

2.5.4.3 Combustion Rate in Room Temperature Airflow

The relations between the surface temperature and combustion rate are shown in Figure 2.129a and Figure 2.129b.[68] The parameter is velocity gradient. Data points are based on the test results, and the solid line indicates the results of calculation. With the velocity gradient at 200 and 640 s^{-1}, the combustion rate increased with surface temperature. However, the combustion rate dropped sharply at a certain temperature and thereafter it again increased. The sharp drop of the combustion rate

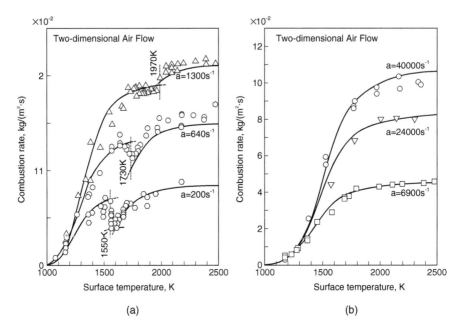

FIGURE 2.129 Influence of the surface temperature on the combustion rate of a solid carbon in room temperature airflow. (A. Makino, *Combust. Flame*, 81(2):166, 1990). The parameter is velocity gradient. Data points are based on the test results, and the solid line indicates the results of calculation.

was closely associated with the formation of a CO flame above the surface.[66,67] With the velocity gradient at 1300 s^{-1}, the combustion rate did not show a sharp drop; however, the rate of increase in the combustion rate fluctuated at a certain temperature. The discontinuity also resulted from the formation of a CO flame. When the velocity gradient was further increased (Figure 2.129b), unlike in the case of Figure 2.129a, the combustion rate increased homogeneously and continuously with the surface temperature, thus getting closer to a certain value (diffusion-dominated combustion rate). The reason no discontinuous change occurred in this case was that the formation of CO flame had been suppressed due to its high velocity gradient.[66,67]

2.5.4.4 Combustion Rate in High Temperature Airflow

The relationship between the surface temperature and combustion rate in a test using high temperature airflow (at the temperature of 1280 K and velocity gradient of 40,000 s^{-1}) is shown in Figure 2.130. In the same way as shown in Figure 2.129b, a CO flame was not formed above the surface due to high velocity gradient and the combustion rate increased homogeneously and continuously with surface temperature toward a certain value.

The figure also shows the results of a case (for the velocity gradient of 10,000 s^{-1}) in which the amount of supplied air was kept at a fixed level and the high temperature air generator was kept out of operation. It is obvious that the combustion

Combustion Phenomena of High Temperature Air Combustion

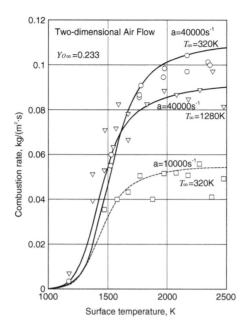

FIGURE 2.130 Influence of the surface temperature on the combustion rate of a solid carbon in high temperature airflow. The parameters are airflow temperature and velocity gradient. Data points are based on the test results, and the solid and dotted lines indicate the results of calculation.

rate was almost doubled due to increases in the flow rate of the oxidizing agent due to thermal expansion when the high temperature air generator was operated. The figure also includes the results of the test using room temperature airflow in which the velocity gradient was maintained at 40,000 s^{-1}. The results show that the combustion rate in high temperature airflow is lower than that in room temperature airflow by about 15%.

2.5.4.5 Dynamic Analysis of Reactive Gas

2.5.4.5.1 Combustion Rate

The combustion of solid carbon (graphite) involves a surface reaction ($2C + O_2 \rightarrow 2CO$ and $C + CO_2 \rightarrow 2CO$, for example) and a gas-phase reaction ($2CO + O_2 \rightarrow 2CO_2$). These reactions influence each other and also influence the combustion rate.[68] In addition, it is also known that the formation of a CO flame is influenced by the velocity gradient, the surface temperature, and oxygen concentration [69] and that the combustion rate at the same surface temperature changes when a CO flame is formed.[66] Other known facts are that the gas-phase reaction has a minimal effect on the combustion rate before the formation of a CO flame[69] and that the formed CO flame can be treated as a flame front after the formation of a CO flame. The analytical expression[66] of the combustion rate (dimensional) in connection with the above is shown below:

$$m = (\rho v)_s \approx \rho_\infty \frac{\sqrt{2^j a \frac{\mu_\infty}{\rho_\infty}}}{K} \ln(1+\beta) \qquad (2.34)$$

where, β is mass transfer number and K is the complement number of material transfer factor. Further, the mass transfer number β is expressed as shown below.

1. When a CO flame is not present:

$$\beta \approx \left(\frac{K A_{S,0}}{1+K A_{S,0}}\right)\left(\frac{2W_C}{W_0} Y_{0,\infty}\right) + \left(\frac{K A_{S,p}}{1+K A_{S,p}}\right)\left(\frac{W_C}{W_P} Y_{p,\infty}\right) \qquad (2.35)$$

2. When the CO flame can be treated as a flame front:

$$\beta \approx \left(\frac{K A_{S,p}}{1+K A_{S,p}}\right)\left(\frac{2W_C}{W_0} Y_{0,\infty} + \frac{W_C}{W_P} Y_{p,\infty}\right) \qquad (2.36)$$

3. When the CO flame remains in contact with the surface of a solid:

$$\beta \approx \left(\frac{K A_{S,0}}{1+2K A_{S,0} - K A_{S,p}}\right)\left(\frac{2W_C}{W_0} Y_{0,\infty}\right) + \left(\frac{K A_{S,p}}{1+2K A_{S,0} - K A_{S,p}}\right)\left(\frac{W_C}{W_P} Y_{p,\infty}\right) \qquad (2.37)$$

The complement number of the material transfer factor K for combustion in two-dimensional stagnation flow is expressed as shown below.

In the cases of (1) and (3):

$$K \approx \left(\frac{T_\infty}{T_s}\right)\left(1 - \frac{T_\infty}{2T_s}\right) + \sqrt{\frac{\pi}{2}} \qquad (2.38)$$

In the case of (2):

$$K \approx \left(\frac{T_\infty}{T_s}\right)\left(1 - \frac{T_\infty}{2T_s}\right) - 0.05\left(1 + \frac{4W_P}{W_0} Y_{0,\infty}\right) + \sqrt{\frac{\pi}{2}} \qquad (2.39)$$

The symbols used in the expressions above represent the following values: a is velocity gradient, ρ is concentration, μ is coefficient of viscosity. Index j refers to the two-dimensional stagnation flow ($j = 0$) or axisymmetric stagnation flow ($j = 1$). Y = mass rate, W = molecular weight. Subscripts C, O, and P mean carbon, oxygen,

Combustion Phenomena of High Temperature Air Combustion

and carbon dioxide, respectively. $A_{S,O}$ and $A_{S,P}$ are surface Damköhler numbers defined in the following equation:

$$A_{S,i} \equiv \frac{B_{S,i}}{\sqrt{2^j a(\mu_\infty/\rho_\infty)}} \left(\frac{T_\infty}{T_s}\right) \exp\left(-\frac{Ta_{s,i}}{T_s}\right) \quad (i = 0, P) \tag{2.40}$$

where, $B_{S,i}$ is the frequency factor of surface reaction, $Ta_{S,i}$ is the activation temperature of surface reaction. Subscripts O and P refer to the C–O_2 surface reaction and C–CO_2 surface reaction, respectively.

The solid carbon test specimen used in this test has the same properties as those used in past studies.[66,67] The frequency factor of the surface reaction and activation energy of the specimen are as described below:

C–O_2 reaction: $B_{S,0} = 4.1 \times 10^6$ m/s, $E_{S,0} = 179$ kj/mol
C–CO_2 reaction: $B_{S,P} = 1.1 \times 10^8$ m/s, $E_{S,P} = 270$ kj/mol

The solid and dotted curves shown in Figures 2.129 and 2.130 represent the combustion rates calculated from the values indicated above. The results of tests correspond comparatively well with the results of analysis. As for the air concentration and coefficient of viscosity used in the calculation of combustion rates, ρ_∞ = 1.09 kg/m^3 and μ_∞ = 1.95 × 10^{-5} Pa·s at 320 K were used since the analysis was based on the condition that $\rho\mu$ is constant.

Since the mass transfer number β that determines the nondimensional combustion rate $[(-fs) = \{\ln(1 + \beta)\}/K]$ is of almost the same value for both room temperature airflow and high temperature airflow, the difference in the combustion rate can be attributed to the concentration or coefficient of viscosity, or both, which are physical property values. With the mass flow rate of the air fixed at the same level, the flow velocity increases in high temperature airflow as a result of thermal expansion. This then increases the velocity gradient to finally raise the combustion rate. With the velocity gradient fixed at the same level, the concentration is lowered in high temperature airflow as a result of thermal expansion and the mass flow rate of oxygen to be transported to the surface decreases, thus lowering the combustion rate.

2.5.4.6 Lower Limit of Oxygen Concentration

With the analytical expression for the combustion rate made available, the influence of the concentration of the oxygen in the high temperature oxidizing agent (at a temperature of 1280 K) on the combustion rate was studied (see Figure 2.131). The results show that a combustion rate almost equal to that available in room temperature airflow can be obtained even if the concentration of the oxygen in high temperature oxidizing agent is lowered to the mass rate of 0.15. In the combustion of a solid carbon, the oxygen concentration can be further reduced in connection with the combustion rate on the condition that the concentration of carbon dioxide is increased since carbon dioxide can serve also as the oxidizing agent in surface reactions. Figure 2.132 shows

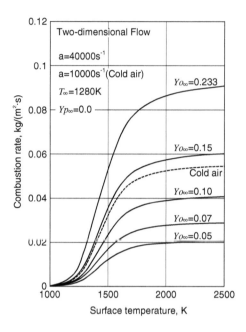

FIGURE 2.131 Influence of the concentration of the oxygen in oxidizing agent flow on the combustion rate of a solid carbon in high temperature airflow.

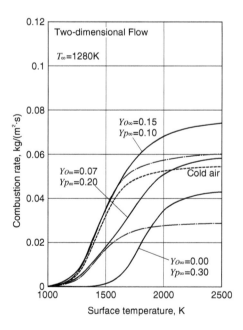

FIGURE 2.132 Influence of the concentration of the carbon dioxide in oxidizing agent flow on the combustion rate of a solid carbon in high temperature airflow.

the combustion rate with the densities of oxygen and carbon dioxide changed while the mole fraction of nitrogen is kept at 0.79. It shows that, if the surface temperature of a solid carbon is 2000 K or higher, a high temperature oxidizing agent with the mass rates of oxygen at 0.07 and carbon dioxide at 0.20 results in a combustion rate almost equal to that in room temperature airflow.

The relationship is expressed by the following expression:

$$2\frac{\ln\{1+\beta_{HO}\}}{K_{HO}} \geq \frac{\ln\{1+\beta_{RA}\}}{K_{RA}} \tag{2.41}$$

where subscript HO refers to the high temperature oxidizing agent and subscript RA refers to room temperature air. The relationship between the oxygen mass rate and the carbon dioxide mass rate is expressed by the following expression:

$$\left[\frac{KA_{S,0}}{1+KA_{S,0}}\right]\left[\frac{2W_C}{W_0}Y_{0,\infty}\right] + \left[\frac{KA_{S,P}}{1+KA_{S,P}}\right]\left[\frac{W_C}{W_P}Y_{P,\infty}\right] \geq (1+\beta_{RA})^{K_{HO}/(2K_{RA})} - 1 \tag{2.42}$$

The relationship is shown in Figure 2.133. The region higher than the descending lines satisfies the condition.

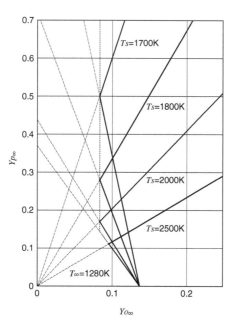

FIGURE 2.133 Combinations of oxygen concentration and carbon dioxide concentration in high temperature airflow to exceed the combustion rate in room temperature air.

Since the C–CO_2 surface reaction is an endothermic reaction, the oxygen concentration has a lower limit. The condition to make the heat generation resulting from the positive surface reaction, with the exothermic value of the C–O_2 surface reaction at Q_O (= 111.4 kJ/mol) and endothermic value of the C–CO_2 surface reaction at Q_P (= 170.7 kJ/mol), is defined by the following expression:

$$Q_O A_{S,O}\left[\frac{2W_C}{W_O}Y_{O,S}\right] + Q_P A_{S,P}\left[\frac{W_C}{W_P}Y_{P,S}\right] \geq 0 \qquad (2.43)$$

The region below the ascending lines satisfies the condition expressed by expression (2.43) above. This means that a combustion rate higher than that in room temperature airflow is obtained from the combination of the oxygen mass rate and the carbon dioxide mass rate on the right side of the thick solid line. It should be noted that the lower limit of the oxygen mass rate is on the order of 0.085 if the surface temperature is 2000 K or lower.

2.5.4.7 Surface Temperature When a CO Flame Is Formed

It has been reported that the combustion rate is drastically lowered by the formation of a CO flame when the velocity gradient is about 1000 s^{-1} or less in the case of combustion in room temperature airflow.[66,67] With a similar case expected in high temperature air combustion, we examined the conditions of CO flame formation by asymptotic analysis method. As a result it was revealed that the conditional expression for ignition by Linan [70] can be applied. The conditional expression for ignition contains velocity gradient and surface temperature, which are represented in Figure 2.134. The abscissa shows the velocity gradient, while the ordinate shows the surface temperature at which a CO flame is formed. The parameter is a surface Damköhler number.

The surface temperature when a CO flame is formed increases with the velocity gradient. A certain amount of CO on the surface of the solid carbon must be secured for the formation of a CO flame. Since the amount of CO largely depends on the flow field and more CO must be supplied for higher-velocity gradients, a high surface temperature upon formation of a CO flame is indispensable. However, if the surface Damköhler number is high and the reactivity on the surface is favorable, a CO flame can be formed at lower surface temperatures. Regarding the conditions of CO flame formation in high temperature airflow, it is revealed that the surface temperature is lower by 100 to 200 K than the conditions of CO flame formation in room temperature airflow.[69] This means that the conditions do not differ from the CO flame formation in room temperature airflow if the velocity gradient is at about 1000 s^{-1} or higher.

2.5.4.8 Combustion Rate in High Temperature Airflow

The relationship between the surface temperature and combustion rate when CO flame formation is expected was examined (see Figure 2.135a). With the velocity

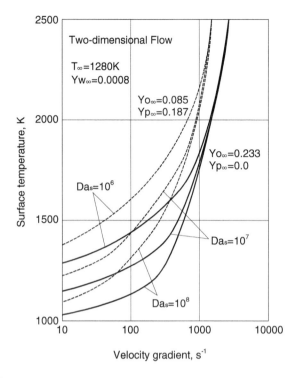

FIGURE 2.134 Relationship between velocity gradient and surface temperature when a CO flame is formed.

gradient at 200 s^{-1}, a CO flame was formed at a surface temperature that is almost equal to that in high temperature airflow. Because the combustion occurred with the CO flame adhered to the surface of the solid, the combustion rate would not increase. To increase the combustion rate in this case, the surface temperature must be higher than 1700 K. With the velocity gradient at 640 s^{-1}, the combustion rate increased with the surface temperature to 1550 K. It however decreased by half with the surface temperature exceeding 1550 K as a CO flame was formed thereafter. To recover the thus half-reduced combustion rate, the surface temperature must be lowered slightly or increased to a level higher than 1800 K. With the velocity gradient at 1300 s^{-1}, the surface temperature at which a CO flame is formed is as high as 1950 K and the formed CO flame is separated from the surface during combustion. In this case the combustion rate is larger, although by a minimal margin, than that in the case where a CO flame is not present and an increase in the combustion rate is expected.

With the velocity gradient at 6900 s^{-1} or higher, the surface temperature at which a CO flame is formed is 2500 K or higher, and the formed CO flame cannot be present in the gaseous phase during combustion. Figure 2.135b shows the relationship between the surface temperature and combustion rate under the above-mentioned conditions.

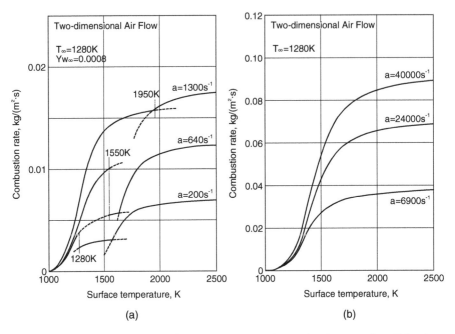

FIGURE 2.135 Influence of the surface temperature on the combustion rate of a solid carbon in high temperature airflow. The parameter is velocity gradient; (a) for the velocity gradient at 1300 s^{-1} or lower and (b) for the velocity gradient at 6900 s^{-1} or higher.

2.5.4.9 Summary

The behavior of the combustion of a solid carbon in high temperature air stagnation flow was investigated. The study proved empirically that an increase in the flow velocity resulted from thermal expansion increases in the velocity gradient to finally raise the combustion rate when the mass flow rate of air is kept at the same level. It was also found that the combustion rate in high temperature airflow is lower than that in room temperature airflow. This was because the lowered concentration as a result of thermal expansion reduces the mass flow rate of oxygen transported to the surface of the carbon when the velocity gradient is kept at the same level.

Analytical study on the influence of the concentration of the oxygen in a high temperature oxidizing agent on the combustion rate was also conducted. It revealed that a combustion rate almost equivalent to that in room temperature airflow was obtained even if the concentration was reduced to the mass rate of approximately 0.085. The test on the formation of a CO flame showed that the combustion rate may be reduced by half depending on the surface temperature with the velocity gradient set at about 1000 s^{-1} or lower.

REFERENCES

1. T. Hirano. *Combustion Science*, p. 31. Kaibundo, 1998 (in Japanese).

2. K. Ito et al. *Trans. of JSME, ser. B*, 56:3508, 1998 (in Japanese).
3. J. Kojima, Y. Ikeda, and T. Nakasima. *36th Japan Symposium on Combustion*, p. 560, 1998 (in Japanese).
4. R. Weber et al. *Proceedings of 2nd International Seminar on High Temperature Combustion in Industrial Furnace*, Stockholm, 2000.
5. S. Mochida, T. Hasegawa, and R. Tanaka. *AFRC Fall Int. Symp.*, 1993.
6. A. Sobiesiak, S. Rahbar, and H. A. Becker. *Combust. Flame*, 115:93, 1998.
7. E. W. Grandmaison et al. *Combust. Flame*, 114:381, 1998.
8. L. Yimer and B. H. A. *Can. J. Chem. Eng.*, 74:840, 1996.
9. E. Shigeta et al. *AFRC/JFRC International Conference on Environmental Control of Combustion Processes*, Paper No. 28, 1991.
10. N. Saiki and L. Koizumi. *AFRC/JFRC International Conference on Environmental Control of Combustion Processes*, 1994.
11. T. G. Kreutz and C. K. Law. *Combust. Flame*, 104:157, 1996.
12. T. G. Kreutz and C. K. Law. *Combust. Flame*, 114:436, 1998.
13. J. Sato et al. *Proc. Combustion Instit.*, 21:695, 1986.
14. M. C. Drake and B. R. J. *Combust. Flame*, 76:151, 1989.
15. T. Hasegawa and R. Tanaka. *Combust. Sci. Technol. Japan*, 1:265, 1994.
16. R. Tanaka et al. *Combust. Sci. Technol. Japan*, 1:257, 1994.
17. M. Katsuki et al. *34th Japan Symposium on Combustion*, p. 441, 1996 (in Japanese).
18. J. A. Miller and C. T. Bowman. *Prog. Energ. Combust. Sci.*, 15:287, 1989. 103
19. R. J. Kee, F. M. Rupley, and M. J. A. Technical Report SAND89-8009, Sandia National Lab., 1990.
20. R. J. Kee et al. Technical Report SAND85-8240, Sandia Natonal Lab., 1993.
21. S. Fukutani and N. Kuniyoshi. *Combust. Sci. Technol. Japan*, 2:269, 1994.
22. G. Dixon-Lewis et al.. *Proc. Combust. Instit.*, 20:1893, 1984.
23. M. Katsuki and K. Ebisui. *The First Asia-Pacific Conf. on Combust.*, p. 294, 1997.
24. J. A. Wunning and J. G. Wunning. *Prog. Energ. Combust. Sci.*, 29:81, 1997.
25. Y. Ju and T. Niioka. *Combust. Theory Modelling*, 1:243, 1997.
26. R. J. Kee et al. Technical Report SAND86-8246, Sandia Report, 1986.
27. K. Kishimoto. *The First Asia-Pacific Conf. on Combust.*, p. 468, 1997.
28. N. Kuniyoshi and S. Fukutani. *The First Asia-Pacific Conf. on Combust.*, p. 456, 1997.
29. S. Fukutani, H. Oike, and N. Kunioshi. *Trans. of JSME, ser. B*, 64(242), 1998 (in Japanese).
30. T. Hasegawa. Personal communication, 1998.
31. H. Richter, W. J. Grieco, and J. B. Howard. *Combust. Flame*, 119(1), 1999.
32. C. T. Bowman et al. *http://www.me.berkeley.edu/gri mech/*, 1995.
33. Y. Hidaka et al. *Int. J. Chem. Kinet.*, 21:643, 1989.
34. S. E. Stein et al. *Proc. Combust. Instit.*, 23:85, 1990.
35. S. D. Thomas, F. Communal, and P. R. Westmoreland. *Prep. Div. Fuel Chem.*, 36:1448, 1991.
36. L. D. Pfefferle, G. Bermudez, and J. Boyle, H. Bockhorn, ed., *Soot Formation in Combustion*, pp. 25–49. Springer-Verlag, Berlin, 1994.
37. T. R. I. of Energy and C. Engineering. *Coal Committee: Review of Coal Engineering 95*. Denryoku Symposia, 1993 (in Japanese).
38. L. D. Smoot. *Fundamentals of Coal Combustion for Clean and Efficient Use, Coal Science and Technology*. Elsevier, 1993.
39. K. Ohtomo and T. Fujiwara. *Combustion Engineering*, p. 106. Korona Co., 1992 (in Japanese).

40. P. R. Solomon and M. A. Serio. *1986 NATO Workshop on Fundamental of Physical-Chemistry of Pulverized Coal Combustion,* Vol. 1, Les Arces, France, 1986.
41. S. K. Ubhayakar et al. *Proc. Combust. Instit.*, 16:427, 1977.
42. G. J. Pitt. *Fuel*, 41:267, 1962.
43. H. Kobayashi et al. *Proc. Combust. Instit.*, 16, 1977.
44. E. J. Niska and A. R. Kerstein. *Energy Fuels*, 5:647, 1991.
45. K. Okazaki et al. *JSME Int. J., ser. 2*, 34(4):533, 1991.
46. E. J. Kansa and H. E. Perlee. *Combust. Flame*, 38:17, 1980.
47. I. Naruse and et al. *Proc. Combust. Instit.*, 24:1415, 1992.
48. S. W. Kang et al. *Proc. Combust. Instit.*, 22:145, 1988.
49. Y. Yamamoto et al. *Trans. of JSME, ser. B*, 60(570):649, 1994. in Japanese.
50. R. H. Essenhigh. *Proc. Combust. Instit.*, 16:353, 1977.
51. W. Nusselt. *V. D. I.*, 68:124, 1924.
52. I. Naruse. *Fundamental Technology Coal Engineering*, p. 146. 1998.
53. T. Tanigawa, T. Hasegawa, and M. Morita. *12th IFRF Members Conference*, Leeuwenhorst Congress Center, Noorwijkerhout, The Netherlands., May 1998. IFRF.
54. T. Hasegawa, R. Tanaka, and T. Niioka. *The First Asia-Pacific Conf. on Combst.*, Osaka, Japan, p. 290, 1997.
55. D. Hardesty and F. Weinberg. *Proc. Combust. Instit.*, 8:201, 1974.
56. A. L. Verlaan et al. Technical Report Document F46/y/1, IFRF, IJmuiden, March 1998.
57. R. Weber et al. *J. Inst. Energy*, 72:77, 1999.
58. A. L. Verlaan et al. Technical Report Document F46/y/2, IFRF, IJmuiden, February 1999.
59. J. Chedaille and Y. Braud. *Measurements in Flames,* Vol. 1. Edward Arnold, 1972.
60. N. Lallemant and J. Locquet. Technical Report Document C73/y/8, IFRF, IJmuiden, January 1998.
61. N. Lallemant, A. Sayre, and R. Weber. *Prog. Energy Combust. Sci.*, 22:543, 1996.
62. S. Chan and C. Tien. *J. Quant. Spectrosc. Radiat. Transfer*, 9:1261, 1969.
63. P. Stabat and N. Lallemant. Technical Report Document G08/y/11, IFRF, IJmuiden, October 1999.
64. J. Haas, M. Tamura., and W. van de Kamp. Technical Report Document F37/y/41, IFRF, IJmuiden, March 1999.
65. K. Yoshikawa. High efficiency power generation from coal and wastes utilizing high temperature air combustion technology part 1: System and overview of the 5 years demonstration project, 1999.
66. A. Makino, N. Arai, and Y. Mihara. *Combust. Flame*, 96(3):261, 1994.
67. A. Makino et al. *Proc. Combust. Instit.*, 26:2067, 1996.
68. A. Makino. *Combust. Flame*, 81(2):166, 1990.
69. A. Makino and C. K. Law. *Combust. Sci. Technol.*, 73(4–6):589, 1990.
70. A. Liñán. *Acta Astronautica*, 1:1007, 1974.

3 Simulation Models for High Temperature Air Combustion

3.1 PRESENT STATE OF COMBUSTION SIMULATION IN FURNACES

3.1.1 INTRODUCTION

It is easy to imagine how difficult it is to calculate even the distribution of one variable, such as temperature, or concentration in a furnace, with so much taking place there. Combustion usually proceeds in the recirculating turbulent flow sometimes associated with a swirling motion. To calculate the performance of the furnace, we must solve a group of partial differential equations governing the balance of mass, momentum, energy, and species together with submodels, such as a turbulence model, phase-interaction model, heat-transfer model, and so on. Regarding turbulence modeling, direct numerical simulation (DNS) of turbulent flows is carried out in the scientific field, and it is unrealistic to apply it to engineering simulation of practical systems with large, three-dimensional geometry. So, presently the use of k–ε model is a prerequisite for the practical engineering calculation of furnaces. This is because examples and experience of the k–ε model have been demonstrated in the past more than any others. Thus, this discussion is limited to combustion and heat-transfer models specific to HiTAC. The other models used for simulation show almost no difference between HiTAC and ordinary combustion.

High temperature air (over 1273 K), far above the temperature level obtainable by conventional preheating methods, is easily realized in industrial furnaces using the high cycle regenerator system. Experience with the combustion method for this system has not been accumulated, and a practical database is not available for such an extreme condition. For effectively devising optimum design for an industrial furnace using high cycle regenerative combustion, it is essential to combine analyses utilizing numerical simulations with appropriate experiments.

Combustion in HiTAC proceeds in an atmosphere of low oxygen concentration as well as at high temperature, mostly above the autoignition temperature of the fuel. Accordingly, reaction zones are relatively distributed to yield somewhat wide and mild heat release, hence a uniform temperature distribution. A pilot burner or conventional flame stabilizer for HiTAC is not needed, and the onset of combustion entirely depends on the mixing processes that result in the encounter of fuel with preheated oxidizer somewhere in the furnace.

Recently, computer software for analyzing thermofluid dynamics including chemical reactions has become commercially available. However, the software cannot predict whether combustion will start or not; it only predicts the burned properties when combustion is ensured by forced ignition or a pilot flame. The onset of combustion in HiTAC depends on the balance between chemical reactions and physical processes, particularly mixing processes for the case. No available models are suitable for use in HiTAC simulations. If they were used for this purpose, accurate predictions could not be expected.

The following reviews the combustion models used for furnace simulations, and considers a new direction of models that reflects the characteristics of HiTAC.

3.1.2 Problems of Existing Combustion Models

This section describes characteristics of combustion models incorporated in the generic commercial software for thermofluid dynamics and their problems when they are applied to HiTAC.

3.1.2.1 Arrhenius Type One-Step Global Reaction Model

In the combustion processes of ordinary hydrocarbon fuels, various intermediate species appear and disappear within a very short time, and correspondingly all these take place within a very thin zone called a flame front or a flamelet. Thus, to simulate precise distribution of chemical species and temperature inside the thin flame, reaction rates have to be calculated regarding each of the species appearing during the combustion processes, coupled with the flow field and the energy transport. It is known that tens of chemical species and more than 200 elementary reactions are involved in the combustion reaction scheme even for methane, the hydrocarbon fuel of the simplest structure. The computers at present available are incapable of performing all these calculations with the conditions for industrial furnaces of complicated three-dimensional geometry.

Since we are interested in estimates of temperature distribution and flame length in combustion equipment, it is appropriate to regard the entire combustion process simply as a one-step reaction between fuel and oxygen to generate heat and combustion product. The reaction when understood in this manner is called a one-step global reaction. Its reaction rate is commonly expressed in the following Arrhenius type formula.

$$R_{fu} = -Fp^2 m_{fu} m_{ox} \exp\left(\frac{-E}{RT}\right) \qquad (3.1)$$

where T is temperature, m_{fu} and m_{ox} are mass fractions of fuel and oxygen, respectively, F is a frequency coefficient, p pressure, E activation energy, and R the universal gas constant. Here, a stoichiometric relationship holds that 1 kg of fuel and r kg of oxygen react to generate $(1 + r)$ kg of burned product. Accordingly, a similar algebraic relationship is valid between the reaction rate of fuel and that of oxygen or burned product.

This model is often combined with the generic software commercially available and is used for defining an upper limit of reaction rate in combination with the eddy-dissipation model, which will be explained later. Although the one-step global reaction model can easily handle complex reactions of combustion through simplification, no consideration is given to any of the intermediates emerging during the combustion processes, and hence generation and emission of CO and NO_x cannot be estimated in practical terms by the model.

It is difficult to introduce the effects of fluctuating properties of turbulent combustion into a chemically controlled reaction model, such as the Arrhenius model. Since fluctuations of temperature and species concentrations in HiTAC are relatively small compared with ordinary combustion, as described before, the use of a chemically controlled reaction model for HiTAC seems reasonable from a practical point of view. However, the validity of recommended empirical constants in the Arrhenius expression has not been examined for HiTAC.

3.1.2.2 Mixing-Is-Reacted Model

The time required for combustion reactions is often far shorter than the time for mixing between fuel and air. This holds true with diffusion combustion where fuel and air are supplied separately and turbulence is relatively weak. In this case, it can be assumed that combustion takes place at the instant of mixing, that is, at an infinite reaction rate. Thus, a simulation is possible simply by calculating the mixing processes, without considering calculation of reaction rate. As shown in Figure 3.1, the mixture fraction f at any point in a furnace is defined by the following equation:

$$f = \frac{m_{fu} - m_{fu,0}}{m_{fu,1} - m_{fu,0}} \tag{3.2}$$

where subscripts 1 and 0 indicate fuel mass fractions at each inlet of fuel and of air, respectively. Thus defined, the value of f for any point in the furnace expresses the mass fraction of fuel existing there in a non-combusting case. Once combustion

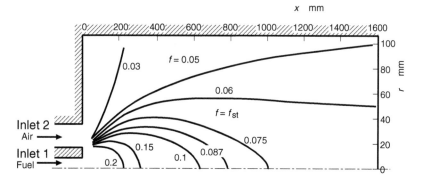

FIGURE 3.1 Distribution of mixture fraction f (f_{st}: stoichiometric f).

occurs at an infinite reaction rate, it proceeds according to the stoichiometric relationship until one of the deficit species, fuel or oxygen, is consumed. Accordingly, as shown in Figure 3.2, the flame front corresponds to the position where the value of mixture fraction conforms to the stoichiometric ratio, f_{st}, and no fuel and oxygen will remain there. Figure 3.2 is also a schematic illustration of adiabatic temperature of combustion in terms of f, assuming complete combustion. By application of the model, the theoretical adiabatic temperature in complete combustion is calculated based on the enthalpy balance. The estimated combustion temperature, therefore, implicitly infers the use of ambient air, hence 21% oxygen content, and tends to be considerably higher than that usually measured in real furnaces. Also, whereas combustion temperature can easily be calculated in an adiabatic system, it is hard to define the temperature under a condition where there is heat transfer or heat loss.

The effect of turbulence is usually considered by the introduction of prescribed probability density function (PDF) for the fluctuations of mixture fraction, that is, clipped Gaussian distribution or β-function, for example. However complicated the reaction models may be, the basic quantitative correlation between mixture fraction and temperature is still dependent on the diagram in Figure 3.2, in which the assumptions of adiabatic conditions and of ambient air are essential. Accordingly, the combustion in high temperature air diluted with flue gas in the furnace, the most important and influential factor of HiTAC, cannot be expressed by the diagram. The local reaction rate of HiTAC differs depending on the local concentration of oxygen as well as the local temperature lowered owing to heat transfer and heat loss.

3.1.2.3 Eddy-Break-Up Model

When combustion occurs in a homogeneous mixture, it can be assumed that combustion is governed by the reaction rate. However, in non-premixed combustion in real furnaces where fuel and air are supplied separately, it is thought that the

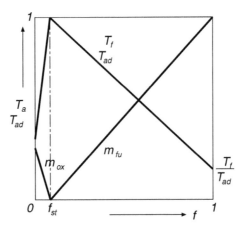

FIGURE 3.2 Adiabatic temperature and mass fractions of remaining fuel and oxygen in terms of mixture fraction based on the mixing-is-reacted model.

combustion reaction proceeds in a turbulent flow where eddies of fuel and air meet and mix with each other down to the molecular level. Since reaction occurs only at the interface of eddies of fuel and air, the gas inside the eddies is not involved in the reaction. Therefore the combustion rate is proportional to the dissipation rate of turbulent energy because interfaces where fuel and air come into contact appear one after another during the process of dissipation of turbulence, namely, the process of large eddies collapsing into small ones. When the k–ε model is adopted as a turbulence model, the dissipation rate of turbulence is proportional to ε/k. When the similarity of the concept is applied to species concentration, which is a scalar quantity, the dissipation of turbulent eddies is considered to correspond to the decay of concentration fluctuation. Regarding a scalar quantity ϕ, its fluctuation intensity g (secondary moment around a mean value $\tilde{\phi}$) is defined as follows:

$$g = \overline{\left(\phi - \tilde{\phi}\right)^2} \tag{3.3}$$

The governing equation for g can be obtained by the analogy of turbulence energy, k, and its dissipation rate, ε, from the Navier-Stokes equation. In the first eddy-breakup model originally proposed by Spalding, the time-averaged reaction rate was modeled to be proportional to the product of turbulence dissipation rate and intensity of concentration fluctuations as given below:[1]

$$\overline{R_{fu,EBU}} = -C_{EBU}\overline{\rho}(\varepsilon/k)g^{1/2} \tag{3.4}$$

where ρ is density and C_{EBU} is an empirical constant. In the eddy dissipation model of Magnussen and Hjertager,[2] which is also based on the same thinking as the above model, concentration itself is used instead of its fluctuation intensity. They considered that combustion occurs only when burned product is involved as a heat source in addition to fuel and oxygen, and thus the combustion reaction rate is regulated by the smallest amount among the three. Then the reaction rate is expressed as follows:

$$\overline{R_{fu}} = -\overline{\rho}(\varepsilon/k)\min\left[A\tilde{m}_{fu},\ A\tilde{m}_{ox}/i,\ A'\tilde{m}_{pr}/(1+i)\right] \tag{3.5}$$

where i is stoichiometric oxygen/fuel ratio and $A = 4$ and $A' = 2$, both being empirical constants. The symbol min[ ] means that the smallest values between the brackets is taken.

It is a common practice to use the smaller value of the combustion reaction rates given by Equations 3.1 and 3.5, considering that the value given by Equation 3.5 is based on a mixing controlled rate and, as such, cannot exceed the value based on a homogeneous mixture.

A problem with this model is that the reaction rate is largely determined by the state of turbulence, and hence the influence of chemical reactions does not appear explicitly. If concentrations and state of turbulence are the same at different points of a furnace, the reaction rates at those points will be considered the same by this model, even when temperatures are different. For this reason, the single use of this model is not appropriate for simulating HiTAC, and it should be combined with the chemical reaction rate model in which reaction characteristics of HiTAC are taken into account.

3.1.2.4 Problems in Temperature Calculation

Among the numerous step reactions forming the complicated combustion reaction scheme, it is in chain terminating reactions where radicals and monatomic molecules combine to form stable species. In the temperature range above 2000 K, the endothermic reverse reactions of the terminal reactions become significant. These endothermic reactions are called thermal dissociation. However, as the temperature falls, the forward reactions still proceed. When the mixture ratio is close to stoichiometric, the combustion temperature in real flames tends to decrease considerably more, due to the thermal dissociation, than the theoretical combustion temperature based on the assumption of complete combustion.

Since most combustion models such as one-step global reaction model do not predict precise concentrations of intermediate species in the flame, as stated before, they are incapable of estimating the correct combustion temperature caused by the thermal dissociation in the high-temperature range. Since it is important in the numerical simulations of practical furnaces to estimate the heat transfer rate to the material to be heated as precisely as possible, the accuracy of flame temperature greatly influences the radiation heat transfer rate, which is proportional to the flame temperature to the fourth power. Therefore, it is necessary to work out some means to raise the accuracy of temperature prediction, even when we use a combustion model that does not deal with intermediate species.

3.2 COMBUSTION MODEL FOR HIGH TEMPERATURE AIR COMBUSTION

3.2.1 CHARACTERISTICS OF HIGH TEMPERATURE AIR COMBUSTION

When highly preheated air (1100 to 1600 K) is used, combustion by autoignition takes place immediately after the mixing of fuel and air; thus stabilizing flames by a pilot flame or a flame holder is not required. Combustion occurs with any type of conventional burners using highly preheated air, but in a HiTAC furnace combustion takes place as a lifted flame relatively free in space. The incoming preheated air is first diluted with burned gas recirculating inside the furnace, then it makes contact with the fuel. Flames formed in such a flow field have totally different features from those of conventional burner flames. The flame has low luminosity, and almost transparent reaction zones resembling a spreading mist in the furnace rather than a flame. In this situation NO_x emission and flame temperature are extremely low despite the high-temperature preheated air.

The above characteristics have not been seen with ordinary combustion using ambient temperature air, because they are not reflected in the combustion models adopted in the generic software available in the market. The mixing-is-reacted model, for example, is suitable for predicting what will happen after combustion has taken place and not for judging whether or not combustion will take place. Accordingly, the model is incapable of calculating whether or where combustion takes place in the furnace. The combustion model to be used for HiTAC simulation must be a model capable of expressing precise reaction rates in a low oxygen concentration atmosphere. The dimensions of the reaction zone tend to be thickened in low oxygen combustion, which is quite different from what common flame sheet models can predict. Therefore, the one-step global reaction model is not suitable for this new type of combustion.

Since, as mentioned above, there may be cool material to be heated in a real furnace, it is important for numerical simulation of practical furnaces to estimate the heat transfer to the material to be heated as precisely as possible. In a thermal field where heat loss through the wall or heat transfer to the material to be heated is involved, the system is no longer taken to be adiabatic. A theoretically complete combustion temperature based on the adiabatic assumption cannot be applied to the temperature of burned gas actually recirculating inside the furnace.

3.2.2 Proposed Improvements

As mentioned above, since HiTAC is a new combustion method, its characteristics cannot be expressed using the combustion models incorporated in the existing generic software for thermofluid calculation. The following are improvements required for making software capable of the numerical simulation of HiTAC.

1. Simulation results have to reflect temperature defects caused by thermal dissociation in the high temperature range above 2000 K, when the full reaction mechanism is not adopted into the combustion reaction model. We can consider the following improvements. One is the introduction of large virtual specific heat into one-step global reaction models to yield the temperature drop resulting from the actual thermal dissociation in the high temperature range. Another way to correct the calculated high temperatures caused by the assumption of perfect combustion is to place a limiting value inferred by the local chemical equilibrium.
2. The rate constants of a reaction model, such as one-step global reaction models incorporated in the existing generic software for thermofluid simulations, are not always appropriate for directly applying to HiTAC, since those constants are mostly adjusted for ordinary combustion using ambient temperature air. It is incorrect to think there will be no problem, even when we use a full reaction mechanism, since the accuracy of all the associated rate constants has not been confirmed, even for the elementary reactions. Uncertainty about the reaction rate constants increases further when it comes to composing global reactions or reduced mechanisms. This is because the reaction rate constants are solely empirical. Therefore, the constants have to be optimized on the assumption that air temperature and oxygen concentration are changeable.

3. A reaction model has to be worked out regarding which model is capable of calculating ignition phenomena inside the furnace space (lifted flames) instead of flames developing from a flame holder of a burner. Conventional combustion simulation models form a flame on the assumptions that there is a high temperature gas jet from the burner rim or that the high temperature boundary conditions are set at a burner solid wall and so on. These numerical techniques are based on the idea in which a stable flame is always formed, and hence they cannot be applied to the prediction of blow-off limit of a flame holder. Thus to calculate flames formed by autoignition inside the furnace, conventional models are incapable of predicting where the flames start to form. We must calculate the process leading to ignition as a result of the coupled effects of mixing and reaction in the furnace. Consequently, the selection of reaction mechanisms and their rate constants is important in modeling of HiTAC flames.
4. Use of the full reaction mechanism is indispensable for taking all the intermediates into account, but a practical calculation including a three-dimensional flow with full reaction mechanism is far beyond the capability of computers today. Hence, the most realistic solution would be to adopt a set of greatly simplified reaction mechanisms covering some intermediates. For expressing the above-mentioned autoignition, a one-step global reaction model is not appropriate and several steps of reduced reaction mechanisms would be required.

3.2.3 Temperature Correction for Thermal Dissociation

As the flame temperature rises, thermal dissociation becomes significant, particularly above 2000 K. Therefore, we examined the effects of thermal dissociation in a high temperature range by use of numerical simulation. The maximum flame temperatures in a counterflow diffusion flame were compared among three reaction models: the full mechanism reaction model, the chemical equilibrium model, and theoretically complete combustion model. The GRI (Gas Research Institute) reaction mechanism comprising 48 chemical species and 275 elementary reactions was selected as the full reaction mechanism for methane–air mixtures.[3] Temperature of combustion air was changed from 300 to 1400 K. The stretch rate was kept constant at 45.5 s^{-1}.

Figure 3.3 shows the maximum temperatures of counterflow diffusion flames in terms of air preheat temperature. Here, the shown theoretical complete combustion temperature was obtained on the assumption of the complete combustion of stoichiometric methane–air mixture, and the adiabatic equilibrium temperature of stoichiometric methane–air mixture was calculated based on the chemical equilibrium. Precise values of specific heat were used for calculating both theoretical and adiabatic equilibrium temperatures. Consequently, the difference between the two may result from the difference in species concentrations of burned product. The difference between the adiabatic equilibrium temperature and that of counterflow diffusion flames may be ascribed partly to the influence of flame stretch caused by the flow, and partly to the difference in concentration of nonequilibrium intermediates. What is clearly observed here is that the flame temperature is changed markedly by the

Simulation Models for High Temperature Air Combustion

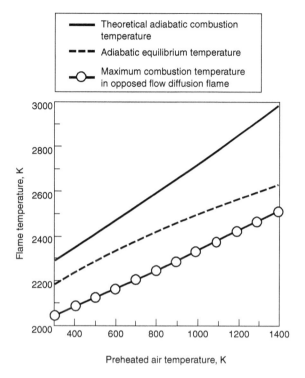

FIGURE 3.3 Comparison of maximum temperature by different combustion models.

effect of stretch. Therefore, we can say that temperature fluctuations in a turbulent flame may be affected not only by fluctuations in equivalence ratio but also by the local stretch rate owing to turbulent motions.

As seen above, it is extremely difficult to predict flame temperature accurately by numerical simulation. The only way to do this is via a simulation using the full reaction mechanism in complex flows with a considerably short time-step, which, however, is an unrealistic operation in engineering calculations. In the simulation of real furnaces where time-averaged temperature and concentrations are calculated, the only practical way is to obtain a reasonable value of combustion temperature by correcting calculated combustion temperature, or to introduce a limit so that reactions do not overshoot chemical equilibrium.

When reverse reactions are not included in the reaction mechanism, the combustion proceeds until the concentration of one of the reactants becomes zero. But, in reality, the chemical reaction does not go beyond chemical equilibrium, and the real maximum flame temperature must be lower than the case where complete reaction is assumed. The difference between the two is too large to be neglected, especially in the high temperature range above 2000 K. For this reason we studied a method for correcting this influence in numerical simulations where a reaction model not including reverse reactions is used.

Temperature obtained in simulation must be corrected using the relationship between the theoretical complete combustion temperatures and equilibrium temperatures calculated based on the relationship between initial temperature and equivalence ratio of the mixture. Figure 3.4 shows the theoretical complete combustion temperatures and the chemical equilibrium temperatures for air preheated to the temperature range of 300 to 1400 K and the equivalence ratio range of 0 to 2. Figure 3.5 shows these results plotted on a plane, the coordinates of which are the theoretical complete combustion temperature and the equilibrium temperature. We can see a simple relationship between them for the fuel-lean side and stoichiometric mixtures, regardless of the initial temperature of the mixture. For the fuel-rich mixtures, in contrast, the complete combustion temperature and the equilibrium temperature are not correlated by a single relationship since the generation of intermediates having high calorific values, such as CO and H_2, is significant. From these results, there are two conclusions. One is that the virtual

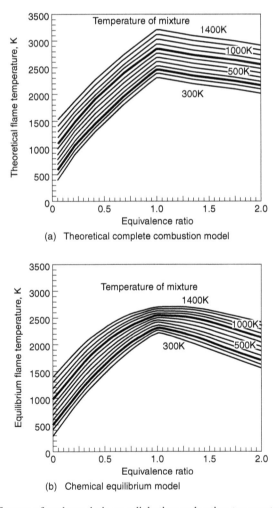

FIGURE 3.4 Influence of preheated air on adiabatic combustion temperature.

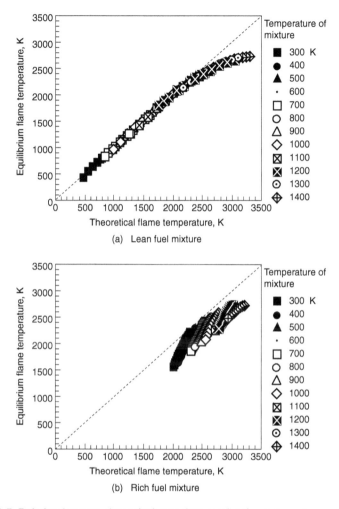

FIGURE 3.5 Relation between theoretical complete combustion temperature and chemical equilibrium temperature.

specific heat taking account of thermal dissociation can compensate for the temperature overshoot resulting from the complete combustion in the fuel-lean mixtures if the global one-step reaction model is adopted. The other is that the global one-step reaction model is not appropriate to be applied to fuel-rich mixtures, and we need to use combustion models predicting, at least, CO and H_2 concentrations.

When an elementary reaction including forward and reverse reactions is at chemical equilibrium, the ratio of forward and reverse reaction rates is equal to the equilibrium constant, and it looks as if the reaction is staying on hold. But, since the reduced reaction mechanisms, including several reaction steps, used in the simulations are not the real elementary reactions, they do not necessarily include reverse reactions. If the following two reactions, for example, that do not include reverse reactions, are included in the reaction mechanism used in a simulation:

$$CO + \frac{1}{2}O_2 \rightarrow CO_2 \qquad (R1)$$

$$H_2 + \frac{1}{2}O_2 \rightarrow H_2O \qquad (R2)$$

and an ordinary Arrhenius reaction rate expression is applied to the forward reactions, they will proceed beyond the chemical equilibrium. Hence, to prevent this from happening, a correction has to be made to the model, so that the rate of those two forward reactions may become zero when the chemical equilibrium is reached.

3.2.4 REACTION MODEL FOR HIGH TEMPERATURE AIR COMBUSTION

What is the most suitable combustion model for HiTAC simulation? Since HiTAC is usually initiated by autoignition, it is totally a chemical controlled phenomenon and it takes place in an atmosphere of high temperature and low oxygen concentration. So, the mixing-is-reacted model and the eddy-break-up model seem inappropriate for the purpose, because these models emphasize that the reaction rate is totally affected by the mixing processes, particularly in turbulence. Therefore, it is necessary here to discuss the potential of reaction models in predicting the balance between reaction and flow as stated before.

If a simulation on a realistic furnace of complex geometry is carried out, only the resultant temperature and concentration distributions in the furnace can be obtained. Thus, it will be difficult to verify the validity of the model, except by comparison with distribution results from experiments as they become available. To examine the potential of reaction models and their applicability to numerical simulation of combustion, a simpler flow system is preferable.

There are reports[4,5] of observations of lifted flames formed in preheated air or diluted airflows in a square (7 × 7 cm) duct with a high speed fuel jet issuing from a fuel nozzle at the center of the duct. A lifted flame will often be formed depending on flow velocity, oxygen concentration, and temperature. The type of flame and its stability limit vary with even minor changes in the influencing factors. Thus, this seems the best geometry to compare the potential of reaction models. As representative reaction models applicable to engineering numerical simulations, the following three models are discussed: a one-step global reaction model and two types of reduced reaction model. The reaction rates are given in Arrhenius type expressions, and the related empirical constants are retained here as originally proposed.

3.2.4.1 One-Step Global Reaction Model (Coffee)

This type of model is one of the most commonly used for numerical simulations of combustion, in which the reaction rate [kg/(m³s)] of fuel is generally expressed as follows:

Simulation Models for High Temperature Air Combustion

$$R_{fu} = -Fp^2 m_{fu} m_{ox} \exp\left(\frac{-E}{RT}\right) \qquad (3.6)$$

The frequency coefficient F and the activation energy E are constants to be defined through experiments. Many researchers have tried to define the values for different kinds of fuel. Among them, Coffee[6] has examined the values to represent the burning velocity in terms of equivalence ratio of methane–air mixtures.

Instead of mass fractions in Equation 3.6, he used volumetric concentrations as follows.

$$\frac{d[CH_4]}{dt} = 2.5 \times 10^{12} [CH_4]^{0.2} [O_2]^{1.3} \exp\left(\frac{-24,355}{T}\right) \qquad (3.7)$$

where [ ] denotes molar concentration, mol/cm^3.

This model was originally proposed for representing the burning velocity of methane–air mixtures at room temperature. Of course, although the expression is not always suitable for mixtures at any temperature, it provides a guideline for the reaction model for HiTAC.

3.2.4.2 Four-Step Reaction Model (Jones and Lindstedt)

In a reduced mechanism, a number of minor elementary reactions and intermediate chemical species are eliminated to create an apparent simplified reaction scheme either through extracting dominant reactions from the detailed chemistry or combining several elementary reactions by assuming partial equilibrium among them. For this reason a single reduced reaction model cannot always be effective for a variety of objects. Note that the accuracy of obtained simulation results might not be reliable unless the most suitable model for the present object is selected.

This four-step model was built by extracting six species and simplifying them into four steps of reactions from detailed chemistry. It has been reported that the model is applicable either to diffusion or premixed laminar flames and that the obtained burning velocity and flame structure, i.e., temperature and species distributions inside the flame, conform well to actual measurements.[7] The original four-step reaction mechanism was proposed for generic saturated lower-hydrocarbon fuels, and in the case of methane, they can be written as follows:

$$CH_4 + H_2O \rightarrow CO + 3H_2 \qquad (R3)$$

$$CH_4 + \frac{1}{2}O_2 \rightarrow CO + 2H_2 \qquad (R4)$$

$$H_2 + \frac{1}{2}O_2 \Leftrightarrow H_2O \qquad (R5)$$

$$CO + H_2O \Leftrightarrow CO_2 + H_2 \quad (R6)$$

The reaction rate is given for each of the four reactions as follows:

$$\frac{d[CH_4]}{dt} = -AT^d[CH_4]^a[H_2O]^b \exp\left(\frac{-E}{RT}\right) \quad (3.8)$$

$$\frac{d[CH_4]}{dt} = -AT^d[CH_4]^a[O_2]^b \exp\left(\frac{-E}{RT}\right) \quad (3.9)$$

$$\frac{d[H_2]}{dt} = -AT^d[H_2]^a[O_2]^b \exp\left(\frac{-E}{RT}\right) \quad (3.10)$$

$$\frac{d[CO]}{dt} = -AT^d[CO]^a[H_2O]^b \exp\left(\frac{-E}{RT}\right) \quad (3.11)$$

where [ ] indicates molar concentration mol/cm³. The reaction rate constants for each of the above reactions are listed in Table 3.1.

This reaction mechanism is divided into two parts: the first half is composed of reactions to turn methane into intermediate compounds and the second half represents combustion of intermediates including comparatively slow reaction of CO oxidization. Since the first-half reactions depend on concentrations of methane and oxygen, they are not much different from the one-step global reaction. The difference lies in that the intermediates, CO and H_2, are also generated from reverse reactions in the four-step reaction model. This model is quite different from the one-step global reaction model in the later steps of combustion.

3.2.4.3 Four-Step Reaction Model (Srivatsa)

This model was originally used for simulating gas turbine combustors and is applicable to combustion under an extreme condition where the characteristic time of mixing is very short, nearly as short as that of chemical reactions, i.e., a combustion with a low Damköhler number. To bring low NO_x combustion into effect using high

TABLE 3.1
Reaction Rate Constants for Jones's Reaction Mechanism

Reaction Rate Formula	A	d	a	b	E cal/mol
3.8	0.30×10^9	0	1.00	1.00	30,000
3.9	0.44×10^{12}	0	0.50	1.25	30,000
3.10	0.68×10^{16}	−1	0.25	1.50	40,000
3.11	0.275×10^{10}	0	1.00	1.00	20,000

temperature combustion air, it is necessary to mix preheated air with burned product inside the furnace by a high-momentum jet. Since the Damköhler number of combustion is considered to be comparatively low under these conditions, this model is suitable for the present purpose in comparison with other models.[8]

$$2CH_4 \rightarrow 2CH_3 + H_2 \quad \text{(R7)}$$

$$2CH_3 + O_2 \rightarrow 2CO + 3H_2 \quad \text{(R8)}$$

$$CO + \frac{1}{2}O_2 \rightarrow CO_2 \quad \text{(R1)}$$

$$H_2 + \frac{1}{2}O_2 \rightarrow H_2O \quad \text{(R2)}$$

No reverse reactions are considered in these reactions. It is likely, therefore, that the forward reactions advance beyond chemical equilibrium in combustion inside a furnace with a long residence time. The reaction rate is given for each of the four reactions as follows:

$$\frac{d[CH_4]}{dt} = -10^x [CH_4]^a [O_2]^b [CH_3]^c \exp\left(\frac{-E}{RT}\right) \quad (3.12)$$

$$\frac{d[CH_3]}{dt} = -10^x S [CH_3]^a [O_2]^b [CH_4]^c \exp\left(\frac{-E}{RT}\right) \quad (3.13)$$

$$\frac{d[CO]}{dt} = -10^x [CO]^a [O_2]^b [H_2O]^c \exp\left(\frac{-E}{RT}\right) \quad (3.14)$$

$$\frac{d[H_2]}{dt} = -10^x [H_2]^a [O_2]^b [CH_3]^c \exp\left(\frac{-E}{RT}\right) \quad (3.15)$$

where [ ] indicates mol concentration [mol/cm^3], $S = 7.93 \exp(-2.48\phi)$, and ϕ is the initial global equivalence ratio, whose value cannot be unity or larger. Values of the empirical constants used in the above formulae are listed in Table 3.2.

3.2.5 COMPARISON OF REACTION MODELS

A 1-mm-square fuel nozzle was placed at the center of a square duct (80 mm × 80 mm), and a region of 600 mm from the nozzle tip was considered as the flow field of simulation. Flow velocity of air u_a in the duct was set to be 6 m/s, its standard temperature T_a 1300 K, and the fuel temperature T_f 300 K. With regard to the oxygen

TABLE 3.2
Reaction Rate Constants for Srivatsa's Reaction Mechanism

Reaction Rate Formula	x	a	b	c	E cal/mol
3.12	17.32	0.50	1.07	0.40	49,600
3.13	14.70	0.90	1.18	−0.37	50,000
3.14	14.60	1.00	0.25	0.50	40,000
3.15	13.53	0.85	1.42	−0.56	41,000

concentration, which is the key to NO_x formation in HiTAC, atmospheric oxygen concentration was adopted as the standard and lower-concentration cases were set forth by dilution with nitrogen.

In the first place, Figure 3.6 shows temperature distributions predicted by the three reaction models: Coffee's one-step global reaction model, Jones' four-step model, and Srivatsa's four-step model, respectively. Since Coffee's model was formulated for the purpose of calculating the burning velocity, it is modeled focusing on the exothermic reactions in relatively early stages among various elementary reactions. Jones' model is divided into two groups of reactions, in which fuel turns into intermediate species in the first half and oxidation-terminating reactions in the second half, and thus the intermediates, such as CO and H_2, exist due to serial occurrence of these reactions. Judging from the simulation results, the reaction rate of the first half of Jones' model is slightly faster than the one-step global reaction model and the oxygen consumption is faster as a consequence. Its oxidation-terminating reactions of CO in the second half are also faster than Coffee's model. Srivatsa's model is characterized by the fact that the time required for starting the first-half reactions to turn fuel into intermediate species is longer than the other models. Since concentration of CH_3 is involved in those reactions, a small amount of CH_3 continues to appear at considerably downstream positions.

3.2.5.1 Comparison of Flame Lifted Height by Different Reaction Models

The height of the flame lift calculated by the three reaction models is shown in Figure 3.7. The definition of lifted height here is the distance from the nozzle tip to the lower edge of flame, not on the center line, even though the H_2O concentration becomes 0.4%. By Jones' model the lifted height does not change much in terms of oxygen concentration, and it is predicted to be as small as 20 mm when oxygen is 5%. In contrast, by Srivatsa's four-step reaction model the flame lifted height is predicted to be comparatively large. The H_2O concentration of 0.4% was adopted here as the definition of the flame lift for the convenience of a comparison with that for the one-step reaction model.

Simulation Models for High Temperature Air Combustion

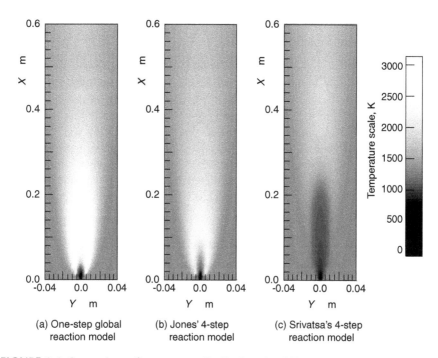

FIGURE 3.6 Comparison of temperature distributions by different combustion models.

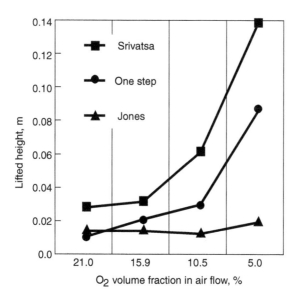

FIGURE 3.7 Comparison of lifted height of flames.

3.2.5.2 Comparison of Maximum Flame Temperature by Different Reaction Models

Figure 3.8 shows a comparison of the maximum flame temperature predicted by each model. If we do not take thermal dissociation into account, Coffee's one-step global reaction model based on the theoretical complete reaction predicts a flame temperature over 2800 K for preheated air of 1400 K reacting with methane (using normal oxidant). The temperature corresponding to this condition was 2292 K by Jones' model, and 1834 K by Srivatsa's model, which seems more realistic. Examining the change of maximum flame temperature in terms of oxygen concentration, a flame temperature over 2000 K was predicted even when oxygen concentration is 10.5% by Coffee's model and Jones' model, if the influence of thermal dissociation was not taken into account.

3.2.5.3 Influence of Jet Velocity on Flame Lift Height

In the numerical simulations, the rim thickness of the fuel nozzle was assumed to be zero, which is different from the real experimental equipment. So, the fluid dynamic influence of the burner rim on the flame stability is somewhat different from the actual condition, and the transition from laminar to turbulent in relation to the increase in fuel ejection velocity cannot be seen exactly by use of k–ε model. However, the influence of reaction model can be compared and discussed, undisturbed by other factors. Figure 3.9 shows the flame lifted height in terms of ejection velocity of fuel flow. Since the definition of the flame lifted height used here is the distance from the nozzle tip to the point where the H_2O concentration is 0.4%, as

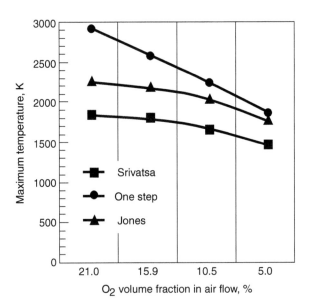

FIGURE 3.8 Comparison of maximum flame temperatures.

Simulation Models for High Temperature Air Combustion

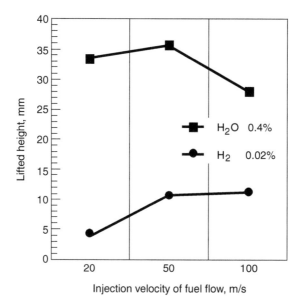

FIGURE 3.9 Influence of fuel ejection velocity on lifted height of flames by different definition.

stated above, the fuel flow rate inevitably affects the results. However, if the different definition 0.02% of H_2 concentration is used, the lifted height and its variation show considerably different results.

It should be noted that the lifted height in experiments would be defined by the visible or detectable intensity of optical emissions, which must be related to a certain amount of apparition of intermediate species. It would be ideal if a quantity reflecting the visually observed lifted height were used for the definition of lifted height in simulations. But, as visually observed flame emission is mainly related to the luminescence of radicals in the flame, it is difficult to predict the intensity of emission using simplified reaction mechanisms alone.

Regarding the desirable characteristics of combustion model to be used in the numerical simulation of HiTAC, Srivatsa's model has shown better features among the three. However, study of the empirical constants relevant to the proposed mechanism is still in progress. At present the following conclusions regarding the combustion model for HiTAC have been reached:

1. The model should provide an appropriate combustion rate in accordance with the changes of air temperature and oxygen concentration, and be capable of predicting flame lift caused by the balance between flow fields and combustion reactions. Accordingly, a chemically controlled reaction model seems suitable.
2. When we use a one-step global reaction model or a reduced mechanism without reverse reactions, it should be combined with virtual specific heat taking into account the influences of thermal dissociation.

3. To raise the accuracy of temperature predicted for any equivalence ratio, the reaction mechanism should at least include reduced reactions relevant to CO and H_2.
4. The influence of turbulence on the reaction rate seems less significant in HiTAC, because the fluctuations of temperature and species concentration in HiTAC are much smaller than in ordinary combustion. So, we may use Arrhenius rate expressions without introducing any influences of turbulence.

3.3 HEAT TRANSFER MODEL FOR HIGH TEMPERATURE AIR COMBUSTION

A large volume reaction zone associated with a mild temperature rise having low luminosity is a typical feature of HiTAC, which is completely different from ordinary combustion observed with luminous flames. When considering the numerical simulation of heat transfer in HiTAC furnaces, we must take this low-luminosity flame into account. In addition, most of the heat transferred to the material to be heated is dominated by radiative heat transfer from the walls, which have been heated by the convective heat transfer of the nonluminous combustion gas. Therefore, we need an appropriate heat transfer model for HiTAC that can deal with heat transfer from both gaseous and solid media to the materials to be heated.

3.3.1 HEAT TRANSFER MODELS

3.3.1.1 Gray Model

If the furnace wall material is a blackbody, radiation in all wavelengths is emitted from its surface at any temperature. The dependency of monochromatic emissive power of a blackbody, $E_{b\lambda}$, on temperature is shown in Figure 3.10. As shown in the figure, the dominant wavelength of radiation emitted from a blackbody moves to the short wavelength range as the temperature of the blackbody rises. The emissive power, E_b, from unit area of the blackbody surface during unit time can be expressed as follows.

$$E_b = \int_0^\infty E_{b\lambda} = \sigma T^4 \quad (3.16)$$

where σ is the Stefan–Boltzmann constant and T is surface temperature.

Fractions of the incident radiant energy on a surface are absorbed, reflected, and transmitted, respectively, and the ratio of respective energy against the incident energy is called absorptivity, a_λ, reflectivity, τ_λ, and transmissivity, d_λ. The interrelation among them can be written as

$$a_\lambda + \tau_\lambda + d_\lambda = 1 \quad (3.17)$$

Simulation Models for High Temperature Air Combustion

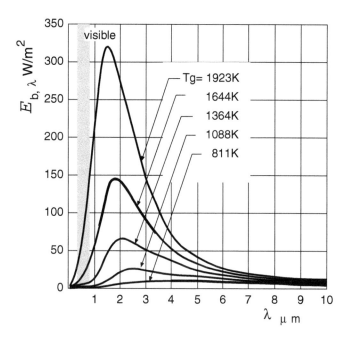

FIGURE 3.10 Radiation intensity of a blackbody.

As a blackbody absorbs all of the incident radiant energy, we have $a_\lambda = 1$, $\tau_\lambda = d_\lambda = 0$. Therefore, we can calculate radiative heat transfer between blackbodies using only their surface temperature, because we need not take into account the wavelength dependency of radiation energy.

When the absorptivity, a_λ, of a material is constant and smaller than unity, it is called a graybody. The monochromatic emissive power, E_λ, of a graybody is proportional to the monochromatic emissive power of a blackbody, and the ratio between the two defines the monochromatic emissivity, ε_λ.

$$E_\lambda = \varepsilon_\lambda E_{b\lambda} \qquad (3.18)$$

The monochromatic absorptivity of the material is also proportional to that of a blackbody and the monochromatic absorptivity a is equal to the monochromatic emissivity ε for graybodies of the same temperature.

$$a = \varepsilon \qquad (3.19)$$

A similar relation is valid for the emissivity and absorptivity,

$$E = \varepsilon E_b \qquad (3.20)$$

Therefore, it is sufficient to calculate the balance of radiant emission and absorption of the wall, when we can adopt a gray model in estimating the radiative heat transfer within an enclosure filled with clear gases.

3.3.1.2 Weighted-Sum-of-Gray-Gases Model

Like a solid medium, gases can absorb and emit radiative energy. Radiation from gas typically exhibits discrete band emission/absorption in terms of wavelength, since the interactions between molecules are relatively weak in gases. Therefore, even if a gas is clear at a certain wavelength, it may behave nonclearly at another wavelength. A mixed gas generally has different emission/absorption characteristics depending on the concentration of nonclear gases. Figure 3.11 shows an example of the distribution of absorptivity of mixed gas containing CO_2 and N_2,[9] which varies with gas temperature and the temperature of radiation source. Thus, it is erroneous to neglect these features when estimating the radiative heat transfer in a furnace filled with burned product by using a constant emissivity.

The relationship between the emissivity–path length for an unclear gas can be expressed by

$$\varepsilon_g = 1 - \exp(-kL) \qquad (3.21)$$

where k is the absorption coefficient and L is the path length of radiation. Hottel[10,11] showed that the emissivity of a real gas can be represented by the weighted sum of gray gases, and the emissivity–path length relationship may be approximated by a series as:

$$\varepsilon_g = \sum_n a_{g,n}\{1 - \exp(-k_n L)\} \qquad (3.22)$$

where

$$\sum_n a_{g,n} = 1 \qquad (3.23)$$

where $a_{g,n}$ is the fractional amount of energy in the spectrum regions where the gray gas of absorption coefficient, k_n. With this expression, the absorption coefficient, k_n, can be made independent of temperature and the temperature dependence of emissivity taken by the weighting factor $a_{g,n}$.

A luminous flame has a continuous spectrum emitted by the small particulates generated during combustion processes. It means that we can assume a luminous flame including soot particles as a gray gas. Therefore, it is convenient to approximate the variable absorptivity of combusting or burned gases, even containing soot particles, with the weighted sum of gray gases. Johnson and Beér[12] have proposed an expression similar to Equation 3.22 taking account of the effect of continuous emission of soot cloud. This is illustrated in Figure 3.12 using three terms representing emissivity bands and the corresponding weighting factor $a_{g,n}$ can be found in the literature; for example, see reference 13.

Simulation Models for High Temperature Air Combustion

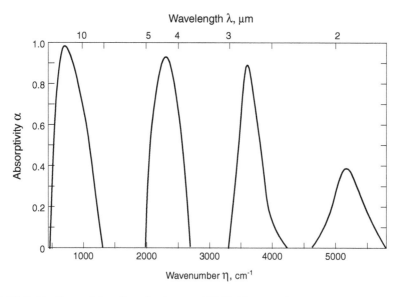

FIGURE 3.11 Absorptivity of a mixed gas at 833 K, 10 atmosphere.

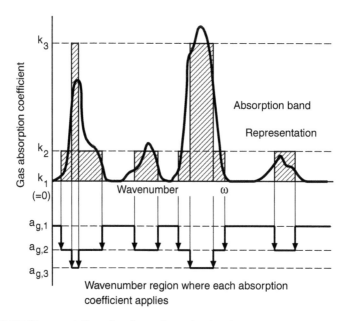

FIGURE 3.12 Representation of real gas absorption band.

3.3.1.3 Nongray Models

As shown in Figure 3.11, radiative heat energy of a gas molecule is transferred through the bands corresponding to respective vibrational energy levels. However, careful observation of these bands reveals that they consist of a series of line spectra corresponding to rotational levels as schematically explained in Figure 3.13.[14] A precise prediction of exact radiative heat transfer cannot be expected unless we calculate the net balance of radiation energy in each wavelength (or wave number), since real radiative heat transfer of gaseous media is carried out through these line spectra. A number of models have been proposed intending to estimate the integrated absorption coefficient taking account of distribution of the line spectrum. The narrowband models, such as the Elsasser model[15] and the statistical model,[16] are in this category. However, it is still difficult to determine their line intensity and overlapping due to broadening, because there are a number of line spectra in a limited wavelength range. Therefore, the accuracy of the obtained results is not always high enough, despite a complex calculation that is necessary in actual execution of numerical simulations. Consequently, practical simulation of heat transfer in furnaces using the narrowband model seems doubtful.

In practical cases, heat transfer engineers are usually interested in estimating radiative heat flux integrated over the spectrum. Thus, it is desirable to have models that can readily predict the total absorption or emission through each band appearing in Figure 3.11. These models are known as wideband models. Edwards and Balakrishnan proposed a model called exponential wideband model (EWBM),[17] in which they assumed three types of line intensity distribution, since it is known from quantum mechanics that the line intensity decreases exponentially in the band wings. The absorption coefficient S/d is represented by one of the following three functions (as illustrated in Figure 3.14):

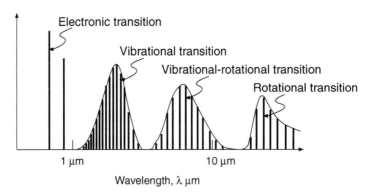

FIGURE 3.13 Spectral lines due to electronic, vibrational, and rotational energy changes in a gas molecule.

Simulation Models for High Temperature Air Combustion

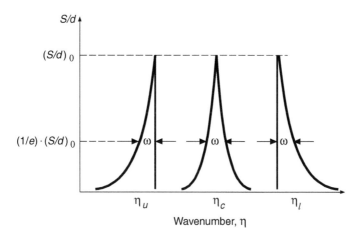

FIGURE 3.14 Exponential wideband model.

with upper limit head $\quad S/d = (\alpha/\omega)\exp\{-(\eta_u - \eta)/\omega\}$
with symmetric head $\quad S/d = (\alpha/\omega)\exp\{-2|\eta_c - \eta|/\omega\}$ (3.24)
with lower limit head $\quad S/d = (\alpha/\omega)\exp\{-(\eta - \eta_l)/\omega\}$

where

$$\alpha = \int_0^\infty (S/d)_\eta d\eta \quad (3.25)$$

is the integrated absorption coefficient and ω is the bandwidth parameter, defined by the bandwidth at $1/e$ value of maximum intensity. The accuracy of the correlation between experimental data and predictions demonstrated in reference 17 was approximately ±20%.

3.3.2 Radiative Heat Transfer Using Nongray Property of Radiation

To discuss the applicability of heat transfer models to HiTAC furnaces, we compared experiments and predictions on characteristics of radiative heat transfer in a small model furnace. The exponential wideband model was adopted for analyzing the radiative heat transfer, in which we used the physical properties of nongray radiation that can take the dependency of radiation of burned gas on wavelengths into account. We measured the radiant heat flux in a propane–air diffusion flame and the total heat flux incident to the surface of furnace wall. The outline of the experimental apparatus is shown in Figure 3.15; the flow rate of propane and air was 0.407 and 40.7 ℓ/min, respectively, and the corresponding flow velocity at the inlet was 0.051

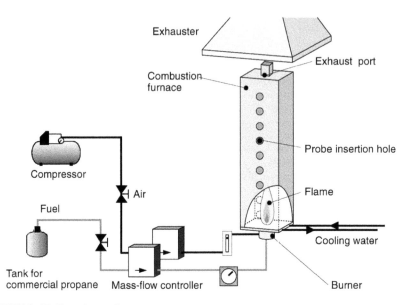

FIGURE 3.15 Experimental apparatus.

and 5.23 m/s, respectively. These experiments were carried out under nonluminous flame conditions, without emitting any soot. Both the radiant heat flux and the total heat flux were measured at several inspection holes prepared in the furnace wall (inside dimensions: 190 × 190 mm^2, height: 1000 mm) at intervals of 100 mm.

Calculating the radiant heat flux, numerical simulations of turbulent combusting flow were carried out using the k–ε turbulence model and the eddy dissipation combustion model including the swirling effect. In representing physical properties of a nongray radiation spectrum, we used the exponential wideband model (EWBM) taking the dependency of both CO_2 and H_2O emission/absorption on the wavelength into account. We assumed line intensity vs. line spacing is continuously and exponentially attenuated as the wavelength is displaced from the band center in this band model. To calculate the radiation transport, we adopted the discrete ordinates method. In the model, the equation of radiation transport is formulated in each scattering direction of radiation, assuming that the whole view angle of a radiant element is split into a specified number of solid angles and the radiation intensity in each solid angle is uniform.

Figure 3.16 shows a comparison of distributions of the radiant heat flux incident upon the wall surface of the furnace obtained by the experiment and the predictions by gray analysis (GRA) and nongray analysis (NGRA), respectively. The distributions of the incident total heat flux are compared in Figure 3.17 for the same cases mentioned above. The results obtained by the nongray model are in good agreement with the values measured. We understand through the present comparison that it is important to take the characteristics of nongray radiative heat transfer into consideration in performing a heat-transfer simulation of HiTAC nonluminous combustion.

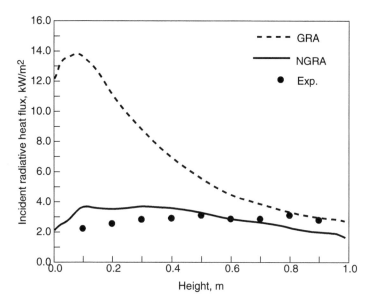

FIGURE 3.16 Comparison of radiant heat fluxes.

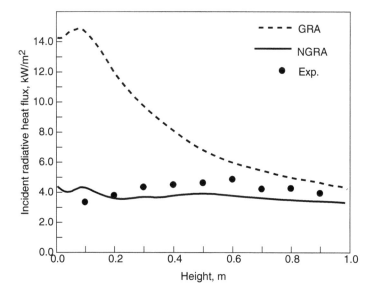

FIGURE 3.17 Comparison of total heat fluxes.

3.4 EXAMPLES OF PRACTICAL APPLICATION

Because HiTAC is a new combustion technology, we have merely constructed a database of the technology and some of its current application for practical design method. Although we have so far discussed the combustion and radiative heat transfer models for HiTAC, we are still at the primitive stage in examining their applicability in all applications with full confidence. Further research is required before we can recommend numerical models for HiTAC. However, some trials have been conducted that allow us to simulate the important characteristics of HiTAC. This is accomplished by combining the existing numerical models and methods. The following examples provide the limits of practical application to numerically simulate the technology utilizing the experience gained so far.

3.4.1 Nitric Oxide Emission

The NO concentration usually observed in furnaces is relatively low and the kinetics of its formation reaction is much slower than the main hydrocarbon oxidation rate. Therefore, it is possible to assume that the reactions involved in the NO chemistry can be decoupled from the main combustion reaction mechanism. Because of the slow formation of NO, equilibrium calculations are not suitable to calculate realistic NO concentrations. Kinetic calculations are essential to obtain reasonable results. Therefore, the mean NO concentration is obtained by solving its transport equation based on the flow field and combustion solution from the main combustion simulations.

$$\bar{\rho}\tilde{u}_i \frac{\partial \tilde{Y}_{NO}}{\partial x_i} = \frac{\partial}{\partial x_i}\left(\rho D \frac{\partial \tilde{Y}_{NO}}{\partial x_i}\right) + \tilde{S}_{NO} \qquad (3.26)$$

where the source term, $\tilde{S}_{NO}$, is determined from different NO mechanisms. Three chemical-kinetic mechanisms for the NO formation/depletion, i.e., thermal NO, prompt NO, and NO reburning, are included in this study.

3.4.1.1 Thermal NO

The formation of thermal NO is determined by the extended Zel'dovich mechanism.

$$O + N_2 \xleftrightarrow{k_{\pm 1}} N + NO \qquad \text{(R3)}$$

$$N + O_2 \xleftrightarrow{k_{\pm 2}} O + NO \qquad \text{(R4)}$$

$$N + OH \xleftrightarrow{k_{\pm 3}} N + NO \qquad \text{(R5)}$$

Based on the quasi-steady-state assumption for N atom concentration, the net rate of NO formation via the above reactions can be determined by

$$\frac{d[NO]_T}{dt} = 2k_1[O][N_2]\left(1 - \frac{k_{-1}k_{-2}[NO]^2}{k_1[N_2]k_2[O_2]}\right) \bigg/ \left(1 + \frac{k_{-1}[NO]}{k_2[O_2] + k_3[OH]}\right) \qquad (3.27)$$

Simulation Models for High Temperature Air Combustion

where $k_i = A_i T^{B_i} \exp\left(\frac{-C_i}{T}\right)$.

The reaction constants, A_i, B_i, and C_i, were taken from Hanson and Salimian.[18]

The OH radical concentration was taken from the main combustion calculation. The O atom concentration was determined by the partial equilibrium approach given as:[19]

$$[O] = AT^B [O_2]^{\frac{1}{2}} \exp\left(\frac{-C}{T}\right) \qquad (3.28)$$

where $A = 36.64$, $B = 0.5$, and $C = 27123$.

3.4.1.2 Prompt NO

The model proposed by De Soete[20] was used to determine the prompt NO formation rate.

$$\frac{d[NO]_P}{dt} = A\eta[O_2]^\alpha [N_2][\text{Fuel}] \exp\left(\frac{-E}{RT}\right) \qquad (3.29)$$

where $\eta = 4.75 + 0.0819 N_c - 23.2\varphi^2 - 12.2\varphi^3$.

The values of A and E were selected from Dupont et al.[21] The oxygen reaction order, α, is related to the oxygen mole fraction in the flame and is given by De Soete.[20]

3.4.1.3 NO Reduction Mechanism (Reburning)

NO reduction is a process where NO reacts with hydrocarbons, CH_i, to form HCN and other products. The NO reduction mechanism proposed by Bowman et al.[22] was employed.

$$CH + NO \xrightarrow{k_4} HCN + O \qquad (R6)$$

$$CH_2 + NO \xrightarrow{k_5} HCN + OH \qquad (R7)$$

$$CH_3 + NO \xrightarrow{k_6} HCN + H_2O \qquad (R8)$$

where $k_i = A_i T^{B_i} \exp\left(\frac{-C_i}{T}\right)$.

The values of A_i, B_i, and C_i are from Bowman et al.[22] The rate of destruction for NO is given by

$$\frac{d[NO]_R}{dt} = k_4 [CH][NO] + k_5 [CH_2][NO] + k_6 [CH_3][NO] \qquad (3.30)$$

The NO source term due to the formation/destruction of the thermal NO, prompt NO, and NO reburning can be calculated by

$$S_{NO} = M_{NO} \left\{ \frac{d[NO]_T}{dt} + \frac{d[NO]_P}{dt} - \frac{d[NO]_R}{dt} \right\} \quad (3.31)$$

To calculate the mean NO formation rate, $\tilde{S}_{NO}$, which is required in the mean NO concentration equation, a PDF is employed to take into account temperature and composition fluctuations. NO is formed mainly by the thermal mechanism in gas-fired furnaces. The rate of thermal NO formation depends on the temperature and oxygen concentration. Therefore, a joint two-variable PDF in terms of temperature and oxygen concentrations was used. The mean formation rate, $\tilde{S}_{NO}$, can be determined by

$$\tilde{S}_{NO} = \iint S_{NO}(T, Y_{O_2}) P_1(T) P_2(Y_{O_2}) dT dY_{O_2} \quad (3.32)$$

The shape of $P_1(T)$ and $P_2(Y_{O_2})$ is approximated by a β-function since the β-PDF is widely used in turbulent combustion simulations.

The governing equations for turbulent non-premixed combustion include the conservation equations of mass, momentum, energy, turbulence kinetic energy, dissipation rate of turbulent kinetic energy, mixture fraction, and its variance. Density-weighted (Favre) averaging was utilized to account for the effects of density change.

The steady three-dimensional Favre-averaged conservation equations in Cartesian coordinates were solved. The numerical simulations were carried out on a nonuniform grid of 89 × 25 × 31 nodes along the x-, y-, and z-coordinates to accommodate the locations and sizes of the slab, air, and fuel injection nozzles, and auxiliary exhaust tube of the experimental furnace. The furnace had inner dimensions of 8 × 4 × 3 m. The grid dependence test in the former work[23] indicated that the grid arrangement described above can yield essentially grid-independent results.

The turbulent combustion model adopted in the simulation was the mixed-is-reacted model combined with the prescribed PDF (β-function) of mixture fraction fluctuations. The radiative heat transfer model used to estimate the source term in the energy conservation equation was the P-1 model by Cheng[24] associated with the weighted sum of gray-gas model. In the numerical simulations, the top and side refractory walls of the furnace were treated as convection/radiation walls with an overall external heat transfer coefficient of 0.58 J/m²sK, which was based on the thickness and thermal properties of refractory walls as well as the surrounding conditions. The bottom wall of the furnace was considered as a wall with a constant heat flux of 8.47 kW/m². This value was estimated from the heat-extraction rate by the cooling water flowing through the bottom of the furnace.

3.4.1.4 Results and Discussion

The temperature distributions on the horizontal plane, including the central plane, of injection ports of combustion air ($z = 1.5$ m) and the slab are depicted in Figure 3.18. This figure shows the main reaction regions are characterized by a high level of temperature. The centers of two main reaction regions in Zone 1 have slightly higher temperatures than the centers of the other two main reaction regions in Zone 2. This may be attributed to the fact that the fuel and combustion air input rates to Zone 1 were higher than Zone 2. The combustion thus occurs at a higher rate, which results in a higher temperature. On the other hand, the temperature outside the main reaction regions in Zone 2 is slightly higher than that in Zone 1 as the slab temperature is higher in Zone 2. Since the low-NO_x burners used in this furnace limit the mixing of the fuel with the combustion air in the initial stage, it can be seen, based on the temperature distribution, that the primary reactions occur downstream from the burners and long flames are produced. This results in a more uniform temperature distribution, which corresponds to a lower NO_x production rate.

The measured NO concentrations at the burner outlets in Zone 1, Zone 2, and the auxiliary outlet were 66, 57, and 76 ppm (dry), respectively. However, the magnitude of predicted NO concentrations was lower than the experimental data. The lack of agreement between the predicted and measured NO concentrations may be attributed to the deficiencies in the NO submodel, which cannot adequately predict the governing flame structure at the current combustion temperature. In the gas-fired

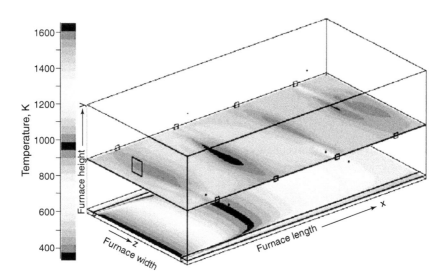

FIGURE 3.18 Temperature distribution in the prototype furnace.

furnace, NO is formed mainly by the thermal mechanism. The thermal NO mechanism used in this study is the Zel'dovich NO formation model, which is the widely used model for NO simulations. However, as indicated by Warnatz,[19] the thermal NO formation is limited to the temperature range above 1800 K. Under the selected operating conditions, the maximum predicted and measured flue gas temperatures were about 1600 and 1550 K, respectively. Therefore, the Zel'dovich NO formation model cannot correctly predict the NO concentration at low temperatures.

3.4.2 Transient Behavior of Furnaces

The highly preheated combustion air for HiTAC is generated by high cycle switching operation of regenerators attached to a furnace. To simulate transient behavior of HiTAC furnaces, we must solve the governing equations including an unsteady time derivative term for mass, momentum, energy, and species mass fractions. This type of simulation is quite different from those described so far. The point of interest is transient behavior of flow and combustion interaction and its resultant emissions. For this purpose, we carried out a fluid-dynamic simulation of the furnace by using a numerical code named PROHEAT. A detailed radiation heat transfer model including gas radiation and a simplified combustion heat model are included in the code.

3.4.2.1 Fluid Dynamics Model

The governing equations for mass, momentum, and energy expressed in nondimensional form were solved by PROHEAT. We assumed the fluid in question was incompressible, since the flow is in low Mach number regime.

Mass conservation equation:

$$\frac{\partial u_j}{\partial x_j} = 0 \tag{3.33}$$

Momentum conservation equation:

$$\frac{\partial u_i}{\partial t} + \frac{\partial}{\partial x_j}\left(u_j u_i\right) = -\frac{\partial p}{\partial x_i} + \frac{\partial}{\partial x_j}\left\{\frac{1}{\text{Re}}\left(\frac{\partial u_i}{\partial x_j} + \frac{\partial u_j}{\partial x_i}\right)\right\} + \delta_{ij}\frac{\text{Gr}}{\text{Re}^2}T \tag{3.34}$$

Energy conservation equation:

$$\frac{\partial T}{\partial t} + \frac{\partial}{\partial x_j}\left(u_j T\right) = \frac{\partial}{\partial x_j}\left\{\frac{1}{\text{Re}\,\text{Pr}}\frac{\partial T}{\partial x_j}\right\} + \frac{Q_{\text{rad}}}{\rho c_p} \tag{3.35}$$

where u_j = velocity, x_j = horizontal axis, t = time, p = pressure, Re = Reynolds number, Gr = Grashof number, T = absolute temperature, Pr = Prandtl number,

Q_{rad} = radiation heat transfer term [W/m³], ρ = density [kg/m³], c_p = specific heat [J/kgK], δ_{ij} = Kronecker's delta.

As a numerical method for the turbulent flow, the large eddy simulation (LES) by the top-hat-type filter function was used. LES seemed the best to describe the time-dependent behavior of the flow compared with the conventional standard Reynolds-average-type turbulent models. The Deardroff–Smagorinsky model was used as a sub-grid-scale model.[25,26]

In the governing equations above, the density change is not included, and thus the effects of thermal expansion and compression are not taken into consideration. The governing equations initially formulated include the density changes due to the temperature variation. The computations took hundreds of hours to finish the calculations. For the first approximation of the transient behavior of the HiTAC furnace, the density change is neglected. To carry out more accurate simulations, the introduction of density variation is essential.

The models and the methods used for the fluid dynamic stimulation are summarized in Table 3.3.

3.4.2.2 Radiation Heat Transfer Model

Here, the READ (radiative energy absorption distribution) method was used in the present calculation code as a radiative heat transfer model. It is a discrete transfer method that calculates the absorption distribution of radiant energy along a ray of radiation. The concept and calculation method of READ can be found in the literature.[27]

This method can deal with radiation and absorption of gas media, and can be applied to nongray gas where the dependence of radiant heat on wavelength is important. It can also handle diffuse reflection and specular reflection on solid surfaces and scattering in a gas body. We need to calculate radiative heat transfer between walls through selective absorbing gases in HiTAC furnaces. Therefore, the flexibility of the READ method seems quite favorable in the present case compared with the zone method.

TABLE 3.3
Models and Methods Used in PROHEAT

Turbulent flow model	Large eddy simulation (SGS model: Deardroff–Smagorinsky model)
Digitization of space	Finite difference method (FDM)
Handling of convection term	Second-order central difference scheme or donor-cell method (second-order central difference and first-order upwind difference schemes)
Making time dispersed	Fluid equation: Adams–Bashforth scheme energy equation: complete implicit solution method
Placement of calculation grid	B-type staggered grid
Algorithm	HS-MAC method

Using the READ method, the radiant heat of a gas element, g_i, per unit time and unit volume can be expressed as follows:

$$Q_{rad} = -4k_{g_i}(1-\alpha_{g_i})\sigma T_{g_i}^4 + \sum_{g_i=1}^{g_I} 4k_{g_i}(1-\alpha_{g_i})\sigma T_{g_i}^4 \frac{V_{gi}}{V_{gj}} R_{dg_i \to gj}$$

$$+ \sum_{w_i=1}^{w_I} \varepsilon_{wi}(1-\alpha_{wi})\sigma T_{wi}^4 \frac{S_{wi}}{V_{gj}} R_{dw_i \to gj} \quad (3.36)$$

Equation 3.36 gives the source term in Equation 3.35, where Q_{rad} = radiant heat [W/m³], T_i = absolute temperature of element i [K], k_{gi} = absorption coefficient of gas element g_i [m⁻¹], ε_{wi} = emissivity of wall element w_i, V_{gi} = volume of gas element g_i [m³], S_{wi} = area of wall element w_i [m²], α_i = self absorptivity of element i, $R_{di \to j}$ = READ value from element i to element j, σ = Stefan–Boltzmann constant.

$R_{di \to j}$ and α_i in Equation 3.36 can be obtained by ray tracing method or Monte Carlo method. In PROHEAT, these values are calculated by the ray tracing method using 10,000 heat rays per element.

3.4.2.3 Combustion Model

The combustion model used in the code is the heat release rate distribution model. We assume that the fuel fed in the furnace shows a one-dimensional distribution in the direction of fuel jet, and the reaction rate is proportional to the concentration of fuel in each cross-section of the jet. These assumptions can express a macroscopic decay of fuel in HiTAC furnaces because fuel and air are fed as high-momentum jets having a strong orientation. The combustion reaction results from the collision between molecules of fuel and oxygen, and the frequency of collision is considered proportional to the number of molecules, that is, the concentration of fuel. Therefore, the present combustion model can be expressed by the following equation:

$$Q_{comb} = Q_0 \frac{d\eta}{dx}$$

$$\approx Q_0 \frac{\eta(x+dx) - \eta(x)}{dx} \quad (3.37)$$

$$\approx Q_0 \frac{\exp(-\kappa x)\{1 - \exp(1-\kappa dx)\}}{dx}$$

where Q_{comb} = local heat release per unit volume [W/m³], Q_0 = combustion rate in the burner [W], η = heat release rate function, x = distance from burner [m], dx = width of calculation grid [m]. In addition, κ is the model constant that can be varied with burner structures or combustion loads, and it must be determined by combustion experiments.

Simulation Models for High Temperature Air Combustion

3.4.2.4 Temperature Distribution during Fuel Changeover

One of the characteristics of a regenerative combustion process is its alternating operation at short intervals. We simulated the time history of the temperature distribution to see the flow behaviors in the furnace at the time of fuel changeover. The serial temperature distributions during the changeover are shown in Figure 3.19, starting at $t = 0$ when the working burner is shut down and the resting burner on the opposite side is ignited simultaneously.

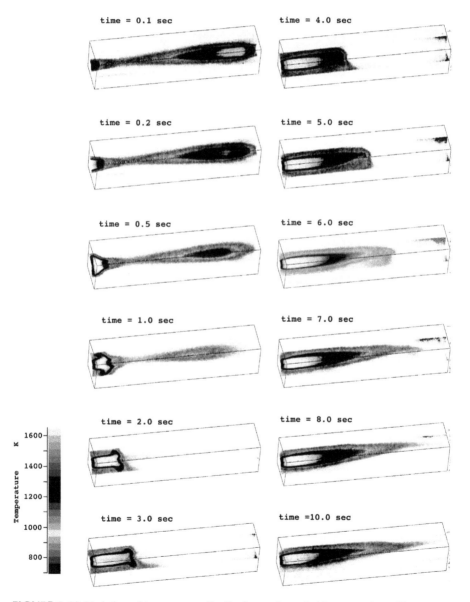

FIGURE 3.19 Variation of temperature distribution at the switching operation of burners.

As shown in the figure, the latest high-temperature region disappears rapidly due to radiation and convection heat transfer. Nevertheless, since the burned gas flow does not change so quickly, we observe that the starting jet flow on the ignition side is pushed back by the remaining inertia of the opposing jet flow. For this reason, a high temperature region is formed in the near-field of the burner in the early period of the start-up, and therefore the possibility of generating NO_x is high.

The present result, not shown here, indicates that there is no problem if the burner is ignited after the air stream has been ejected without fuel for about 3 s. However, this air-ejection period is required to be as short as possible because the heating capability is in proportion to the burning period. As for the optimum ignition time, a delayed ignition by 1 s is considered most favorable. During this ignition time the flow of jet flame switches from one burner to the other. The optimum ignition time depends on the dimensions of the burner and furnace.

3.4.2.5 Comparison with Measured Temperatures by Suction Pyrometer

In the discussion so far, values that fit with heat flux values obtained in a prototype furnace were used. Since burned gas temperatures were made available by measurements obtained by suction pyrometers, we compared the values measured with the pyrometer and the calculated values. We adopted measurements taken at 200 mm from each plane, i.e., the ceiling, floor, and sidewalls, which are considered located away from the combustion regions. In addition, for calculated values, time averages taken from several points because they greatly fluctuate in time and space.

As seen in Figure 3.20, the calculated temperatures agree well with the measured temperatures near the sidewall regions, but the former are estimated slightly below the measured ones near the floor area indicating better needs for the model used. The radiative heat transfer to the water pipe for heat removal is considered to be overestimated.

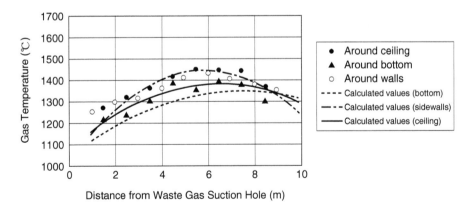

FIGURE 3.20 Comparison of experimental and calculated average temperatures.

3.4.2.6 Calculation on Wide Regenerative Furnace

To verify the performance of the numerical method described above, we carried out a numerical experiment by using the code PROHEAT on one of the largest furnaces in operation. The furnace has the dimensions of length 9.8 m, width 14 m, height 2.6 m, and has four sets of regenerative burners placed at equal distances in its upper zone. It was assumed that there was a heat sink with a constant temperature of 1273 K at the bottom. The upper soaking zone was taken as the calculation domain. It has no infow of gas from the upstream side, but some flow goes out downstream from the gap between the partition of 1 m high and the steel slab.

Two cases were considered regarding the burner arrangement as shown in Table 3.4.

1. Four sets, or eight units, of high-speed regenerative burners were placed in a staggered arrangement.
2. Eight units of conventional preheated air burners were set. By comparing furnace temperature distributions for the cases, we evaluated the effect of the regenerative burners on the uniformity of temperature distribution in furnaces.

Calculations use the code assuming the combustion heat value of $\kappa = 2.0 \text{m}^{-1}$ as described above. The calculation under unsteady state condition was continued for 30 s by using temperatures calculated under steady state condition. The average temperatures mentioned in the following section are obtained through the unsteady calculation results.

Figure 3.21 shows the temperature distribution obtained in the furnace width direction with the temperatures averaged both in the longitudinal and the vertical (height) directions. For both burner arrangements, the temperature at the center of the furnace width was lower than that in other regions.

While the maximum temperatures did not differ greatly in the two cases, 1479 K for the regenerative burner and 1483 K for the conventional burner, the minimum

TABLE 3.4
Calculation Conditions Regarding Upper Soaking Zone of Wide Regenerative Furnace

	Regenerative Furnace	Conventional Burner Furnace
Number of burners	4 sets 8 units	8 units
Burner combustion load (W)	1.50×10^6	0.75×10^6
Preheating air temperature (K)	1273	873
Air ratio	1.05	1.05
Pull-back ratio	0.8	—
Ejection velocity (m/s)	20	10

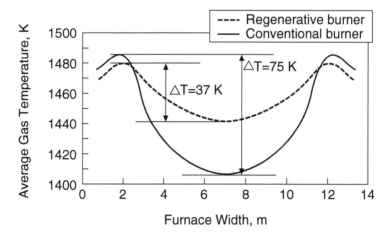

FIGURE 3.21 Comparison of temperature distributions in width direction between regenerative burner and conventional burner.

temperatures differed greatly, 1442 K for the regenerative burner and 1408 K for the conventional burner. Hence, the conventional burner gave much lower temperature at the furnace center in the width direction. This is because the regenerative burner allows twice as much combustion load as the conventional burner. Therefore, it had a high jet velocity generating a long burner flame. The regenerative burner showed a temperature difference of 37 K between maximum and minimum, whereas the conventional burner showed 75 K, about twice that for the regenerative burner. Temperature uniformity can be realized in the furnace with the regenerative burner. The temperature distribution in the furnace becomes more uniform with the higher Reynolds number of the combustion air.

By demonstrating some examples, it is shown that the transient behavior in switching operation of regenerative burners can be simulated using LES turbulence model, but some time is necessary before the method can be used as a practical design tool.

REFERENCES

1. D. B. Spalding, *Proc. Combust. Instit.*, 13:649, 1970.
2. B. F. Magnussen and H. Hjertager, *Proc. Combust. Instit.*, 16:719, 1976.
3. C. T. Bowman et al., http://www.me.berkeley.edu/gri mech/, 1995.
4. T. Fujimori, D. Riechelmann, and J. Sato, The First Asia-Pacific Conf. on Combust., p. 298, 1997.
5. T. Fujimori et al., *Proc. Combust. Instit.*, 34:447, 1996.
6. T. P. Coffee, *Combust. Sci. Technol.*, 43:333, 1985.
7. W. P. Jones and R. P. Lindstedt, *Combust. Flame*, 73:233, 1988.
8. S. K. Srivatsa, Technical Report CR-167930, NASA Report, 1982.
9. D. K. Edwards, ASME, *J. Heat Transfer*, 84C:1, 1962.
10. H. C. Hottel, *Heat Transmission*, McGraw-Hill, New York, 1954.

11. H. C. Hottel and A. F. Sarofim, *Radiative Transfer*, McGraw-Hill, New York, 1967.
12. T. R. Johnson and J. M. Beér, *Fourth Symposium on Flames and Industry*. British Flame Research Committee and Institute of Fuel, Imperial College, 1972.
13. J. S. Truelove, Technical Report R-8494, AERE Harwell Report, 1976.
14. M. F. Modest, *Radiative Heat Transfer*, McGraw-Hill, New York, 1993.
15. C. L. Tíen, *Advances in Heat Transfer* 5, Academic Press, New York, 1968.
16. S. S. Penner, *Quantitative Molecular Spectroscopy and Gas Emissions*, Addison-Wesley, Reading, MA, 1959.
17. D. K. Edwards and A. Balakrishnan, *Int. J. Heat Mass Transfer*, 16:25, 1973.
18. R. K. Hanson and S. Salimian, W. C. Gardiner, ed., *Combustion Chemistry*, 1984.
19. J. Warnatz, *Proc. European Gas Conference*, p. 303, 1991.
20. G. G. DeSoete, *Proc. Combust. Instit.*, 15:1093, 1975.
21. V. Dupont et al., *Fuel*, 72(4):497, 1993.
22. C. T. Bowman, W. W. Bartok, and A. F. Sarofim, eds., *Fossil Fuel Combustion — A Source Book*, Wiley, New York, 1991.
23. C. Zhang, T. Ishii, and S. Sugiyama, Numerical heat transfer, Part A: applications, 32:1662, 1997.
24. P. Cheng, *AIAA J.*, 2:1662, 1962.
25. J. W. Deardroff, *J. Fluid Mech.*, 41(2):453, 1970.
26. J. Smagorinsky, *Mon. Weather Rev.*, 91:99, 1963.
27. H. Hayasaka, *ASME HTD*, 74:59, 1987.

4 Practical Combustion Methods Used in Industries

4.1 HISTORICAL TRANSITION OF INDUSTRIAL FURNACE TECHNOLOGIES

Every avenue of development of energy-saving technologies has been pursued since the oil crises. Discussions on new technologies at the Third Conference of Parties (COP3) promoted further incentives. Even in the field of industrial furnaces, the foundation of industry, a high performance industrial furnace applying high temperature air combustion technology has reached the stage where it can be put to practical use. This innovative high performance industrial furnace is having an impact on both conventional industrial furnaces and the instrumentation control technologies for combustion in terms of reducing energy consumption.

4.1.1 Energy Technologies Discussed at COP3

To meet the problems of carbon dioxide emissions in 2010, many models and results from private and government bodies such as the Environment Agency, the Ministry of International Trade and Industry (now called METI), the Japan Energy and Economy Research Institute, and the Electric Power Central Research Institute were discussed. The models and results are now being compared and evaluated for the basis of the final program. Table 4.1 shows a list of energy technologies considered in the AIM end-use model by Matsuoka[1] and others. It can be seen from this table that recent technical progress has been remarkable in service technologies such as electric automobiles, hybrid cars, fuel cells, solar-utilized systems, repowering of private power generation, latent heat recovering thermal components, and high performance industrial furnaces. With advances in these technologies, the equipment cost and running cost of the technical installations have decreased remarkably. In many cases, these costs were the result of defects related to the new technologies. For example, a cogeneration system for domestic use is being widely disseminated by manufacturers, proving its cost effectiveness. This is along the same lines of technology developments as in fuel cells and micro gas turbines for domestic use.

There is a possibility that the high temperature air combustion technology for industrial furnaces can be applied to all types of combustion and heat transfer technologies. It is considered that in the future this technology will be related to the greater portion of the technical items listed in Table 4.1.

TABLE 4.1
Energy Technologies Discussed at The Third Conference of Parties (COP3)

Sections	Field	Service Technology
Industry	Iron and Steel	Humidistat for coke oven, scrap preheating apparatus, coke oven for the next generation, sintering furnace, blast furnace, converter, smelting reduction furnace, AC electric furnace, DC electric furnace, casting apparatus, continuous casting apparatus, conventional-type heating apparatus, hot direct rolling/hot slab charging, annealing apparatus, continuous annealing apparatus, dry-type coke quenching apparatus, wet-type coke quenching apparatus, wet-type electricity generating apparatus utilizing the top pressure of blast furnace, dry-type electricity generating apparatus utilizing the top pressure of blast furnace, private power generation, repowering of private power generation, combined cycle power generation
	Cement	Tube mill, auxiliary grinding machine, NSP/other than SP, NSP/SP, high-efficiency clinker cooler, vertical-type mill, fluidized bed firing furnace, diesel power generator, waste heat power generator, repowering of private power generation, combined cycle power generation
	Petrochemical	Apparatus for naphtha cracking reaction, apparatus for high efficiency naphtha cracking reaction, low-density polyethylene manufacturing apparatus, high-efficiency low-density polyethylene manufacturing apparatus, high-density polyethylene manufacturing apparatus, polypropylene manufacturing apparatus, high efficiency polypropylene manufacturing apparatus, ethylene oxide manufacturing apparatus, styrene monomer manufacturing apparatus, acetaldehyde manufacturing apparatus, polyethylene manufacturing apparatus, high performance polypropylene manufacturing apparatus, oil burning boiler, low air ratio-type oil burning boiler, acrylonitrile manufacturing apparatus, propylene oxide manufacturing apparatus, BTX manufacturing apparatus, other petrochemistry products manufacturing apparatus, private power generation, repowering of private power generation, combined cycle power generation

TABLE 4.1 (CONTINUED)
Energy Technologies Discussed at The Third Conference of Parties (COP3)

Sections	Field	Service Technology
	Paper and pulp	Conventional-type steam melting apparatus, preliminary penetration-type continuous steam melting apparatus, conventional-type cleaning apparatus, high performance pulp cleaning apparatus, conventional-type delignifying apparatus, oxygen delignifying apparatus, drum bleaching apparatus, diffuser bleaching apparatus, conventional type evaporator, liquid film flowing-down-type evaporator, waste paper pulp manufacturing apparatus, semi-chemical pulp manufacturing apparatus, mechanical pulp manufacturing apparatus, sulfite pulp manufacturing apparatus, conventional-type drier hood apparatus, high-performance drier hood apparatus, conventional-type size pressing apparatus, high-performance size pressing apparatus, machine-made paper and other papers, conventional-type dehydrating apparatus, high-performance bearing dehydrating apparatus, conventional type caustification, direct caustification, coal burning boiler, oil burning boiler, low air ratio-type oil burning boiler, black liquor burning boiler, private power generation, repowering of private power generation, combined cycle power generation
	Industrially cross-sectional technologies	High performance industrial furnace, boiler combustion control, repowering of private power generation, combined-cycle power generation, inverter control, high efficiency motor, solar power generation
Domestic	Cooling	Ordinary air conditioner, energy saving air conditioner
	Cooling/heating	Ordinary air conditioner, energy saving air conditioner, gas-fired air conditioner, oil-fired air conditioner
	Heating	Oil-burning space heater, oil-burning fan forced heater, oil-burning hot air heater, gas-burning fan forced heater (city gas-specified and LP gas-specified), gas-burning hot air heater, electric radiant heater, electric ceramic fan forced heater
	Hot water supply	Gas hot water heater (city gas-specified and LP gas-specified), oil-burning hot water heater, electric water heater, solar system, solar thermal hot water heater, latent heat recovering type hot water supplying appliance (city gas-specified and LP gas-specified)

TABLE 4.1 (CONTINUED)
Energy Technologies Discussed at The Third Conference of Parties (COP3)

Sections	Field	Service Technology
	Cooling/heating and hot water supply	Electric multifunctional heat pump, gas heat pump, oil heat pump
	Illumination	Incandescent lamp, fluorescent lamp, incandescent lamp-type fluorescent lamp
	Motor power, etc.	Television, refrigerator and cold storage, washing machine, cleaning machine, electric oven, others
	Electric power	Solar power generation
	Residence	Highly insulated dwelling
Business	Cooling	Electric cooler and air conditioner
	Cooling/heating	Gas heat pump
	Heating	Electric space heater, oil-burning space heater, gas-burning space heater (city gas-specified and LP gas-specified), coal-burning space heater
	Electric power, cooling/heating, hot-water supply	Co-generation type engine, co-generation type gas turbine, cogeneration type diesel engine, solar electric power generation
	Illumination	Fluorescent lamp, high frequency (HF) inverter illumination, sensor attached illumination
	Illumination for emergency exit	Ordinary type guide light, high-intensity guide light
	Kitchen	Kitchen gas (city gas-specified and LP gas-specified), kitchen coal
	Building	Energy saving type building
	Motor power, etc.	Copying machine, calculator, elevator, other power controlled apparatus
Transportation	Passenger transportation	Gasoline fueled automobile (light/small-sized/ordinary/for commercial use/private bus), direct fuel-injection automobile (small-sized/ordinary/for commercial use), diesel power automobile (small-sized/ordinary/for commercial use/private bus/business bus), LPG automobile (for commercial use), CNG (small-sized/for commercial use/private bus), electric car (light/small-sized/for commercial use), hybrid automobile (business bus), ultrahigh efficiency automobile (light/small-sized/ordinary/for commercial use), railway, passenger ship, aircraft
	Bulk shipment	Gasoline fueled automobile (light/small-sized/ordinary), diesel power automobile (small-sized/ordinary), electric car (light/small-sized), CNG (light/small-sized), hybrid automobile (ordinary), railway, cargo vessel, aircraft
Energy conversion	City gas manufacturing	City gas (LPG), city gas (furnace gas), city gas (natural gas)

TABLE 4.1 (CONTINUED)
Energy Technologies Discussed at The Third Conference of Parties (COP3)

Sections	Field	Service Technology
	Oil refining	Conventional-type oil refining plant, high efficiency oil refining plant
	Electric power enterprise	Coal thermal power generation (traditional, repowering, high efficiency), LNG thermal power generation (traditional, repowering, high-efficiency), oil thermal power generation (traditional, repowering, high-efficiency), hydraulic power generation, solar power generation, wind power generation, fuel cell, waste power generation, biomass power generation, nuclear power generation

4.1.2 CONVENTIONAL TECHNOLOGIES OF ENERGY SAVING AND COMBUSTION CONTROL FOR INDUSTRIAL FURNACES

The oil crises between 1965 and 1975 inspired consideration of counter-measures for reducing energy consumption in all industrial furnaces. As an example of industrial furnaces, consider a heating furnace for manufacturing steel materials; the energy-saving counter-measures for it are shown in Figure 4.1. The expected results are listed in Table 4.2. In this case, it must be noted that the basic units of consumption mentioned in Figure 4.1 and Table 4.2 are different from each other. The 2.1 GJ/t (50×10^4 kcal/t) basis for Table 4.2 differs from the 1.89 GJ/t (45×10^4 kcal/t) basis for Figure 4.1. Furthermore, the technical items for reducing energy consumption are more or less dissimilar, as well.

From the above-cited Figure 4.1 and Table 4.2, the items of energy-saving technologies for the heating furnace in the steel-making process, which were established during the period from 1975 to 1985,[2] are

1. Extending the furnace length
2. Reinforcing the preheating of the air and gas through renewing and rationalization of the recuperator
3. Increasing the thermal insulation (fiber lining in the furnace body and multiple skid insulation)
4. Decreasing the number of openings
5. Controlling the lower air ratio (control of the oxygen)
6. Heating by jet impinging
7. Skid boiler, waste gas boiler
8. Improving the heat pattern
9. Extracting at a lower temperature
10. Operations of direct rolling, hot charging

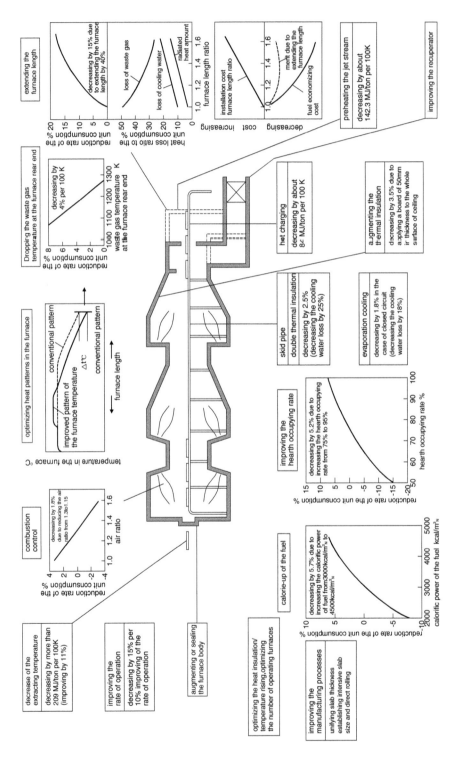

FIGURE 4.1 Energy saving in the heating furnace (reduced rate %) 1.89 GJ/t (45 × 10⁴ kcal/t) base.

TABLE 4.2
Energy Savings and Expected Values of Energy Saved

Classification	Energy Saving	Energy Reduction Objectives				Expected Values of Energy Saved		
		Heat Contained in Steel Materials	Heat Loss to Waste Gas	Heat Loss to Radiation	Heat Loss to Cooling Water	Value for Countermeasures	Value of Energy Saved ($\times 10^3$ kcal/t)	Remarks
Improving of installations	1. Direct rolling	O	O	O	O	100%	500	
	2. Hot charging	O	O			Charging at 500°C	100	
	3. Extending of the furnace length		O	×	×	Extending by 10%	40	
	4. Augmenting of thermal insulation			×		+30%	5	Fiber-veneering
	5. Double insulation of skids			O	O	caster/fiber ≈ 1	35	
	6. Locking of openings			O		$-m^2$	25	Including a shortening of time for opening/closing the doors
	7. Controlling of low air ratio		O			O_2, −10%	10	
Recovering of waste heat	1. Preheating of air		O			Preheating the air at 300°C	75	
	2. Preheating by jet impinging		O			Preheating the steel materials at 300°C	65	
	3. Cooling by evaporation				O	Vaporizing of the total amount of water for cooling skids	60	Evaluation: 800 kcal/kg of steam
	4. Waste heat boiler		O			Increasing the vapor by 0.1 t/t of steel	80	
	5. Preheating of gas		O			Preheating the gas at 300°C	12	

TABLE 4.2 (CONTINUED)
Energy Savings and Expected Values of Energy Saved

Classification	Energy Saving	Energy Reduction Objectives				Expected Values of Energy Saved		
		Heat Contained in Steel Materials	Heat Loss to Waste Gas	Heat Loss to Radiation	Heat Loss to Cooling Water	Value for Countermeasures	Value of Energy Saved ($\times 10^3$ kcal/t)	Remarks
Improving of operation	1. Improvement of heat patterns		○			Decreasing the waste gas temperature by 100°C	15	
	2. Extracting at a lower temperature	○				Decreasing the extracting temperature by 50°C	22	
	3. Improvement of manufacturing efficiency		○	○	○	Improving by 1%	50	Including a reduction in electric power
	4. Improvement of the rate of operation		○	○	○			
	5. Carefully thought-out controlling of the operation (heat insulation, temperature rising, furnace pressure, cooling water control, etc.)		○					
	6. Unifying the sizes of steel material					Unifying the thickness of all steel materials	~50	

Representative examples of conventional combustion control systems for industrial furnaces are shown in Figure 4.2. From the figure, the general items to be controlled are shown below.

1. Furnace temperature control
   - Temperature in the zones
   - Cascades in both the upper and lower zones
   - Burner thinning
   - Steel slab discharging and discharging temperature
2. Flow rate control
   - Fuel flow
   - Airflow
3. In-furnace oxygen control
4. Combustion safety control with shutdown conditions
   - Fuel supply pressure-down
   - Combustion air pressure-down
   - Cooling water pressure-down
   - Instrumentation fluid pressure-down
   - Power breakdown
   - Atomizing air pressure-down for heavy oil burning
5. In-furnace pressure control
6. Recuperator protection control
7. Discharging slab tracking with photo detector and ITV camera system

4.1.3 Development of High Performance Industrial Furnaces

With the heating furnace for manufacturing steel materials as representative of industrial furnaces, the results of the development of high performance industrial furnaces are reviewed by tracing the changes in such heating furnaces since the end of World War II.

Regarding the transition of the furnace shape, the focus of discussion was on the burning technologies such as axial-flow burning, side burning, and top burning, to ensure the uniformity of temperature in heating furnaces with wide widths. Table 4.3 shows the characteristics of these three types of burning. Taking the arrangement of skids into consideration, side burning was to be adopted. However, a problem arose with regard to the flame spreading to more than half the furnace width. If the flame could not spread completely, axial-flow burning should be adopted and a fine control at the final adjustment was to be carried out by top burning. The problem was which type of burning should be adopted because the outer shape of the furnace might be very different if this type of burning was adopted. Much discussion ensued regarding which type of burner layout should be adopted because the dimensions of the furnace might significantly affect combustion performance depending largely on the burner layout.

On the basis of this concept, the length of flame is designed to be 6 to 7 m at most in the case of a furnace having the burning capacity of about 3.48 MW (3 × 10^6 kcal/h). Some of the furnace shapes are shown in Figure 4.3, and the actual

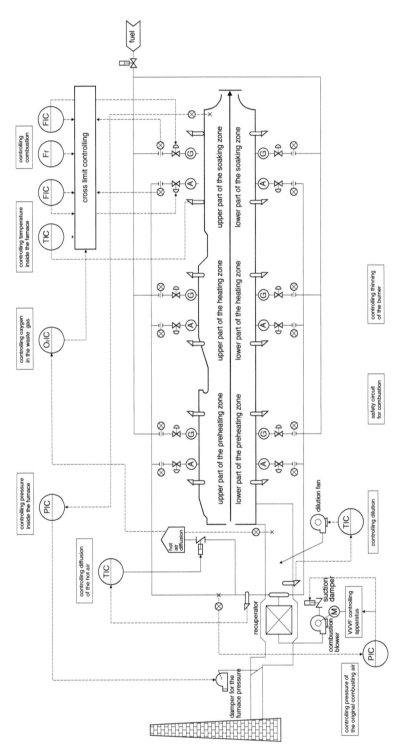

FIGURE 4.2 An example of conventional instrumentation for heating furnace.

examples of the shape of a slab heating furnace in the hot rolling process is shown in Figure 4.4. A furnace 57 m long (maximum) and 14 m wide (maximum) was developed. On average, it had four zones of preheating, first heating, second heating, and soaking together with ten zones that made up the upper and lower sides of the soaking zone, and was parted into halves by a roof. Later, a furnace with an even more finely divided 13-zone construction was developed. However, while there is the axial flow burning type regenerating furnace under a line of regeneration as shown in Figure 4.5, the side-burning type furnace (box model) has become the basic heating furnace. Since burning takes place between respective paired-burners in this type of furnace, actual zones have little meaning. Even the existence of a partition wall, which was indispensable in the past, makes little difference in the characteristics of the heating furnace. The heating furnace has taken on a new meaning as a heat-treatment furnace. When this type of furnace is compared with that of Figure 4.3, it looks as if the furnace dimension has reverted to the all-side burning type. But, in fact, the furnace height, furnace length, and the partition wall have been changed greatly. Temperatures at the furnace end should be lowered to attain energy savings. Therefore, a preheating zone within the furnace should be as long as possible within the limits of budget. The preheating zone is known as a "deep charger" or a "matching zone." It also resulted in creating additional space within the furnace. Additional burners added at the preheating zone would improve the production capacity of the facility.

In designing the regenerating furnace, extending the furnace length as adopted above only increases radiant heat loss. This results in increased energy consumption. Furnace length is therefore very important in the critical design of the regenerating furnace and is greatly shortened in comparison with conventional furnace lengths. However, such extension of the furnace length has some significance. Figures 4.6 and 4.7, respectively, show the fuel unit consumptions of both a high performance industrial furnace and the conventional recuperator type furnace that are designed under the same conditions. These figures show the relation between the fuel unit consumption and the furnace length when a slab (250 mm thick) is extracted at 250 ton/h at respective temperatures of 1150, 1200, and 1250°C. These fuel unit consumptions are the results calculated by the so-called heat balance model, where the heat balance by zone is maintained together with the heat balance in the furnace at large.

Figure 4.8 shows the relation between the heat loss (radiated heat from the furnace body plus heat loss to the cooling water) and the furnace length. In the conventional design of heating furnaces, a load of 600 to 700 kg/m²h to the furnace hearth is considered rational from the viewpoint of energy savings. But in the case of the design of the regeneration-type furnace, the load to the furnace hearth is now designed to be in the region of 1000 kg/m²h. In addition, it can be concluded that the optimum value from the viewpoint of energy savings that results in minimum value of energy consumption per unit basis is now being designed. The design basis for the specifications is entirely different from the conventional specifications for the recuperator type furnace.

The typical furnace height (the height of the top portion from the top of a skid, in this case, is equivalent to about half the total furnace height) is around 3 m. It is reported that the thickness of a gas layer should be about ten times as great as the

TABLE 4.3
Comparison of Characteristics of Various Burners in the Heating Furnace

	Items	Axial-Flow Burner	Side Burner	Roof Burner
1.	Burner position against the furnace	see Figure 4.3d.	see Figure 4.3b	see Figure 4.3c
2.	Fuel	Heavy oil, C gas, M gas, etc.	Heavy oil, C gas, M gas, etc.	Kerosene, C gas, M gas, etc.
3.	Type of burner flame	Long flame type	Short flame type (variable flame type)	High-intensity combustion radiating type
4.	Controllability of specific combustion	Adjusting range is wide	Adjusting range is narrow (about 1/2); thinning control is required according to the specific combustion	Adjusting range is relatively wide
5.	Heating load	Heating load can be secured by a larger burner	Heating load can be secured by a larger burner	Heating load cannot be secured by a smaller burner
6.	Restrictions by furnace inner size	Length per one zone in the furnace direction is restricted	Furnace width is restricted	No restrictions; however, combustion at the lower part is regarded as wrong
7.	Combustion gas flow in the furnace	The combustion gas flows in the furnace direction without trouble	A polarized gas flow occurs easily, as the burner direction and the furnace length direction cross at right angles to each other	The combustion gas flows in the furnace without trouble, since combustion is carried out mostly in the burner tile
8.	Burner mounting structure to the furnace	The structure of furnace body is made complicated, as nose parts are required	The furnace body structure is made simple, since nose parts are not required	Although the furnace body structure is simple, there are many burners, and therefore the pipe arrangement is made complicated
9.	Uniformity of heating (heating pattern)	Heating in the furnace-width direction is easily made uniform Temperature at nose parts in the furnace direction is apt to drop down	Uniformity of heating in the furnace-width direction is rather inferior; heating in the furnace-length direction can be easily made uniform	Uniformity of heating in both the furnace-width direction and the furnace-length direction can be easily achieved

TABLE 4.3 (CONTINUED)
Comparison of Characteristics of Various Burners in the Heating Furnace

Items	Axial-Flow Burner	Side Burner	Roof Burner
10. Workability	Workability is comparatively good; however, since the environment surrounding the burners in the lower zone is at a higher temperature, the workability is not good	Working environment is good	Workability is slightly inferior, since the environment surrounding the roof burners is relatively at a higher temperature and there are many burners used

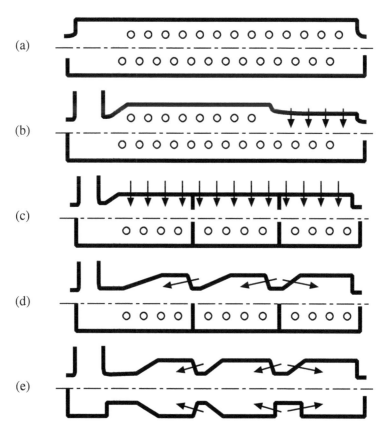

FIGURE 4.3 Positioning of burners for heating the furnace and furnace shapes. (a) All side burners. (b) Roof burners in the upper part of soaking zone, and all side burners for the others. (c) All roof burners in the upper zone, and all side burners in the lower zone. (d) Axial-flow burners in the upper zone, and side burners in the lower zone. (e) All axial-flow burners. (Circles and arrows indicate the location of burners.)

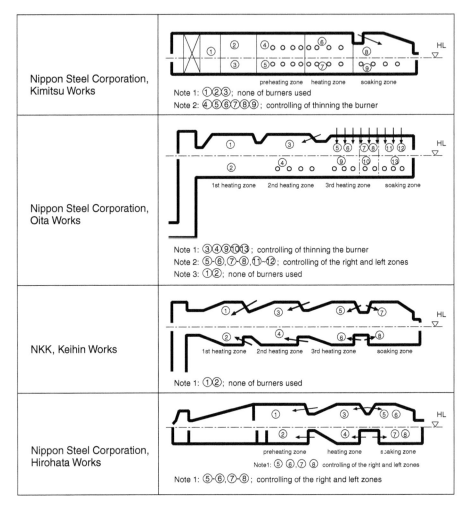

FIGURE 4.4 Examples of the shape of slab heating furnace in the hot rolling process. (Circles and arrows indicate the location of burners.)

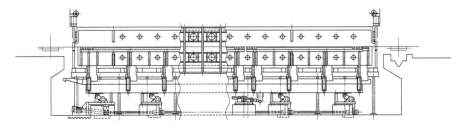

FIGURE 4.5 Examples of trial designing of a high performance heating furnace.

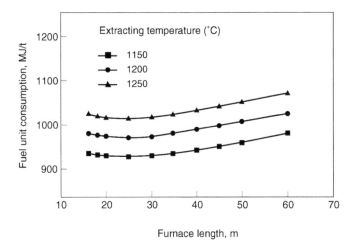

FIGURE 4.6 Effects of extracting temperature of the high performance heating furnace.

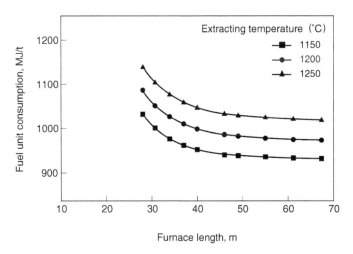

FIGURE 4.7 Effects of extracting temperature of the conventional-type heating furnace.

thickness of a steel slab. This has been interpreted as an empirical factor for determining the furnace height. The results presented in reference 3 have been used in the design caluclation as a sufficient condition. The pitch between burners has been determined by experimental factors such as the size of the tile. Some experimental results[4] showed that the optimum point of the heat flux ratio exists in a pitch ranging between 2.0 to 2.5 m. This is due to a deviation of the heat flux (a temperature gradient in the steel plate) affected by both the furnace height and the burner pitch in the case of combining the furnace height and the burner pitch, as shown in Figures 4.9 and 4.10. These figures come from the results of tests on changing the furnace

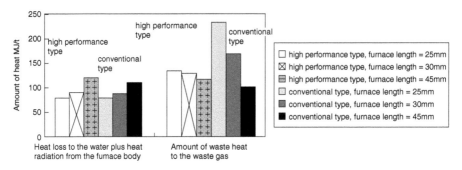

FIGURE 4.8 Comparison of heat losses.

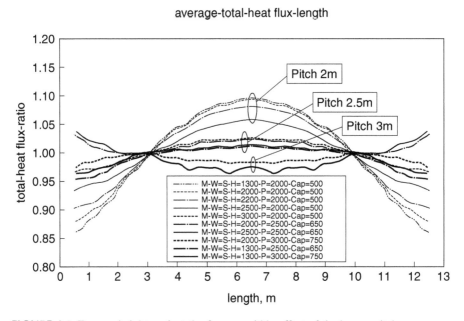

FIGURE 4.9 Furnace height against the furnace width, effect of the burner pitch.

height and the results of analyses through the overall in-furnace model (taking such reactions as heat flow, combustion reaction, and heat transfer into consideration).

It is clear that, even if the furnace height is lowered to 1.2 m, there is no difference in the heating efficiency, although the nonsymmetrical heat may be increased slightly. Therefore, a value for the height of about 2 m is proposed, taking the maintenance of inner parts of the furnace into consideration. However, if maintenance work is performed entirely by a robot for gunning and repairing, and carried out only from the outside of the heating furnace, a heating furnace with a superlow height of less than 1.5 m would be feasible. From the discussion of furnace height, it can be seen that the outer shape of the heating furnace leans toward the shape of the

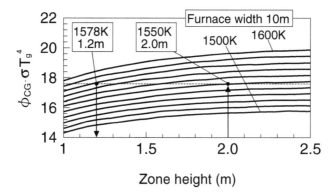

FIGURE 4.10 Examination of the optimum furnace height. Relationship between the zone height and the radiant heat flux to steel materials, taking into account the effect of the furnace wall (numerical analysis solution).

heat-treatment furnace. As for heating furnaces, there have been four types of recuperators as shown in Figure 4.11: the multipipe type, the Hasen type, the radiating type, and the tile type.

In a recently designed recuperator, a very large heat exchanger, which has a heat transferring area in the 10,000 m² class for applying to three heating furnaces, is set between the furnace end and the smoke stack. The heat resistance temperature of this recuperator is about 1050°C, and the maximum temperature obtained of the preheated air is 720 to 730°C. If the temperature of the preheated air is raised to more than 730°C, problems occur, such as the increase of NO_x content, the increase in equipment investment due to the expanded heat transferring area of the recuperator, and the lowering of the durability of the recuperator itself. Between 720 and 730°C should be the limit for the maximum temperature of preheated air. The surplus waste heat is utilized for preheating the gas to about 300°C, which would be the final temperature attained assuming that the gas composition is not transformed.

However, with a regeneration-type furnace, a heat reservoir for exchanging heat is included in the burner itself, so heat exchanging is completed within the paired burner. Thus, the heat recovery is carried out by the heat-exchange method of a distributed type, an entirely different system, not used by the conventional heat exchange of an integrated type. The heat recovery ratio in the case of the conventional-type recuperator is 60% at most, as shown in Figure 4.12. But, in the regeneration-type furnace, up to about 80% of heat can be recovered by a direct heat exchange method of a heat-reserving type. This is almost 100% in terms of temperature efficiency. When the temperature in the heating furnace is 1300°C, the temperature of the preheated air reaches between 1150 and 1250°C, which is 50 to 150°C less than the temperature in the heating furnace. Thus, "realization of the in-system critical heat recovery" is included.

Regarding the burners, there has been a transition to technologies for lengthening the flame against a broader width of the heating furnace and for reducing the NO_x to pass the NO_x regulation values stably. Various NO_x reduction technologies have

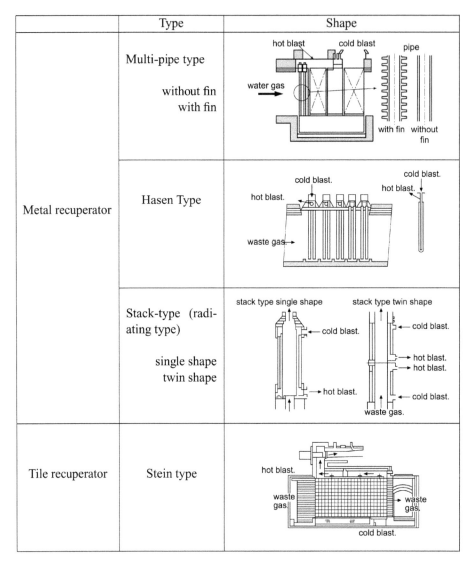

FIGURE 4.11 Types of air recuperator for the slab heating furnace in the hot rolling process.

been developed, such as multistage combustion, dispersion combustion, steam blowing, premixed combustion, lean fuel combustion, and exhaust gas recirculation.

In HiTAC technology applied burner, the effectiveness of waste gas recycling has been improved only by increasing the airflow velocity. The target for reducing NO_x has been realized by the multiplier effects of the dispersion combustion. In the past, it was difficult to suppress the level of 100 ppm of NO_x at preheated air temperature in the range between 500 to 600°C, when the furnace temperature was

Practical Combustion Methods Used in Industries 229

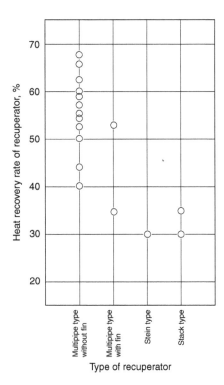

FIGURE 4.12 Types of recuperator and its heat recovery rate.

higher than 1250°C. The value of 100 ppm NO_x can be passed without problems at a preheated air temperature well over 1000°C and at a furnace temperature higher than 1300°C.

The change of driving mechanism from a pusher-type furnace to a walking-beam-type began in 1967. Recently, a multiple-zone-type walking beam furnace has become predominant. Technical measures to counter skid marks have not been solved yet. Various counter-measures have been worked out, such as shifting of the skid beam, heating method for aiming at the skid contacting portions by burners with a long head that project out from the noses standing up through the furnace hearth and from the furnace sidewall, adoption of gate-type posts, adoption of a taller skid-height, and moderate cooling of skids. However, no effective countermeasure has been found to date.

The applied HiTAC burner uses technologies of blowing air at a high velocity to achieve homogeneous combustion entirely inside the furnace. Therefore, it is expected that the gas, which was stagnating in all methods so far, will be able to flow to the surroundings of the skid supports to where the gas did not diffuse. By this effect, it is also expected that homogeneity of ambient atmospheric temperatures in the lower part of the furnace can be improved, resulting in a decrease of skid marks.

4.2 ENERGY CONSERVATION

4.2.1 BASIC APPROACH

By heating combustion air with high temperature gas that has completed heat transfer to the material to be heated, the volume of fuel consumed can be reduced by an amount equal to the recovered heat equivalent of the amount of energy required for the heating. Although it is easily determined from the balance of input and output of the heat, heating the combustion air increases the adiabatic flame temperature and improves the heating efficiency proportionally. Consequently, heating within a specified range becomes possible even if the fuel supply volume is reduced.

In an actual heating application, in-furnace temperature is maintained at a level corresponding to a specified heating velocity (production rate) because a low in-furnace temperature causes underheating and a high in-furnace temperature causes overheating. The volume of heat held by the combustion gas in excess of the in-furnace gas is the key to heating the object and simultaneously maintaining the temperature of the in-furnace gas temperature at a uniform level. Failure to maintain the in-furnace gas temperature means an interruption of normal system operation. In other words, the heating potential of the combustion gas is determined not merely from the heat balance calculated on the basis of the total heat input including recovered heat, but from the amount of heat in excess of the in-furnace gas temperature. For example, it is impossible to heat an object to a temperature higher than the adiabatic flame temperature, no matter how much fuel is fed into the furnace. It is also very difficult to heat the object to a temperature close to the adiabatic flame temperature. However, if an adiabatic flame temperature higher than the required in-furnace gas temperature is realized with the combustion air heated using the same fuel, the type of heating mentioned above can be realized.

4.2.2 EFFECT OF IMPROVEMENT

(1) A fuel saving effect by returning recovered heat to the heating system by preheating air and (2) an enhanced heat-transfer acceleration effect by increased adiabatic flame temperature can be expected from high-temperature-air combustion as shown in Table 4.4. The former can be categorized as an energy-saving effect resulting from the increased heating potential (heating capacity in excess of the in-furnace gas temperature); the latter is aimed at realizing an increased heat transfer rate by utilizing the advantageous condition whereby formation of an extensive high temperature field is readily realized if the adiabatic flame temperature is increased by preheating the high temperature air. It is also possible to lower the maximum flame temperature by averaging the in-furnace gas temperature field with the total amount of transferred heat fixed at a certain level. This characteristic is important in connection with the maintenance of the heating equipment. If the same maximum temperature is allowed, the amount of recovered heat returned to the heating system can be substantially increased. The following part of this section discusses the effects of items 1 and 2 above, with the in-furnace conditions varied into two different cases. The first case is for a furnace where the in-furnace temperature is maintained

TABLE 4.4
Improvement of Heat Transfer by High Temperature Air Combustion

Characteristics	Heating Control Method
Increase in heating potential • Adiabatic flame temperature is increased by returning recovered heat to the heating system	The volume of fuel supply is adjusted so that the furnace temperature maintains at the same level as in the conventional method. ~1/heating potential
Averaging/raising furnace temperature • Maximum flame temperature is lowered through averaging • Substantially more recovered heat can be returned to the heating system until the same maximum temperature is reached	The volume of fuel supply is adjusted so that the air temperature is raised without exceeding the permissible upper limit and the heat transfer rate is fixed
Improved heat transfer rate • A high temperature zone and a low temperature zone are prepared and the furnace temperature is averaged and controlled on a zone-by-zone basis	The volume of fuel supply is adjusted on a zone-by-zone basis so that the high temperature zone is heated to the permissible upper limit and the heat transfer velocity is uniform throughout the furnace

at a uniform level and the second case is for a furnace where the in-furnace space is divided into zones for zone-based heating control.

If the temperature is uniform throughout the furnace, heat exchange between gases does not occur and the temperature of the gas discharged from the furnace is equal to the in-furnace gas temperature. The heat balance of the heating system can be expressed by the following heating potential scenario.

Now, assuming that the calorific value of fuel is H_0, the preheated air sensible heat for burning the unit volume of fuel is H_a, and fuel v_0 is fed when air preheating is not executed, the fuel input volume v_a required to make the total heat input to the heating system uniform is calculated by the following equation:

$$H_0 v_0 = (H_0 + H_a) v_a \tag{4.1}$$

If air preheating is executed, $v_0 > v_a$ is always satisfied to save fuel by $v_0 - v_a$. It is an improvement when the in-furnace gas temperature is at room temperature (atmospheric temperature) and achieves a reduction in the volume of fuel supply by the preheated air sensible heat. However, during actual heating operations, the heat transferred from the in-furnace gas to the material to be heated or the furnace wall must be compensated for and the in-furnace gas temperature must be kept high for the duration of the operation.

Figure 4.13 shows the relationship between the preheated air temperature and the heating potential of the combustion gas (maximum heat that can be transferred to the material to be heated), represented in typical form. The enthalpy region above

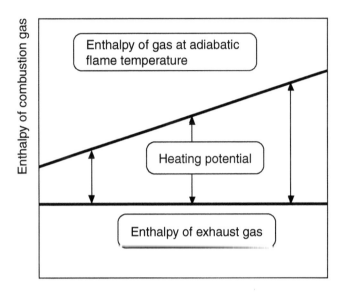

FIGURE 4.13 Heating potential of combustion gas.

the enthalpy of the in-furnace gas can actually contribute to the heating during operation. The heating potential of this combustion gas is equal to the enthalpy of the gas between the adiabatic flame temperature and the in-furnace gas temperature disregarding the heat loss in heating. Figures 4.13 and 4.14 shows the conceptual definition of the heating potential. In the case of an ideal adiabatic system without heat extraction, the in-furnace gas temperature is the adiabatic flame temperature T_{ad} and the adiabatic enthaply of flame gas is the enthalpy H_{ad} equivalent to the temperature. In the case of heating with heat extraction, the in-furnace temperature is the temperature T_g determined by the heat extraction amount and the exhaust gas enthalpy is the enthalpy H_g equivalent to the temperature.

The difference between H_{ad} and H_g is generated by heat extraction, and is proportional to the volume of heat extraction. Disregarding the fixed heat-loss amount of the heating equipment, the volume of heat extraction is basically the amount of heat transferred to the material to be heated. In other words, the difference between H_{ad} and H_g was used for the heating.

It is assumed that heating is executed without preheating the air and the process is referential. We can see by what margin the fuel supply requirement for the process is reduced by introducing an air-preheating step. Assume that the enthalpy of the combustion gas per unit volume of fuel supply, which is equivalent to the in-furnace gas temperature, is H_g. Heating is executed without preheating the air to satisfy the condition $H_0 > H_g$. With the acceleration maintained at a fixed level, the following equation can be established:

$$(H_0 - H_g)V_0 = (H_0 + H_a - H_g)V_a \tag{4.2}$$

Practical Combustion Methods Used in Industries

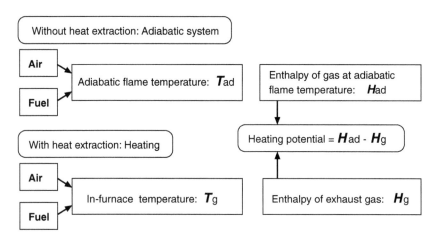

FIGURE 4.14 Conceptual definition of heating potential.

where V_0 and V_a are fuel supply amount when no air preheating is involved, and ($H_0 - H_g$) and ($H_0 + H_a - H_g$) are both heating potentials. This represents the actual phenomenon in progress. The energy saving effect of combustion with air preheating is now obtained from $\Delta V = V_0 - V_a$ in place of $\Delta v = v_0 - v_a$. Under these conditions, the percentage volume of fuel supply obtained by normalizing the fuel supply requirements at respective preheated air temperatures by the fuel supply amount for the heating without air preheating is found to be the reciprocal of the heating potential rate. Further, with $H_0 > H_g > 0$, the following equation can be established:

$$\frac{V_a}{V_0} = \frac{H_0 - H_g}{H_0 + H_a - H_g} < \frac{H_0}{H_0 + H_a} = \frac{v_a}{v_0} \qquad (4.3)$$

Thus, it may be proved that the effect is greater than a mere reduction of fuel supply volume by the preheated air sensible temperature. Although the heating potential increases only by the preheated air sensible temperature, the heating potential rate increases greatly. The larger the enthalpy of the in-furnace gas (or the higher the in-furnace gas temperature), the greater the effect becomes. The left side of the inequality represents the effect of fuel saving from combustion with air preheating, and the right side represents the apparent effect of an increased calorific value with the preheated air sensible heat considered to be an increase in the latent heat of the fuel.

Figure 4.15 shows the relationship between the preheated air temperature and the fuel supply requirement when CH_4 is burned at an air ratio of 1.0 and at an in-furnace gas temperature of 1773 K. The solid line in the figure shows the fuel supply rate and the dotted line indicates the ratio of the latent heat of fuel to the total energy input, which includes the preheated air sensible heat. These values are calculated, respectively, by V_a/V_0 and v_a/v_0 in Equation 4.3 above. The difference between the ordinate axis 1.0 and the solid lines vs. the ordinate axis 1.0 and the dotted line, respectively, indicate the rate of fuel saving and the ratio of preheated air sensible

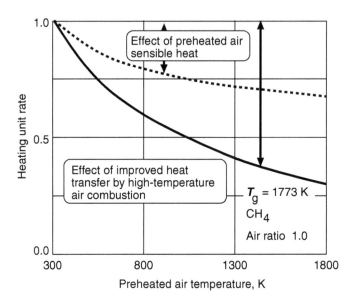

FIGURE 4.15 Effect of improved heat transfer by preheated air combustion.

heat to the total energy input — in other words, the aforementioned fuel saving effect and apparent calorific value improvement. In this case, the fuel-saving effect is more than double the apparent calorific value improvement effect. This implies that estimations on the basis of recovered heat bear little relation to the real effect indicator. With the preheated air sensible heat at 1100 K or higher, the volume of fuel supply will be reduced by half.

Heating Control Applied on a Zone-by-Zone Basis — The same fuel saving effect as mentioned above is applied here, as well. In addition, zone-by-zone heating control in high temperature air combustion can accelerate heat transfer as described below by utilizing the advantageous condition whereby formation of an extensive high temperature field is realized if the adiabatic flame temperature is increased.

The dominant heat transfer inside a furnace takes the form of radiation heat transfer and the heat transfer rate can be defined as a function of the fourth power of gas temperature. The heat transfer rate increases to a considerable extent at high temperatures. Accordingly, high temperature air combustion has the potential to obtain a high heat transfer rate in excess of that resulting from an increase in the flame temperature by increasing the preheated air temperature. Heating potential is now combined with the time-based capacity, or the heat transfer capacity per unit time. It should be noted that the increased amount in the enthalpy corresponding to the increased temperature of the flame is evaluated as the fuel saving effect described above. An effective method of utilizing the potential of the high heat transfer rate is, for example, to divide the inner space of a furnace into zones to control heating. The total heat transfer rate can be further increased by heating a $(T_g + \Delta T_g)$ zone and a $(T_g - \Delta T_g)$ zone at different temperatures. The rate achieved will be greater than that for a single furnace space which is heated at T_g uniformly. For example, a furnace at an in-furnace temperature of 1473 K is divided into a zone of

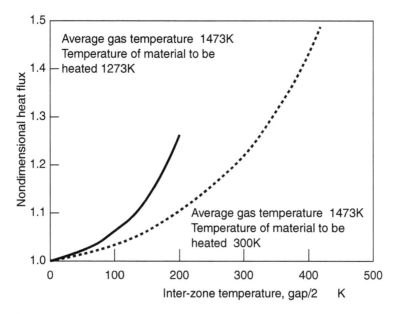

FIGURE 4.16 Example of heat transfer acceleration effect by zone-to-zone heating control method.

$(1473 + \Delta T_g)$ K and another zone of $(1473 - \Delta T_g)$ K. The relationship between ΔT_g and the velocity of heat transfer to the material to be heated under the above conditions, disregarding heat loss, is shown in Figure 4.16.

The ordinate axis indicates the heat transfer rate in the unit of transfer velocity ratio obtained by normalizing the heat transfer rate under each condition, by the heat transfer rate when the in-furnace gas temperature is kept at a uniform level at 1473 K. The figure shows that the heat transfer rate of the material to be heated with a temperature of 1273 K in the range of $\Delta T_g \sim 200$ K (1673 and 1273 K zones) is about 25% higher than in the furnace at 1473 K. This method results in a higher heat transfer and a higher heating velocity (production rate) for the same heat input per unit time. Further, it is possible to lower the average in-furnace temperature with the heat transfer rate maintained at a fixed level, which means a proportional reduction of heat input to the lowered temperature.

As described above, formation of a stable high temperature field is a prerequisite to dividing a furnace space into a high temperature zone and a low temperature zone, which are easily realized in high temperature air combustion.

4.3 POLLUTION REDUCTION

4.3.1 Basic Concept of Low NO_x Combustion

NO_x generation is basically a function of temperature, oxygen concentration, and residence time. In a typical heating furnace, residence time is usually long enough to enable the generation of NO_x and the basic measures for low NO_x combustion are to lower the peak temperature of the flame and to reduce the oxygen concentration. However, since an excessively low oxygen concentration in a combustion field results in unstable combustion, the most typical method of low NO_x combustion employs a reduction of oxygen concentration while increasing the air preheating temperature to secure stable combustion. This method meets the needs of both energy saving and low NO_x combustion.

Figure 4.17 shows the basic concept of low NO_x combustion. Superlow NO_x combustion can be achieved with a flame that forms a flat temperature distribution by controlling the peak temperature of the flame instead of using a conventional flame that forms a temperature distribution with a local high temperature spot (hot spot). High temperature air combustion agrees well with this concept, with reinforced in-furnace recirculation, which lowers the oxygen concentration in the combustion reaction field, and stabilized combustion, which results from the use of extremely hot combustion air.

Figure 4.18 shows the results of a combustion analysis using CFD in accordance with the basic concept described above. The abscissa axis indicates the distance from the burner, which is nondimensionalized by the width of the furnace (4 m). The ordinate axis indicates the temperature. In a conventional combustion method, a conspicuous high temperature peak is present in the temperature distribution and the oxygen concentrations at this peak point at 1600 and 1200 K are both 7%. The temperature distribution of high temperature air combustion in which recirculation of in-furnace gas is reinforced and the mixture of fuel and combustion air is controlled to be entirely flat although there is a peak. The oxygen concentration at the peak point is 1%, far below the level in a conventional combustion method. The peak temperature of the high temperature air combustion at 1600 K is lower than

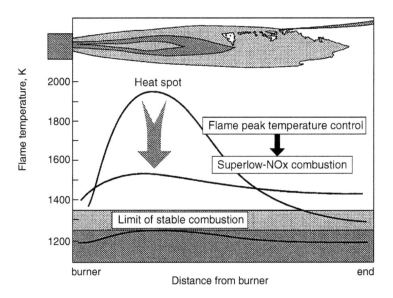

FIGURE 4.17 Basic concept of low NO_x combustion.

Practical Combustion Methods Used in Industries 237

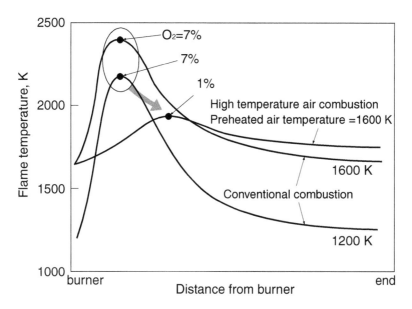

FIGURE 4.18 Oxygen concentration at peak point and distribution of temperature of flame on burner center axis.

that of conventional combustion at 1200 K. It proves that the peak temperature in high temperature air combustion can be controlled even if the preheated air temperature is raised. Further, since the oxygen concentration at the peak point is much lower than that in the conventional combustion, it is possible to realize low NO_x combustion in high temperature air combustion.

4.3.2 RESULTS OF THE TEST

Figure 4.19 shows a series of data including NO_x concentration, furnace temperature, preheated air temperature, and exhaust gas temperature of the exhaust gas obtained from the combustion test executed in the common test facility II. The furnace temperature was measured by a thermocouple at 100 mm below the furnace ceiling. Note that the preheated air temperature was lower than the furnace temperature by only 50 K. Using such high temperature combustion air, NO_x concentration as low as 90 ppm (not converted) resulted. This value is less than 1/3 of NO_x concentration in a conventional low NO_x burner and less than 1/10 of that in a quick-mixing type burner, thus marking a large NO_x reduction in high temperature air combustion. The time-based variations in the preheated air temperature, exhaust gas temperature, and NO_x concentration were caused as a result of the change of combustion behavior of a regenerative burner combustion system using the burner switching time intervals of 30 s.

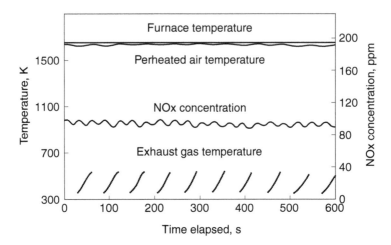

FIGURE 4.19 Test data: preheated air temperature, exhaust gas temperature, and NO_x concentration.

4.3.3 POLLUTION REDUCTION

The combustion phenomenon has been clarified through experiments using small-scale laboratory combustion equipment on the combustion and the fluid flow, and through flame image analysis.

When the combustion air reaches as high as 1273 K during combustion, if a conventional burning method is used, a flame with an extremely high temperature and luminosity is formed, and, therefore, the generation of NO_x will increase quite suddenly. However, even if the same high temperature air is used, if a low oxygen content combustion (diluted combustion) by increasing the burned gas circulation is realized, an entirely new high temperature air combustion will be realized with characteristics of increased flame volume, lowered flame luminosity, and decreased temperature gradient. At the same time a steady burning flame is formed in a large space, and at that moment NO_x generation is decreased rapidly.

The ultimate aim of our research is the commercialization of a furnace using the clarified combustion phenomena described above, diluted and high temperature air combustion in actual industrial furnaces. We have collected combustion and heat transfer data using large industrial scale combustion laboratory equipment.

As a result, it has been verified that we can realize diluted slow burning in a low oxygen content field even in a commercial furnace and can achieve a low concentration of NO_x generation even with high temperature air as high as 1273 K by using the method of a nontile burner structure, high speed injection of combustion air, and separate feeding of fuel and air, by promoting burned gas circulation, and by suppressing the mixing of fuel and air simultaneously. To compare the structural differences in the burners, the structure of the conventional burner and that of the diluted and high temperature air combustion burner are shown in Figures 4.20 and 4.21, respectively.

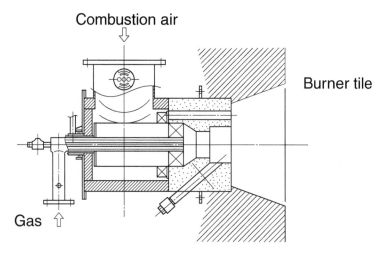

FIGURE 4.20 Structure of conventional burner.

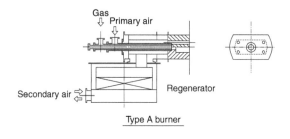

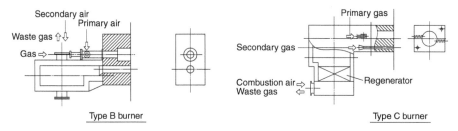

FIGURE 4.21 Structure of diluted and high temperature air combustion burner.

The diluted and high temperature air combustion burner has no burner tile to hold the flame and has a longer feeding distance between fuel and combustion. Its structure may appear to make it impossible to maintain a steady flame in the conventional combustion concept. However, it is considered that steady combustion can be maintained even in this structure by the use of high temperature air well beyond the conventional range of combustion, and that slow combustion is realized by suppression of the rapid combustion reaction, which otherwise occurs as the result of an introduction of high temperature air.

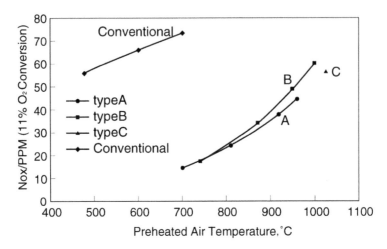

FIGURE 4.22 Example of data on NO_x generation measurement.

Figure 4.22 shows an example of measurement data on NO_x generation in the combustion tests conducted using large-scale combustion equipment. The horizontal axis shows the combustion air temperature, and the vertical axis shows the NO_x generation equivalent to 11% O_2. In addition, Types A, B, and C in the figure correspond to those burner types shown in Figure 4.21. All data were obtained from use of the diluted and high temperature air combustion burner.

LNG or LPG was used as fuel. By using the diluted combustion burner, the concentration of NO_x is suppressed to a level of 60 ppm even at a combustion air temperature as high as 1273 K and 30 ppm at 973 K, whereas with the conventional burner, it is 70 ppm at the same temperature of 973 K. As stated above, the degree of reduction in NO_x generation with this combustion has been verified. Furthermore, it is easily estimated from this figure that combustion with the diluted combustion burner does not exist in the extended line of that of conventional combustion burners, and that a completely different combustion phenomenon occurs in HiTAC.

Based on the achievements described above, the construction of continuous reheating furnaces, incorporating the new concept of high temperature air combustion system, has been steadily increasing. NO_x generation data in actual operations are beginning to be collected as well (Table 4.5). According to these data collected, the total NO_x generation for an operation temperature of 1323 to 1623 K is in a range of 30 to 90 ppm (equivalent to 11% O_2). It is assumed that the test data from the combustion laboratory equipment are valid.

Furthermore, this fact shows that the application of the diluted and high temperature air combustion technology to industrial furnaces can greatly contribute to reducing NO_x generation.

TABLE 4.5
Examples of NO_x in Continuous Reheating Furnace

No.	Furnace Type	Fuel	Operating Temperature, K(°C)	NO_x Generation (Equivalent to 11% O_2), ppm	Remarks
1.	Continuous WBF	LNG	1373 (1100)	89	Regeneration in all zones
2.	Continuous WBF	Mix Gas	1423 (1150)	77	Regeneration in preheating zone
3.	Continuous WBF	LNG	1323 (1050)	82	Regeneration in all zones
4.	Continuous WBF	C-Heavy Oil	1323 (1050)	36	Regeneration in all zones
5.	Continuous WBF	LNG	1483 (1210)	72	Regeneration in preheating zone
6.	Continuous WBF	Mix Gas	1423 (1150)	30	Regeneration in preheating zone
7.	Continuous PF	LNG	1623 (1350)	40	Regeneration in all lower zones
8.	Continuous WBF	COG	1453 (1180)	34	Regeneration in preheating zone

Note: WBF: Walking beam furnace, PF: Pusher furnace.

REFERENCES

1. Y. Matsuoka, Global warming measures valued from energy demand, paper presented to the Science Council of Japan, 1997.
2. T. Iron and S. I. of Japan, Eds., *Recent Hot Strip Production Technology in Japan*, p. 200, 1987.
3. H. C. Hottel and A. F. Sarofim, *Radiative Transfer*, McGraw-Hill, New York, 1967.
4. M. Uede, K. Tanaka, M. Imada, and K. Murakami, *Proc. IJPGC Conf.,* 15082, 2000.

5 Design Guidelines for High Performance Industrial Furnaces

5.1 FLOWCHART ON GENERAL DESIGN

5.1.1 DESIGN CONCEPT OF A HIGH PERFORMANCE INDUSTRIAL FURNACE

In the design phase of an actual industrial furnace, the main specifications are first determined by integrating the test data, corrected data from scale-up modification of the system, calculation results from model analysis with respect to heat transfer, flow dynamics, and reactions aimed at evaluating the effect of the respective variations from the test conditions. Then the details of the design are worked on. Table 5.1 shows the overall configuration image of the technical arrangements and Table 5.2 the design flow of a high performance furnace as an example of an industrial furnace. The basic concept of the design flow is the same as for a conventional furnace. However, the design of a high performance furnace makes use of various databases which have been compiled as a result of development activities, such as (1) data on the height of the furnace, (2) data on overall heat transfer coefficient (total absorptivity) ϕ_{CG}, and (3) data on preheated air temperature. Because these data are different from those of a conventional furnace, the final appearance of a newly designed furnace will be altered accordingly.

Also in the process of defining the control system, the incorporated control software and control devices differ from those for a conventional furnace, reflecting the results of the development in the form of furnace temperature and furnace pressure control data.

5.1.2 OPTIMAL DESIGN FOR FURNACE LENGTH AND HEIGHT

In most cases the height of a furnace has been determined, not according to heat transfer conditions, but by the physical restrictions of the furnace such as the structure and installation dimensions. The optimal furnace height for the highest heating efficiency available can be determined from the balance between gas thickness and heat loss. The greater the height of the furnace, the thicker the gas layer becomes. As a result, the radiation rate of the gas increases as the quantity of heat loss through the furnace wall increases. As long as the increase in the heating efficiency resulting from the increased radiation rate of the gas remains higher than the increase in the

TABLE 5.1
Overall Configuration of Technical Arrangements of High Performance Industrial Furnace Development Project

(I)

Flame characteristics
1. High capacity flame (flame with a high flame capacity); flame with low concentration and temperature gradients
2. Transparent flame; flame without conspicuous luminous flame
3. Lifted flame; stable flame distanced from fuel nozzle or delayed combustion flame

Flames not conforming to the conventional patterns classified by combustion technology; new flame and new combustion
1. Composite flame and composite combustion

$$\text{combustion} \begin{cases} \text{Diffusion} \\ \text{Premixed} \end{cases} \text{flow} \begin{cases} \text{Turbulent} \\ \text{Laminar} \end{cases}$$

   turbulent diffusion flame with the characteristics of a premixed flame
2. Da: Damköhler number
   Da = reaction rate/transfer rate
         transfer time/reaction time
   Low Damköhler number combustion
3. Well-stirred reactor; complete mixture system flame

Formation of radicals
1. Green flame generation of CH-component fuel by C_2 radical
2. Quantization of C_2, CH, OH radicals
3. Reforming to polynuclear aromatic hydrocarbon (PAH), precursor of soot by fuel pyrolysis

Characteristics of high temperature air combustion:
  Spontaneous ignition
  Flame holding mechanism is not required
  Flame patterns differ substantially depending on the method of mixing air and fuel
  Instantaneous peak flame temperature is low despite using high temperature preheated air
  Temperature variation of the flame is very small
  Diluted combustion or low-oxygen combustion results from recirculated burned gas
  Reaction zone becomes thick in low oxygen combustion

(II)

Air / Gas	High speed injection	
Dispersed combustion Multistage combustion	Fuel dispersion Diffusion · Suppression of mixing Effect of premixing Air dispersion (oxidizing agent)	
Strong recirculation of exhaust gas Hard mixing, hard diffusion		
Balanced combustion · Spontaneous combustion-combustion in low oxygen field higher than autoignition temperature after-burning, reburning different from spatial position and time axis		

TABLE 5.1 (CONTINUED)
Overall Configuration of Technical Arrangements of High Performance Industrial Furnace Development Project

Combustion devices	Hardware condition:
Regenerative burner	Position of air and gas nozzles, air and fuel piping
	Angle of air and gas nozzles
Gun ⎧ Center-gun system	Shape and number of air and gas nozzles
⎨ outer-gun (accentric gun) system	Discharge flow rate of air and gas
⎪ single-axis type (rotary type)	Position and pitch of burners, paring,
⎩ double-axis type (fixed type)	gas-circulating face, position, shape, and structure of partition
	Escape balance
regenerative type ⎧ honeycomb	Vertical heat balance
⎩ ball	Effective gas convection method
Gas duct and smoke stack	Improvement in furnace bottom ϕ_{CG} and shielding
⎧ Duct stack type Escape type	In-furnace pressure balance and furnace pressure control
⎩ Single stack type Zero-escape type	
	Software condition:
	ϕ_{CG} and calorific value distribution
	Optimum furnace type (furnace length, width, and height)
	Waste heat recovery condition, heat balance
	Gas recirculation conditions, partition, type, structure, and pairing configuration
	Zone devision
	Combustion control: heat pattern control, zone control, and startup control
	Furnace pressure control

heat loss through the furnace wall, the heating efficiency of the furnace increases with the height of the furnace.

In the case of a regenerative burner furnace that features high waste heat recovery, sufficient heat can be recovered from the exhaust gas even if the in-furnace gas temperature is increased, where the heating efficiency does not change significantly with the gas layer thickness. The optimal furnace height can be small. The relationship between the furnace height and unit fuel consumption was determined by heat transfer analysis using a basic heat transfer model in the same manner as with the examination of schedule-free heating. The heat loss was set according to the area ratio based on the data obtained from an actual furnace. Calculations were made for both the upper region and the lower region in the furnace.

Table 5.3 shows the calculation conditions and Figure 5.1 the results of the calculations. The furnace height is at its optimum level. The optimal height is reduced by about 1 m for both lower and upper regions of the furnace height side in the case

TABLE 5.2
High Performance Furnace Design Flow

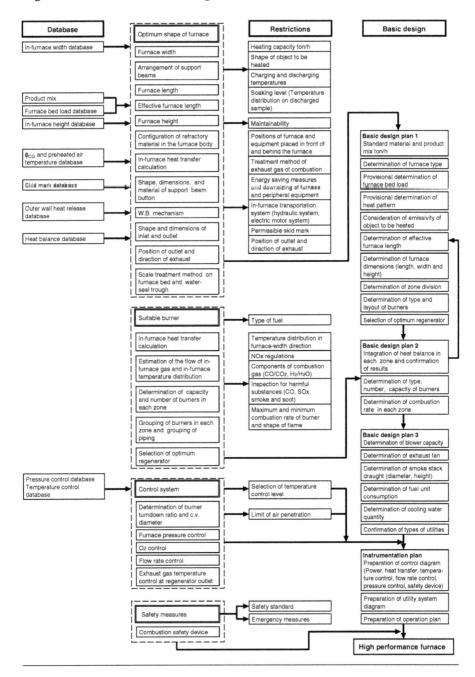

TABLE 5.3
Calculation Conditions

Furnace length	36 m
Heating time	2.5 h
Fuel	M gas
Heating temperature	20–1200°C
Slab thickness	220 mm

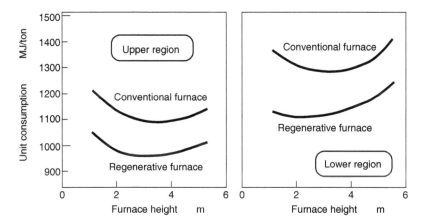

FIGURE 5.1 Relationship between furnace height and unit fuel consumption: estimation of optimal furnace height.

of high temperature air combustion as compared to a conventional furnace. Further, the gradient in the vicinity of the lowest value of unit fuel consumption is small, indicating that high thermal efficiency is available across a wide range of furnace heights. However, it must be noted here that only the amount of heat transfer by radiation is taken into consideration in the calculations. It can be estimated that the optimal furnace height will be much lower when heat transfer due to convection cannot be ignored.

Figure 5.2 shows the results of the test conducted to examine the total heating efficiency using the common test furnace II with the furnace height changed from 2.5 to 3.0 m and further to 3.5 m on the basis of the results of the above analysis. The lower the height of the furnace, the higher the heating efficiency becomes. However, the efficiency does not increase significantly at a height below 2.5 m. It can therefore be assumed that the optimal furnace height that maximizes the heating efficiency lies in this region of furnace height. This result represents a much lower optimal furnace height than the results shown in Figure 5.1, presumably because heat transfer by convection is not considered in the case of Figure 5.1. In short, lower furnace heights will result in greater amounts of heat transfer due to convection for the same amount of combustion.

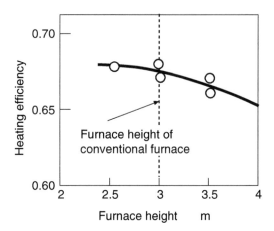

FIGURE 5.2 Furnace height and heating efficiency.

5.1.3 OPTIMAL DESIGN FOR OTHER FURNACE CONFIGURATIONS

5.1.3.1 Pitch and Capacity of Burner

In the design of a high performance furnace, not only the dimensions, but also the capacity and installation pitch (the number of burners installed) of the burners are crucial. The effect of burner pitch on the heating efficiency was tested with a doubled burner pitch by using each of four pairs of burners installed in the furnace length direction. As shown in Figure 5.3, the heating efficiency and combustion load were found to be negatively correlated. Presumably this is because the high temperature region expanded in the furnace width direction as the combustion load of burners increased to finally increase the temperature of the zone close to the wall on the other side of the burners. It is also shown that there is an optimal burner capacity with respect to a given furnace width, because the heating efficiency was lowered when the combustion load was too small and the in-furnace temperature in the center position became high. The in-furnace temperature distribution in high temperature air combustion was flattened more than in the conventional method. However, it is also true that a gap in temperature was generated between the burner center axis and the burners. So it is necessary to determine the burner capacity with these findings taken into consideration. According to the results of calculations, the optimal burner pitch can be assumed to be about 18% of the furnace width.

5.1.3.2 Partition Wall

The effect of the partition wall was examined by placing a partition between the first zone and the second zone in such a manner that a combustion pattern with a higher load occurred in the second step and was set up in the common test facility II. With a partition wall used to separate the high temperature zone from the low temperature zone and to block the transfer of gas and radiation of heat from gas between the zones, the temperature of the high temperature zone can be considerably raised with corresponding decreases in the low temperature zone. The results of the

Design Guidelines for High Performance Industrial Furnaces 249

Condition	First Zone Combustion Amount kW(Mcal/h)		Second Zone Combustion Amount KW(Mcal/h)		Heating Efficiency
	No. 1 Pair	No. 2 Pair	No. 3 Pair	No. 4 Pair	
Doubled pitch 1	0	1175(1010)	0	1163(1000)	0.598
Doubled pitch 2	0	1420(1220)	0	1420(1220)	0.590
Usual pitch	698(600)	698(600)	698(600)	698(600)	0.652

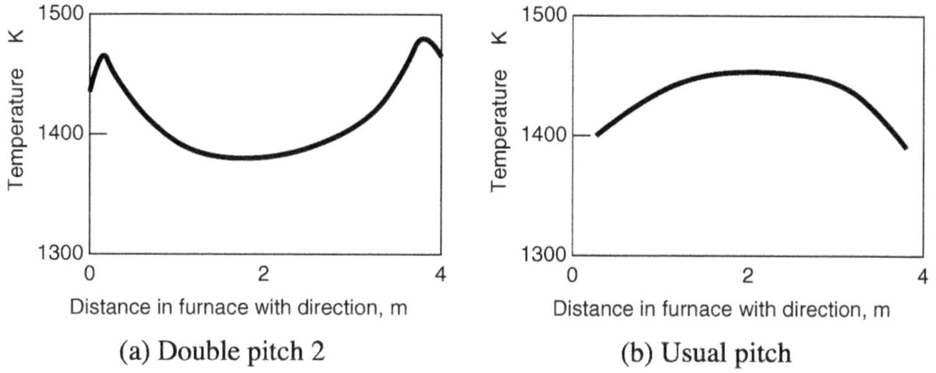

(a) Double pitch 2

(b) Usual pitch

FIGURE 5.3 Changes of gas temperature profiles with burner pitches.

test show that a temperature variation of 300°C was caused between the first zone and the second zone and the heating efficiency increased by 4% as shown in Figure 5.4, with the temperature deviation of the slab in the thickness direction increased from 65 to 90°C. For industrial heating applications, the quality of products is very important. Therefore, the balance between efficiency and uniform-heating condition must be taken into consideration in the design.

5.1.3.3 Analytical Study of the Effect of a Partition Wall

To identify the effect of a partition wall, studies were made on (1) the measurement of wind velocity using air at normal temperature, (2) fluid analysis using air at normal temperature, (3) measurement of temperature distribution associated with combustion, (4) analysis of thermal fluid associated with combustion, and (5) optimization of the length of the intermediate partition wall.

The measurement of the wind velocity using air at normal temperature was carried out using the testing setup shown in Figure 5.5, and the length of intermediate wall was varied to four lengths of 0, 1.3, 2.2, and 3 m (or expressed in the percentages of effective furnace length, 0, 43, 73, and 100%). Fluid analyses were performed using a total analysis model of reheating furnaces and with the mesh having the number of approximately 38,000 as shown in Figure 5.6. The results of these measurements and analyses are shown in Figures 5.7 and 5.8, respectively. The results from the actual measurements and the analyses are similar. Regarding the

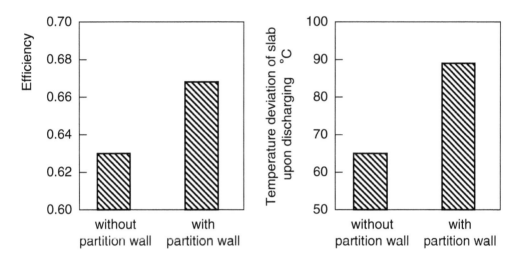

FIGURE 5.4 Partition wall effect on heating efficiency and temperature deviation.

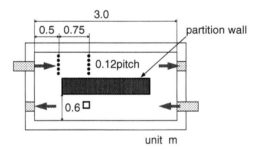

FIGURE 5.5 Measurement locations.

effect of the intermediate wall to the gas recycling inside the furnace, the wind velocity along the line of burner axis is measured and the volume of wind that passes through is calculated, to obtain the recirculation ratio. This ratio is shown in Figure 5.9, and the direction of the gas flows is shown in Figure 5.10.

The actual measurement of temperature distribution was made using the setup shown in Figure 5.11. The gas burned was city gas 13A; the combustion capacity was 230 kW, and the air ratio 1.1. The results of these actual measurements and the analyses are shown in Figure 5.12. The difference of absolute temperature values between the two is fairly large, but the general trend is similar. The results of analyses are summarized in the furnace temperature distribution, heat flux distribution, heat flux improvement ratio, and heat flux deviation, and are shown in Figures 5.13 and 5.14.

The results showed that the deviation of the temperature inside the furnace is reduced by the intermediate partition wall. A longer partition wall makes the average furnace temperature higher. A longer partition wall also causes an increase in the

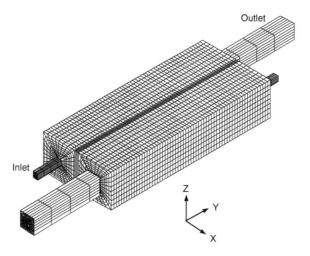

FIGURE 5.6 Analysis model.

furnace temperature distribution deviation; it is especially noticeable when the wall is longer than 80%. A longer partition wall also results in a larger heat flux in the material to be heated, and the heat flux to the direction of burner axis changes from concave distribution curves (curves less than 60%) to convex curves, and for curves over 80%, the deviation increases. It also means that the longer the partition wall is, the smaller the convection heat flux, and for curves over 80%, it becomes smaller by around 15%. To summarize, the optimum length of the partition wall is 70% ± 10% of the effective furnace length (in the direction of the longer side of the heating material), and in this range the furnace temperatures are higher and their deviation is smaller.

5.1.3.4 Lower Part of Furnace

A number of tests using the model test furnace shown in Figure 5.15 yielded several significant results. In one experiment where the heating pattern of the furnace remained constant, the comparison of the furnace temperature distribution was made between combustion with air at normal temperature and combustion with air at a high temperature. The result showed that the distribution was more uniform during combustion with high temperature air than combustion with normal temperature air as shown in Figures 5.16 and 5.17. Where heat input was constant at 930 kW, the time required for the surface temperature of steel at the burner side to rise from normal temperature to 1100°C was compared with conditions where normal temperature air was used. The result showed that temperature rise from combustion with high temperature air was quicker, as shown in Figure 5.18.

In the comparison with 1200°C of the furnace temperature with 1.2 MW of combustion, it was found that in the case of combustion with high temperature air, the temperature of steel near the skids was higher and the difference of temperatures between the portion of the steel directly above the skids and the portion around the middle of the steel was smaller than with normal combustion (Figure 5.19). By

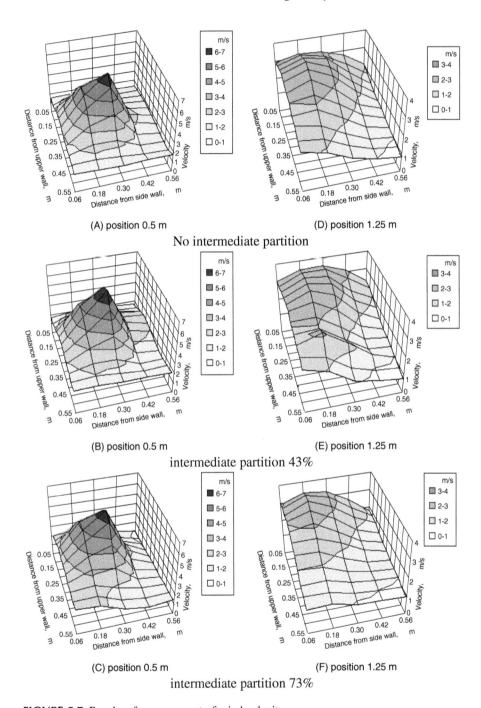

FIGURE 5.7 Results of measurement of wind velocity.

Design Guidelines for High Performance Industrial Furnaces 253

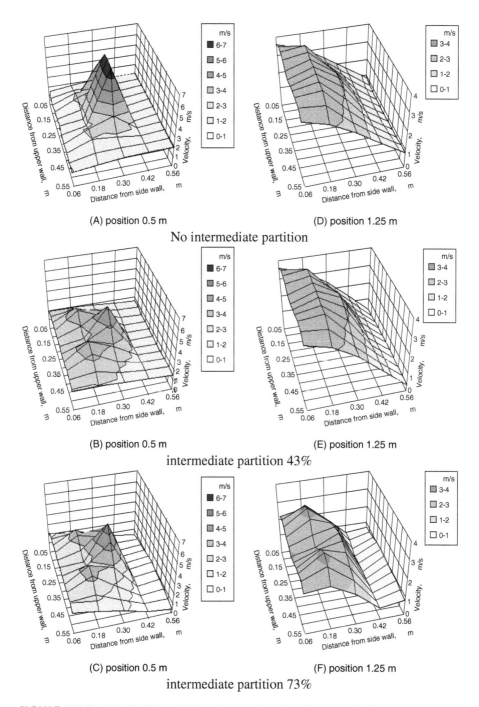

FIGURE 5.8 Results of wind velocity analyses when there is no combustion.

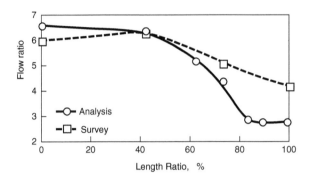

FIGURE 5.9 Recirculating ratio and the lengths of an intermediate partition.

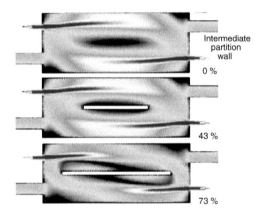

FIGURE 5.10 Effect of the lengths of an intermediate partition on velocity distribution.

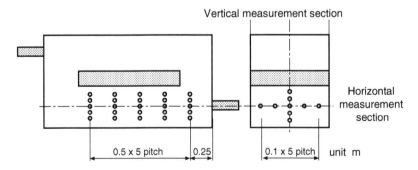

FIGURE 5.11 Measurement locations.

Design Guidelines for High Performance Industrial Furnaces

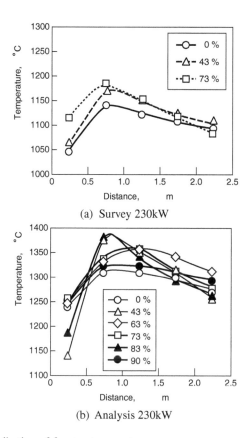

FIGURE 5.12 Distribution of furnace temperature.

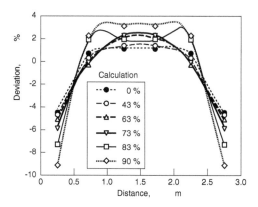

FIGURE 5.13 Distribution of furnace temperature after additive treatment.

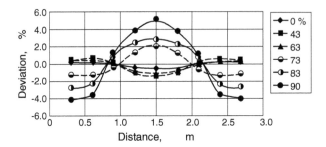

FIGURE 5.14 Deviation of heat flux.

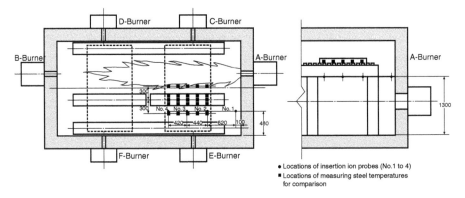

FIGURE 5.15 The lower part of heating furnace structure for model testing and the locations of inserted ion probes and of measurement steel temperature for comparison (units = mm).

measuring the ion current near the water-cooled skids, it was found that the improvement of the steel temperature uniformity was mainly attributable to the increase of the flame volume caused by HiTAC (Figure 5.20).

For the purpose of heating the material uniformly, it is possible to spread the flames with the high temperature air to the entire area inside the furnace with HiTAC, but at the same time it will increase the risk of releasing unburned components outside the furnace. Perfect combustion inside the furnace is necessary for the method to be applied to furnaces for industrial operation. The following experimental studies were made to improve the uniformity of steel temperature satisfying this condition: one study on the effect of the uniform heating of steel if the flames are burned in the shorter direction of a furnace to change the locations where the flames are formed; another on the effect of the heat transfer to the parts of steel near the skids if a method is adopted that feeds fuel and air in such a way as to dilute them near burners by the furnace gas, but under the condition of burning in the longer direction of the furnace; and, last, on the effect of the heat transfer if the furnace gas near the skids is agitated by some external force.

Design Guidelines for High Performance Industrial Furnaces 257

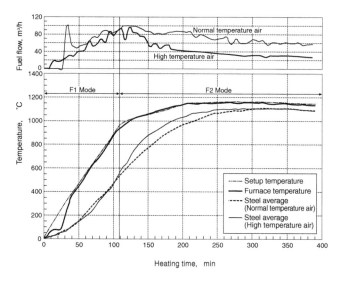

FIGURE 5.16 Change of the furnace temperature and the average temperature of steel during the process of heating.

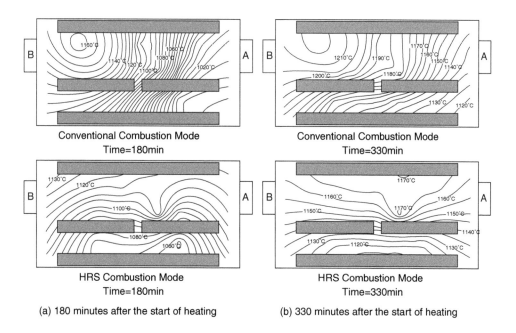

(a) 180 minutes after the start of heating   (b) 330 minutes after the start of heating

FIGURE 5.17 Distribution of the temperatures of furnace atmosphere in the cross section beneath the skids.

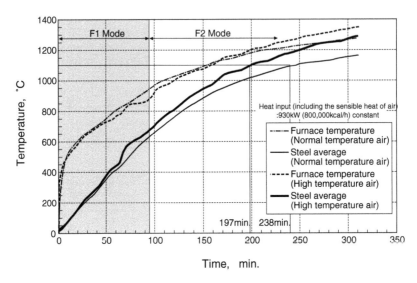

FIGURE 5.18 Comparison of the rate of temperature rise under the constant heat input of 930 kW.

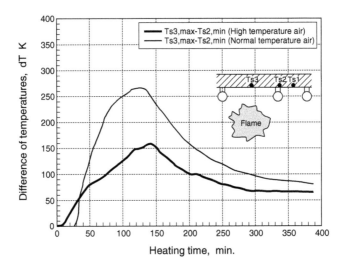

FIGURE 5.19 Change of steel surface temperature with time.

1. *Effect of flame locations*

To change the locations where the flames were formed, combustion in the direction of the furnace width (burners C, D, E, and F) was established so that the flames were formed beneath the skids. As a result, the steel temperature near the skids rose even under normal combustion, and the ion current that shows the existence of flames also increased, but ΔT was larger compared with that of the combustion with high temperature air (Figures 5.21 and 5.22). To enhance the dilution of air jet by the

Design Guidelines for High Performance Industrial Furnaces 259

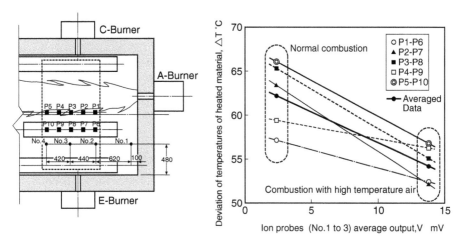

FIGURE 5.20 Output voltage of average ion current and the difference of temperatures on the steel surface (units = mm).

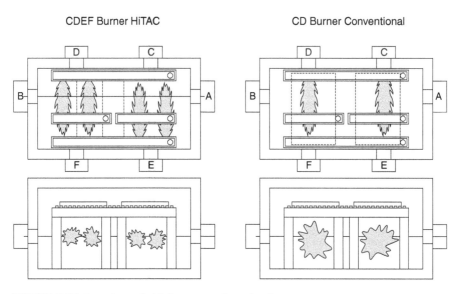

FIGURE 5.21 Locations of side burners and combustion.

furnace gas, a rectangular jet nozzle was used, and by shortening the distance to the confluence point of fuel and air, their mixing near the burner was improved. This was because the steel temperature near the skids and the ion current value were almost the same as they were when burners of original shape were used. When a device to circulate the exhaust gas by force (shown in Figure 5.23) was used under and near the skids, the steel temperature where it was in direct contact with the skids rose both for the cases of combustion with normal air and with hot air, and the temperature difference became smaller. It was also noted that there was a slight

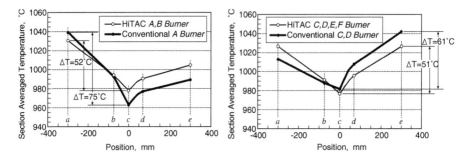

FIGURE 5.22 Comparison of the temperatures on the steel surface at the locations on two sides centering skids. (Conditions: furnace capacity = 40 ton/ch; fuel = A-type heavy oil; escape ratio = 15%.)

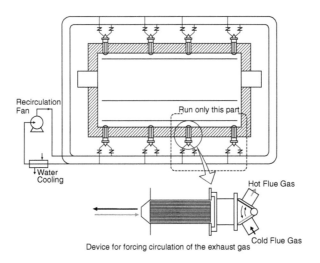

FIGURE 5.23 Device for forced circulation of the furnace gas.

increase of the ion current near the skids in the case of combustion with high temperature air.

2. Temperature difference near the skids

When combustion flames with normal air were formed near the water-cooled skids, the temperature of steel at the location in direct contact with the skids rose higher than in the case of normal combustion where the flames were not formed near them. But the temperature of steel at the location away from the skids also rose and, as a result, the reduction of ΔT of the steel was small. This is probably explained by the flames locally existing near the skids as indicated by the ion current. The potential core of a jet depends on the diameter of the nozzle, and the length of the potential core was shortened making the jet two dimensional by using an air jet nozzle shaped as an elliptic slit. The diluted condition near the burners was enhanced by the increased inducement of the furnace gas. The flow volume and the velocity were unchanged, but instead of the original round port of 162 mm diameter, a slit-shaped port of 55 mm in width was used, and the length of the potential core was about one third of the core of the round port. The direction of the fuel jet was also changed to 10° inward to the center axis of the burner. The flame thus formed was highly transparent by visual observation but the steel temperature near the skids and the ion current remained similar to those measured in the case of the original burner shape. This can be interpreted to show that such high temperature air combustion, which allows detection of the ion current of 20 mV, is feasible even at locations far from the burners by controlling the feeding method of air and fuel, even when their mixing near the burners is enhanced. Thus, this method is considered effective as a practical method to perform uniform heating, while preventing the escape from the furnace of the gas during the combustion process.

The forced recirculation of exhaust gas near and under the skids improved heating performance near the skids. Because there was a slight increase of ion density compared with the other cases of high temperature air combustion, it is conceivable that, aside from the effect of the increased convection heat transfer due to the greater flow speed of the furnace gas, the increase of ions carried toward the skids by the increased flux near the location where the probe was inserted also contributed to the improvement. However, considering that the temperature difference under normal combustion also decreased, the effect of convection heat transfer was seen to be greater in this test. Quantitative comparison of the effects of each factor was not carried out, but this method showed the possibility of heating control in the domain where it was difficult to control the flow by the flame from burners or the momentum of combustion gas alone (Figures 5.24 and 5.25).

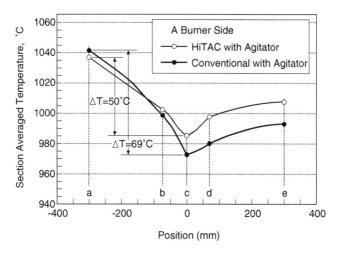

FIGURE 5.24 Comparison of temperatures on the steel surface near skids when the forced circulation of furnace gas is added.

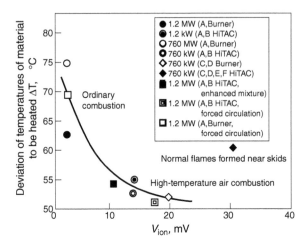

FIGURE 5.25 Comparison of uniform heating performances of steel materials by the change of mixture status near burner and the forced circulation of furnace gas.

3. Improvement in steel heating

Steel-heating tests using a model testing furnace of the lower part of a furnace structure demonstrated several points. It was possible to obtain a uniform distribution of furnace temperature even in a furnace space where heat-absorbing, water-cooled supports exist, if combustion with high temperature air is adopted. The heating time could be shortened by the improvement in the deviation of steel temperature made possible by the high temperature air combustion. The steel temperature at the location near the water-cooled skids could be raised by the flame volume increased from the high temperature combustion, and the localized heating problem caused by the skids could be improved. The locally cooled steel temperature could be raised by forming normal flames near the skids, but combustion with high temperature air was better in terms of obtaining a uniform heating of the whole steel. It was possible to implement high temperature combustion, even with an enhanced mixture of fuel/air nearer the burner by controlling the feeding method. The freedom of heating control increased when a forced recirculation of furnace gas was added.

5.1.3.5 Furnace Width and Maximum Combustion Capacity

HiTAC has the characteristics of diluted slow combustion in low oxygen and, with a narrower furnace width than the combustion space necessary to complete combustion, the problem of ejection of unburned matter through discharge burner arises. It is impossible to obtain the desired temperature distribution and heat transfer quantity. Therefore, it is important to estimate specifically the relationship between furnace width and suitable combustion capacity of burners.

Figure 5.26, with actual results and computational fluid dynamics (CFD) analysis results, shows the relationship between furnace width and maximum combustion capacity of burners to be considered to complete combustion in a furnace. This can be used as a qualitative index to check the burner capacity derived from the heat

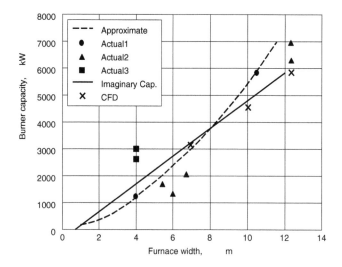

FIGURE 5.26 The relationship between furnace width and maximum combustion capacity.

balance model. Further, when selecting larger-capacity burners than this index to a given furnace width such as the point of Actual3 in Figure 5.26, counter-measures can be adopted to avoid the problem as the burner design concept is changed.

5.2 HEAT BALANCE AND PERFORMANCE ESTIMATION WITH SIMULATION PROGRAM

5.2.1 OUTLINE OF SIMULATION PROGRAM

Recently, many high performance industrial furnaces (regenerative furnaces) that use regenerative heating exchangers have been developed. These furnaces are replacing conventional reheating furnaces that use recuperators (conventional furnaces). Comparison between the high performance industrial furnaces and the conventional furnaces in the case of continuous reheating furnaces for semifinished steel (billets, slabs, etc.) shows a big difference not only in combustion air and the flow of exhaust gas, but also in heat patterns and the rising temperature curves of semifinished steel (Figure 5.27). There may also be a difference in the characteristics of thermal efficiency. If this is true, future designs for saving energy will have to be reconsidered.

For this purpose, it is necessary to make a comparative evaluation of high performance furnaces with regenerative heating exchangers and conventional furnaces with recuperators (Figure 5.28). It then becomes necessary to quantify the relationship among factors including fuel consumption and heat recovery rates, discharging temperatures, processing mass capacities (ton/h), size of steels to be reheated, and charging temperatures. Therefore, a simulator was developed to evaluate the performance of the two types of reheating furnaces: those with regenerative heat exchangers and recuperators. This would also be useful when drawing up basic plans for the furnaces.

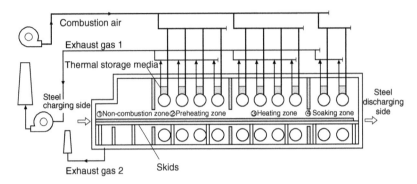

FIGURE 5.27 Continuous reheating furnace for semifinished steel with regenerative heat exchangers.

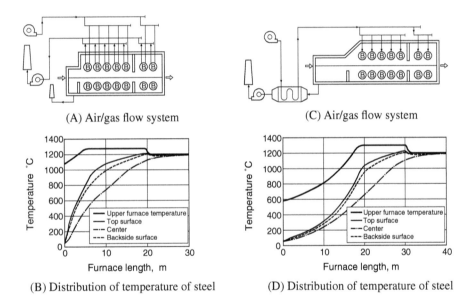

FIGURE 5.28 Distributions of temperature of steel in a conventional furnace (C, D) and a high performance furnace (A, B) with air/gas flow systems.

For the basic functions of the simulator, it was also decided that the simulator would be a process simulator for studying such characteristics of a reheating furnace as fuel consumption and required furnace length. It was then necessary to obtain the temperature of steel in the furnace and to observe the effects of varying other parameters. For calculation method and input data, the simulator is a simplified model and basic data are input to carry out the calculations. The furnace size, zone length, furnace height, and furnace width serve as one factor. The simulator calculates the heat balance (Figure 5.29) and the following are input as operational conditions required for the calculation:

Design Guidelines for High Performance Industrial Furnaces

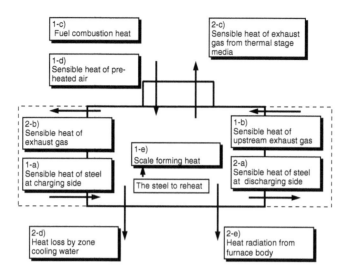

FIGURE 5.29 Heat balance in a zone.

1. ϕ_{CG}
2. The temperatures set for each zone and combustion
3. The heat radiation loss from the furnace and the overall heat transfer coefficient for the loss by cooling water
4. Combustion conditions (gas composition, air ratio, and atmospheric temperature)

This simulator is applicable to semifinished steel in the form of slabs, round billets, and billets and it requires the data input of 22 thermal properties of various types of steel. The following data are necessary on the operational conditions related to each form of steel:

1. Steel material type
2. Discharge temperature and the difference of the temperature
3. Size of the steel, thickness, width, and length, and charging intervals
4. Charging temperature and designated steel type

5.2.2 Basic Functions of the Simulator

A heat balance model of a basic reheating furnace was constructed. The basic features of the reheating furnace used for the present study are explained by using the example of a continuous reheating furnace with regenerative heating exchangers. The continuous reheating furnace is divided into four zones: (1) noncombustion zone, (2) preheating zone, (3) heating zone, and (4) soaking zone. In each zone, the heat balance between heat input and output is obtained using calculation models. These are models of the temperature inside the steel, of combustion based on fuel composition, of preheated air temperature based on a given thermal efficiency, and of the radiation heat from the furnace body. This latter model uses the furnace temperature and the heating surface, plus the heat loss by cooling water.

The required heat volume and steel temperature are calculated based on the heat balance obtainable by combination of the above, the combustion heat, and the sensible heat of the exhaust gas. The difference between the continuous reheating furnace model with regenerative heat exchangers from the conventional continuous reheating furnace model is the difference depending on whether the combustion air to be heated goes through regenerative heat exchangers or recuperators. Furthermore, regarding the high performance model, the program is also applicable to the non-combustion zone at the furnace tail.

5.2.2.1 Estimation Method of Fuel Flow Volume and Exhaust Gas Temperatures Using Heat Balance

The method to calculate the volume of fuel flow and exhaust gas temperature to obtain the heat balance in each divided zone is shown in Figure 5.29. The heat inputs related to the combustion gas are the sensible heat of the semifinished steel coming into the zone from the steel charging side, as well as the sensible heat of preheated air and exhaust gas coming into the zone from the steel discharging side. Scale-forming heat and fuel combustion heat are also heat inputs, even though sensible heat of preheated air and scale-forming heat do not exist in the noncombustion zone.

Heat outputs are sensible heat of a semifinished steel going out of the zone on the discharge side and of the exhaust gas flowing to the zones on the steel charging side (downstream), as well as the sensible heat of exhaust gas going out to the regenerative heating exchangers (no heat exchangers in the noncombustion zone), heat loss by zone cooling water, and heat radiation loss from the furnace wall.

The way to obtain the heat balance between these heat inputs and outputs is to calculate the temperature and the fuel necessary. The zones are split into two groups and the calculation is carried out for each group. For the noncombustion zone, the foregoing heat input and output values of items, a semifinished steel, exhaust gas into the zone, scale forming, cooling water heat loss, and heat radiation loss are given (or hypothetically determined) and the sensible heat of exhaust gas from the zones on the steel charging side (downstream) is calculated. For the preheating zone, heating zone, and soaking zone, the foregoing heat input and output values (which are sensible heat of exhaust gas flowing out to the zones on the steel charging side) are given, in addition to the case of calculation for the noncombustion zone. By using these data input, the fuel combustion heat and its associated airflow rate are calculated. Both calculation methods assume that the temperature of the semifinished steel and its heat input are known, and the calculation required here is a convergence calculation involving a heat balance and a steel-temperature calculation.

5.2.2.2 Calculation Method of the Internal Temperature of the Semifinished Steel

The semifinished steel is approximated to one of the following three forms to calculate the temperature. The forms are one-dimensional infinite plate (slabs), infinite column (round billets), and two-dimensional infinite plate (billets). As shown in Figures 5.30 to 5.32, the heating curve varies significantly depending on the steel form. Therefore, when inputting data, an appropriate form is selected as an approximation, based on which the calculation is carried out. Temperatures along the furnace length are obtained by calculating the temperature for each location of the steel, converting the location to time.

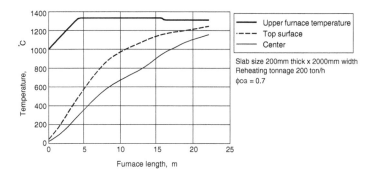

FIGURE 5.30 Slab temperature increase with furnace length.

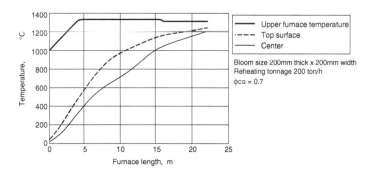

FIGURE 5.31 Square bloom temperature increase with furnace length.

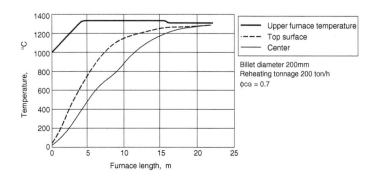

FIGURE 5.32 Round billet temperature increase with furnace length.

Heat transfer inside the steel is calculated using a non-steady finite-difference method (temperature–converted temperature–retained heat method) shown in Figure 5.33 by "Heat transfer testing and its calculation method for a continuous reheating furnace" edited by the Iron and Steel Institute of Japan. Concerning the thermal properties of semifinished steel, the retained heat and the conversion temperature of steel are applicable to the 22 types of steel addressed in "Heat transfer testing and its calculation method for a continuous reheating furnace." The following sections are examples of average specific heat and thermal conductivity (Figures 5.34 and 5.35).

In the case of the one-dimensional heat transfer partial differential equation, the finite-difference method of:

$$H(i)_{t+\Delta t} = H(i)_t + \frac{\lambda_0}{\rho} \frac{\Delta t}{\Delta \chi^2} \{\phi(i-1) - 2\phi(i) + \phi(i+1)\}$$

is used, which is obtained by linear differentiating:

$$\frac{\partial \theta}{\partial t} = \frac{\partial}{\partial x}\left(\frac{\lambda}{\rho c_p} \frac{\partial \theta}{\partial x}\right)$$

by retained heat to conversion temperature method.

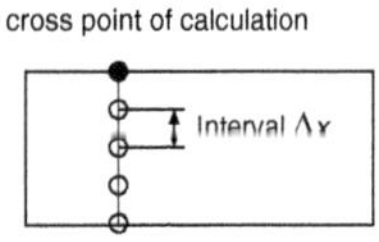

cross point of calculation

Interval Δy

FIGURE 5.33 One-dimensional calculation method of heat conduction in steel.

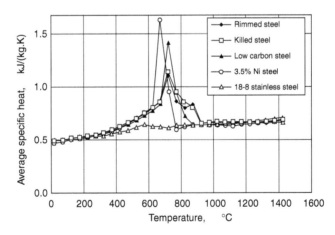

FIGURE 5.34 Examples of the relationship between steel temperature and average specific heat.

Design Guidelines for High Performance Industrial Furnaces

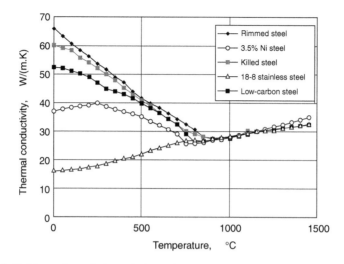

FIGURE 5.35 Examples of the relationship between steel temperature and thermal conductivity.

5.2.3 Calculation of Preheated Air Temperatures and Exhaust Gas Temperatures after Heat Exchange

The regenerative heat exchanger simulator carries out a non-steady convergence calculation requiring extensive computing time. It should also be noted that reheating furnaces have multiple regenerative heat exchangers and it is very difficult to do the calculation combining all the exchangers. Therefore the present reheating furnace simulator does the calculation using as input data the heat recovery rates of heat exchangers obtained by the regenerative heat exchanger simulator. Air and the specific heat of exhaust gas are the nonlinear functions of temperatures. Therefore, a simple multiplication of the exhaust gas temperature by the heat recovery rate does not give the correct temperature of the preheated air.

As Figure 5.36 shows, (a) the sensible heat of exhaust gas is obtained from the exhaust gas temperature, and (b) the sensible heat of the preheated air is obtained from the heat recovery rate from this sensible heat of exhaust gas. Next, (c) the preheated air temperature is obtained from the sensible heat of the preheated air where a convergence calculation is also carried out to obtain the heat balance, taking into consideration also the condition that the exhaust gas temperature is greater than the preheated air temperature. This method is applied both to regenerative heat exchangers and to recuperators.

5.2.4 Radiation Heat from the Furnace Body and Heat Loss by Cooling Water

The radiation heat from the furnace body of this zone (I) is calculated by Formula 5.1. The temperature simulator is programmed in the form of a box to simplify its

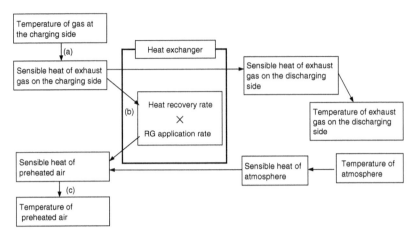

FIGURE 5.36 Calcualtion of preheated air temperatures and exhaust gas temperatures.

inputs and program. Therefore, where the equivalent heat transfer coefficient is concerned, it is necessary to input values that reflect area and adiabatic conditions.

$$Q_{rad} = h_{rad} A_w (T_{fw} - T_{at}) \quad (5.1)$$

Q_{rad}	= radiation heat of furnace body (I), kW
h_{rad}	= equivalent heat transfer coefficient (I)
L	= zone length, m
a	= furnace wall area per unit length, m
A_w	= furnace wall area, m² = La
T_{fw}	= furnace temperature, K
T_{at}	= atmospheric temperature, K

The heat loss by cooling water of this zone (I) is calculated by the following formula. Equation 5.2 is also simplified and a linear heat transfer coefficient (I) that includes the vertical skids has to be input. To obtain the values necessary for Equations 5.1 and 5.2, support tools shown in Table 5.4 are programmed and used in combination.

$$Q_{loss} = h_{rad} Ln (T_{fw} - T_{wat}) \quad (5.2)$$

Q_{loss}	= heat loss by cooling water, kW
n	= number of skids
T_{fw}	= furnace temperature, K
T_{wat}	= water temperature, K

Design Guidelines for High Performance Industrial Furnaces 271

TABLE 5.4
Main Support Tools for the Evaluation Simulator of High Performance Industrial Furnaces

Support programs

The area of opening/calculation program of radiation heat (radiation heat is calculated on the basis of opening area)

Calculation program of the area of the furnace that radiates heat (the area of heat radiation is calculated based on the shape of the furnace)

Calculation program of heat transfer volume through furnace wall (the heat transferred through the furnace wall is calculated by one-dimensional heat transfer calculation)

The calculation program of steady heat transfer in a cylinder (calculation of the heat transfer volume of skids)

Calculation program of the heat transfer coefficient of water tubes (the calculation of the heat transfer coefficient on the water side of skids)

5.2.5 OUTLINES OF SYSTEM OPERATION METHOD AND SIMULATION RESULT

For the input data and the calculated results of this simulator, spreadsheet applications of PCs are used. Therefore, it is easy to reuse the calculated results. The books for the calculated results also contain the input data shown in Figure 5.37. Figure 5.38 shows the processing flow of steel temperature calculation and the heat balance model. This simulation program is applicable to high performance furnaces as well as to conventional furnaces. A sample calculation was performed under conditions shown in Table 5.5; its results are also shown graphically in Table 5.6. Each figure in this table corresponds to one in Table 5.5, and sample data sheets for Examples 1 and 2 are also shown in Tables 5.7 and 5.8. Table 5.9 shows outputs of the calculations based on the fuel consumption and the heat balance table in addition to the aforementioned steel temperatures. In addition, the effects on fuel consumption by different parameters can be calculated by applying two different parameters relative to heat recovery rates and the length of furnace, as shown in Figure 5.39.

5.2.6 COMPARISON OF CALCULATION AND MEASUREMENT

Using the programmed simulator, the calculations made and their results were compared with actual measurements. Tables 5.10 and 5.11 show the results of the comparison; it is demonstrated that the simulation results correspond closely with the figures actually measured.

Comparative studies of the performances of furnaces equivalent in scale with the following example are carried out, selecting samples out of trial designs. Comparisons were made of a high performance reheating furnace and a conventional furnace for slabs of 285 ton/h; the results of a study are shown in Figures 5.40 and

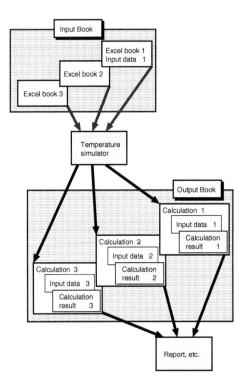

FIGURE 5.37 Flow of input and output data.

5.41. The specifications of these two types of furnaces are shown in Table 5.12, and the increasing temperature curves of slab are shown in Figures 5.42 and 5.43. In Table 5.13, comparison of the heat balances for the two types of furnaces for slabs is presented. A comparison of the results of 285 ton/h for slabs is shown in Figures 5.44 to 5.46.

5.2.7 Effect of Fuel Calorific Value on the Fuel Consumption of Reheating Furnaces

The Mix Gas produced by mixing coke oven gas (COG) and blast furnace gas (BFG) has a low calorific value and generates more exhaust gas (see Figure 5.47), resulting in a low heat recovery rate. This is the reason the effect of calorific value of fuels on the heat recovery rates of regenerative heating exchangers was studied. Based on the results of this research, the way in which calorific value of fuels affected the fuel consumption of a continuous reheating furnace was studied. In combination with calorific values, the effects of several factors were also examined. These were the effect of heat recovery rates, of discharging temperatures, of the processing rate by ton/h, and of slab thickness. Comparison was made of the effect of charging temperatures. Other factors were the effect of heat recovery rates in the case of the charging temperatures of 300 and 600°C.

Design Guidelines for High Performance Industrial Furnaces

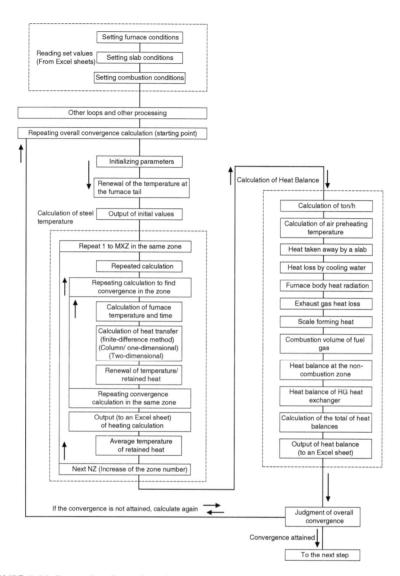

FIGURE 5.38 Processing flow of steel temperature calculation and heat balance model.

The kinds of fuel gas used were LPG, LNG, COG, and MixGas. MixGas has three calorific values of 11.7, 8.4, and 7.3 MJ/m_N^3 (m_N^3 denotes the volumes at the normal state), obtained by mixing COG and BFG with ingredients as shown in Table 5.14. Note that the heat recovery rate differs depending on the calorific value, and for this reason the heat recovery rates of a heat exchanger for each kind of fuel gas are shown in Table 5.15. In addition, facility and operating conditions are shown in Tables 5.16 and 5.17.

From the results obtained through the studies of Sections 5.1 and 5.2, the values of fuel consumption are summarized as shown in Figure 5.48, presuming that the

TABLE 5.5
Applicable Heating Forms

Descriptions	Example 1	Example 2	Example 3	Example 4
Furnace type	Complete RG combustion	Conventional furnace	Noncombustion + RG combustion	Noncombustion + RG combustion
Targeted discharging temperature	Exists	Exists	Exists	Exists
Form of semifinished steel	Slab	Slab	Slab	Round billet

RG = regenerative.

TABLE 5.6
Examples of Heat Patterns and Temperature Rising Curves

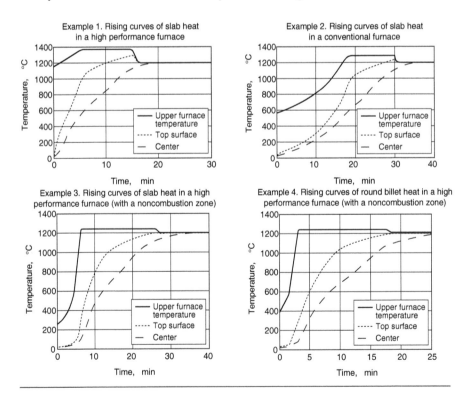

Design Guidelines for High Performance Industrial Furnaces

TABLE 5.7
Input Data Example 1

Input data example 1

Description and Setting		Setting		
1. Furnace type		Complete RG combustion		
2. Targeted discharging temperature		Exists		
3. Temperature difference of the slab at discharge		Exists		
4. Form of semi-finished steel		Slab		
			Line	Column
External loop	Not defined		0	0
Middle loop	RG heat recovery rate		18	35
Inner loop	Zone length		5	4

	1	2	3	4	5	6	7	8	9	10	11	12	13	14
1	Input data													
2														
3	Conditions of Reheating Furnaces												Setting	Used steel type (Enter a letter in the Setting space)
4			<1>	<2>	<3>						Furnace			1. Rimmed steel
5	Zone length	m	500	1000	1000						width	10.00		2. Killed steel
6							Furnace length	25						3. Low carbon steel
7	ϕ_{CG} Upper		0.85	0.85	0.65						Height	2.00		4. Middle carbon steel
8	ϕ_{CG} Lower		0.65	0.65	0.65									5. Middle carbon steel
9											Depth	2.00		6. Eutectoid steel
10	Upper zone furnace temperature at charging	°C	1,100	1,300	1,204									7. Eutectoid steel
11	Upper zone furnace temperature at discharging	°C	1,300	1,300	1,204									8. Carbon tool steel
12	Lower zone furnace temperature at charging	°C	1,100	1,300	1,204									9. 1.5%Mn steel
13	Lower zone furnace temperature at discharging	°C	1,200	1,300	1,204									10. 3.5%Ni steel
14														11. 3.5%Ni steel 1%Cr steel
15	Combustion	1or0	1	1	1									12. 3.5%Ni steel 1%Cr steel
16														13. 3.5%Ni steel 1%Cr-Mo steel
17	RG rate of exhaust gas	%	100	100	100									14. 1%Cr steel
18	RG heat recovery rate	%	60	60	60									15. 1%Cr-Mo steel
19														16. Si-Mn steel
20	Recuperator heat recovery rate	%	0											17. 13%Mn steel
21														18. 28%Ni steel
22														19. 18-8 stainless steel
23	Upper furnace body radiation heat	cal/hm	400	400	400									20. 12%Cr stainless steel
24	Lower furnace body radiation heat	cal/hm	400	400	400									21. 12%Cr tool steel
25														22. 18-4-1 High speed steel
26	Heat loss by cooling water	cal/hm	400	400	400									
27														
28	Conditions of semi-finished													
29	steel	Form	1		← 0: round billets, 1:Slab, 2:Billet									
30	Reheating tonnage	ton/h	200											
31	Discharging temperature(target)	°C	1200		← If set, soaking zone heat pattern is automatically modified									
32	Slab temperature difference at discharging	°C	25		← If set, heating zone heat pattern is automatically modified									
33	Thickness/Diameter of steel	mm	250											
34	Width of steel (0 for round billet)	mm	1950											
35	Length of steel	mm	9600											
36	Intervals of charging steels	mm	100		Slab unit weight 36.7 t									
37	Temperature of steel at charging	°C	25.0		The time in the furnace 134.4									
38														
39	Combustion Conditions													
40	Ingredients of fuel gas		N2	O2	CO	CO2	H2	H2O	CH4	C2H4	C2H6	C3H8		
41			25.0	0.0	10.9	29.6	0.0	15.5	1.2	0.6	0.0			
42	Atmospheric Temperature	°C	25.0											
43	Air ratio		1.10											
44										Heating	steam	steam		
45	for Liquid fuel, minus value		C	H	S	O	N	H2O	Ash	value	kg/kg	temperature		
46	to the air ratio		85.60	13.40	0.10	0.81	0.09	0.09	0	10,260				
47														
48														
49														
50	loop calculation													
51														
52	External loop		17	35	100									
53														
54														
55	Middle loop		18	35	40	60	80							
56														
57			CASE	I	II	III	IV	V	VI	VII	VIII	IX	X	
58	Internal loop		5	4	5	7.5	10	12.5	15	17.5	20	25	30	35
59														

TABLE 5.8
Input Data Example 2

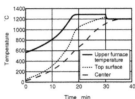

Input data example 2

Description & Setting		Setting
1. Furnace type		Conventional furnace
2. Targeted discharging temperature		Exists
3. Temperature difference of the slab at discharge		Exists
4. Form of semi-finished steel		Slab

		Line	Column
External loop	Not defined	0	0
Middle loop	RG heat recovery rate	20	3
Inner loop	Zone length	5	4

	1	2	3	4	5	6	7	8	9	10	11	12	13	14
1	Input data													
2														
3	Conditions of Reheating Furnaces												Setting	Used steel type (Enter a letter in the Setting space)
4			<1>	<2>	<3>									1. Rimmed steel
5	Zone length	m	8.0	5.0	5.0	12.0	10.0			Furnace width				2. Killed steel
6						Furnace length 40					10.00			3. Low carbon steel
7	ϕ_{co} Upper		0.85	0.85	0.85	0.85	0.85			Height				4. Middle carbon steel
8	ϕ_{co} Lower		0.89	0.89	0.89	0.89	0.89				2.00			5. Middle carbon steel
9										Depth				6. Eutectoid steel
10	Upper zone furnace temperature at charging	°C	500	800	1,100	1,300	1,204				2.00			7. Eutectoid steel
11	Upper zone furnace temperature at discharging	°C	800	1,100	1,300	1,300	1,204							8. Carbon tool steel
12	Lower zone furnace temperature at charging	°C	500	800	1,100	1,300	1,204							9. 1.5%Mn steel
13	Lower zone furnace temperature at discharging	°C	800	1,100	1,300	1,300	1,204							10. 3.5%Ni steel
14														11. 3.5%Ni steel 1%Cr steel
15	Combustion	1 or 0	0	0	0									12. 3.5%Ni steel 1%Cr steel
16														13. 3.5%Ni steel 1%Cr-Mo steel
17	RG rate of exhaust gas	%	0	0	0	0	0							14. 1%Cr steel
18	RG heat recovery rate	%	0	0	0	0	0							15. 1%Cr-Mo steel
19														16. Si-Mn steel
20	Recuperator heat recovery rate	%	50											17. 13%Mn steel
21														18. 28%Ni steel
22														19. 18-8 stainless steel
23	Upper furnace body radiation heat	kcal/hm²	3.00	3.00	3.00	3.00	3.00							20. 12%Cr stainless steel
24	Lower furnace body radiation heat	kcal/hm²	3.00	3.00	3.00	3.00	3.00							21. 12%Cr tool steel
25														22. 18-4-1 High speed steel
26	Heat loss by cooling water	kcal/hm³	0.80	0.80	0.80									
27														
28	Conditions of semi-finished steel													
29	steel	Form	1	← 0: round billets, 1:Slab, 2:Billet										
30	Reheating tonnage	ton/h	200											
31	Discharging temperature(target)	°C	1200	← If set, soaking zone heat pattern is automatically modified										
32	Slab temperature difference at discharging	°C	25	← If set, heatking zone heat pattern is automatically modified										
33	Thickness/Diameter of steel	mm	250											
34	Width of steel (0 for round billet)	mm	1950											
35	Length of steel	mm	9600											
36	Intervals of charging steels	mm	100	Slab unit weight 36.7 t										
37	Temperature of steel at charging	°C	36.0	The time in the furnace 215.1										
38														
39	Combustion Conditions													
40	Ingredients of fuel gas		N2	O2	CO	CO2	H2	H2O	CH4	C2H4	C2H6	C3H8		
41			25.0	0.0	17.1	10.9	29.6	0.0	15.5	1.2	0.6	0.0		
42	Atmospheric Temperature	°C	25.0											
43	Air ratio		1.10											
44										Heating	steam	steam		
45	for Liquid fuel, minus value		C	H	S	O	N	H2O	Ash	value	kg/kg	temperature		
46	to the air ratio		85.60	13.40	0.10	0.81	0.09	0.09	0	10.260				
47														
48														
49														
50	loop calculation													
51														
52	External loop		0	15	100									
53														
54														
55	Middle loop		20	0	40	60	80							
56														
57			CASE	I	II	III	IV	V	VI	VII	VIII	IX	X	
58	Internal loop		5	4	5	75	10	12.5	15	17.5	20	25	30	35
59														

TABLE 5.9
Heat Balance Table

Processing Volume 200 ton/h			Fuel Consumption 11.01×10^2 kJ/ton	Furnace Efficiency 72.7%	
Heat Balance Total					
Heat Volume	8.90×10^4	kW		18.63×10^2	kJ/ton
Ratio of Input Heat	%	MJ/ton	Ratio of Output Heat	%	MJ/ton
Sensible heat of charged steel	1.1	16.87	Sensible heat of discharged steel	51.0	816.62
Fuel combustion heat	68.7	1100.24	Heat loss by cooling air	6.3	100.68
Sensible heat of fuel air	30.3	485.01	Radiation heat from furnace body	1.3	20.13
Scale forming heat	1.0	16.33	Heat loss of exhaust gas	41.5	664.68
	100.0	1602.43		100.0	1602.43
Sensible heat of cool air	0.6	9.56	Sensible heat of exhaust gas	11.8	189.22

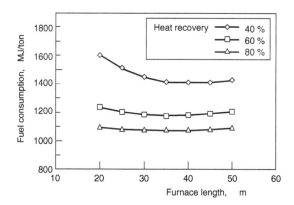

FIGURE 5.39 Relationship between the furnace length and fuel consumption.

TABLE 5.10
Calculation Assumptions

	Company		
	A	B	C
Furnace length, m	36.00	19.00	24.65
Furnace width, m	10.10	8.30	5.60
Reheating tonnage, ton/h	200	100	100
Discharging temperature, °C	1200	1050	1150
Steel thickness, mm	250	130ϕ	120
Steel width, mm	1950	—	120
Steel length, mm	9600	7700	5300

TABLE 5.11
Comparison of Actual Fuel Consumption and Calculated Results

	Company		
	A	B	C
Actual fuel consumption, Mcal/ton	250–259	225	242
Calculated results, Mcal/ton	259	224	238

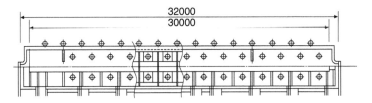

FIGURE 5.40 High performance reheating furnace for slabs of 285 ton/h (unit = mm).

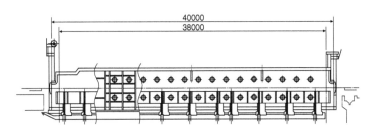

FIGURE 5.41 Conventional reheating furnace for slabs of 285 ton/h (unit = mm).

heat recovery rate is at its maximum for each heating value. Figure 5.49 is a graph obtained by choosing the specific point in Figure 5.48 of the furnace length = 30 m. The tendency for the fuel consumption to increase with a decrease of the calorific value is also seen with the ratio of exhaust gas volume/air shown in Figure 5.47. This is caused by the difference in exhaust gas volume.

The air ratio also has a large effect on the fuel consumption. The effect was studied on a reheating furnace for slabs of 250 ton/h. Five factors were studied in combination, and the results of these studies are shown in Figure 5.50. These factors are the effect of air ratio, furnace length, heat-recovery rates, charging and discharging temperatures, processing capacity, ton/h, and the slab thickness.

Design Guidelines for High Performance Industrial Furnaces

TABLE 5.12
Basic Specifications of the Reheating Furnace for Slabs of 285 ton/h

	Description	High Performance Reheating Furnace	Conventional Reheating Furnace
1.	Heating performance, ton/h	285	285
2.	Slab size		
	Thickness, mm	255	255
	Width, mm	960	960
	Length, mm	9,016	9,016
3.	Charging temperature, °C	20	20
4.	Discharging temperature		
	Thickness direction average minimum temperature, °C	1,200	1,200
	Cross section average, °C	1,220	1,255
5.	Principal figures on the furnace size		
	Effective furnace length, mm	30,000	38,000
	Internal width of furnace, mm	10,000	10,000
	Furnace height upper zone, mm	1,800	2,400
	Lower zone, mm	2,000	2,200
	Load on the furnace floor, kg/m²/h	1,054	832
6.	Number of combustion controlled zones	6	6
7.	Designed furnace temperature		
	Preheating zone °C	1,300	1,300
	Heating zone °C	1,380	1,380
	Soaking zone °C	1,330	1,330
8.	Heating time, h	1.8	2.3

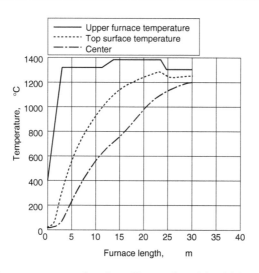

FIGURE 5.42 Slab temperature as a function of furnace length in a high performance furnace.

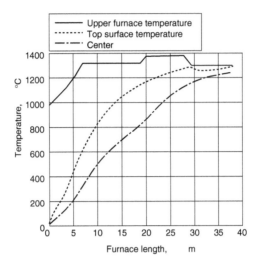

FIGURE 5.43 Slab temperature rising curves of a conventional furnace.

TABLE 5.13
Heat Balance Table of 285 ton/h Slab Furnace

	Description	High Performance Furnace, MJ/ton	Conventional Furnace, MJ/ton
Heat input	Sensible heat of charged steel	9.38	9.38
	Fuel combustion heat	1002.66	1445.85
	Sensible heat of combustion air	490.31	264.34
	Scale forming heat	38.51	38.51
	Total	**1540.86**	**1758.07**
Heat output	Sensible heat of discharged steel	829.71	852.81
	Heat loss by cooling water	68.67	113.81
	Heat radiation from furnace body	51.14	67.03
	Heat loss of exhaust gas	591.34	724.43
	Total	**1540.86**	**1758.07**

5.3 COMBUSTION CONTROL SYSTEM

Regenerative combustion is characterized by alternately switching a pair of burners. This periodic switching consequently has harmful effects on measurement systems and disturbs the feedback control needed for stable combustion. In combustion systems in particular, air-to-fuel ratio control is very delicate and difficult. As a result, the in-furnace pressure varies substantially, making measurement of only

Design Guidelines for High Performance Industrial Furnaces

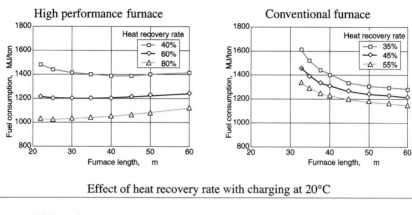

Effect of heat recovery rate with charging at 20°C

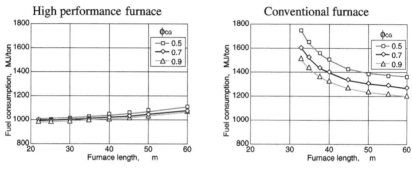

Effect of radiation parameter ϕ_{CG}

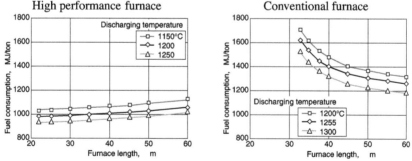

Effect of discharge temperature

FIGURE 5.44 Effects of heat recovery, ϕ_{CG}, discharging temperature to fuel consumption rate in high performance furnace and conventional furnace.

average values possible. It is necessary to analyze the characteristics of these measured signals so that signals can be processed correctly to meet the requirements of conventional continuous control.

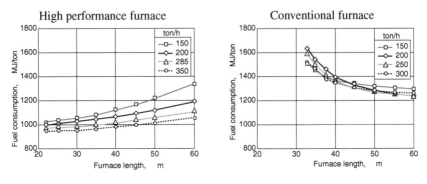

Effect of production rate

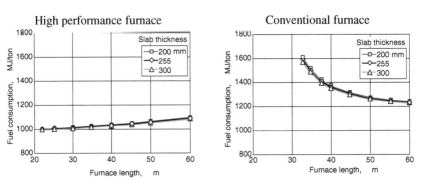

Effect of slab thickness

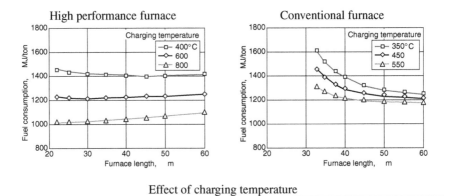

Effect of charging temperature

FIGURE 5.45 Effects of production rate, slab thickness, and charging temperature to fuel consumption rate in high performance furnace and conventional furnace.

5.3.1 BASIC COMBUSTION CONTROL SYSTEM FOR STABLE OPERATION

For regenerative combustion in which burning and extinguishing are repeatedly switched in a few minutes, it is very important to build up operation sequences of

Design Guidelines for High Performance Industrial Furnaces

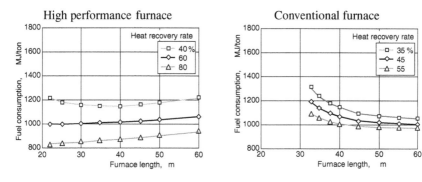

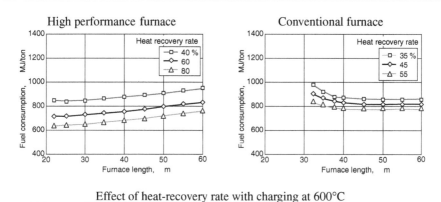

FIGURE 5.46 Effects of heat recovery charging at 300 and 600°C to fuel consumption rate in high performance furnace and conventional furnace.

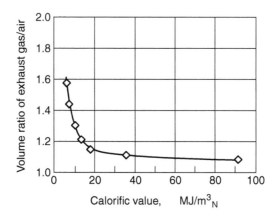

FIGURE 5.47 Relationship between calorific values and exhaust gas/air.

TABLE 5.14
Components of COG and BFG Used for Calculation

	N_2	O_2	CO	CO_2	H_2	H_2O	CH_4	C_2H_4	Calorific Value
COG	6	0.6	7.4	2.9	54.1	0	25.4	2.3	18,167 kJ/m_N^3
BFG	54.5	0.2	21.5	21.2	2.6	0	0	0	3,324 kJ/m_N^3

m_N^3 denotes the volume at the normal state.

TABLE 5.15
Studied Fuels and Heat-Recovery Rates

	LPG	LNG	COG	MG2800	MG2000	MG1750
Heating value kcal/m_N^3	21,620	8,556	4,316	2,804	2,012	1,760
kJ/m_N^3	90,519	35,822	18,070	11,740	8,424	7,369
Heat-recovery rates of heating exchangers, %	83	82	80	74	67	62

m_N^3 denotes the volume at the normal state.

TABLE 5.16
Facility Conditions

Length of noncombustion zone	0–14 m
Length of preheating zone	5.00 m
Length of heating zone	15.00 m
Length of soaking zone	5.00 m
Heat recovery rate	80.0%
Fuel (studied factors)	
Air ratio	1.10
ϕ_{CG}	0.70

changeover valves and regulative flow valves that allow stable combustion operation. For this purpose, two requirements must be satisfied. The first is to prevent unstable condition of flow control of air and exhaust gas resulting from the flow rate reaching the zero level at the moment of changeover. The second is to prevent excessive variation in the flow rate of air and exhaust gas at the moment of changeover so that pressure in the furnace will not be markedly changed and the pilot burner will be

TABLE 5.17
Operational Conditions

Reheating tonnage	250 ton/h
Discharging temperature	1150°C
Difference of temperature at the time of discharge	30°C
Steel thickness	250 mm
Steel width	2000 mm
Steel length	9000 mm
Intervals between steels	100 mm
Temperature of steel at the time of charging	20 and 600°C

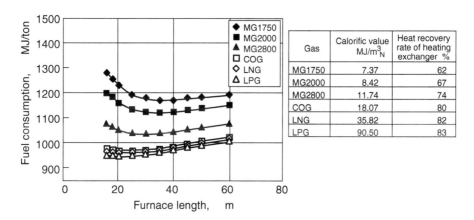

FIGURE 5.48 Effect on fuel consumption of different calorific values and furnace lengths.

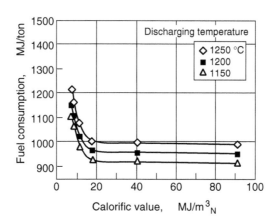

FIGURE 5.49 Calorific values and fuel consumption.

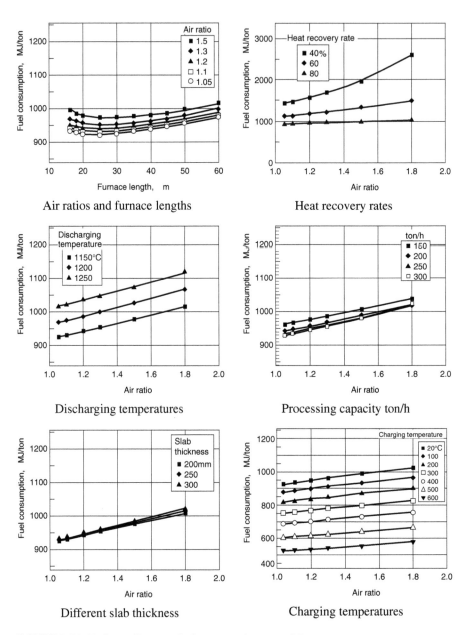

FIGURE 5.50 Various effects on fuel consumption rate of furnaces.

able to continue stable combustion. The basic concept behind these requirements is *to realize timing-optimized aperture control of the flow regulating valves at the moment of burner changeover.* Figure 5.51 shows the concept of the operating sequences.

Design Guidelines for High Performance Industrial Furnaces

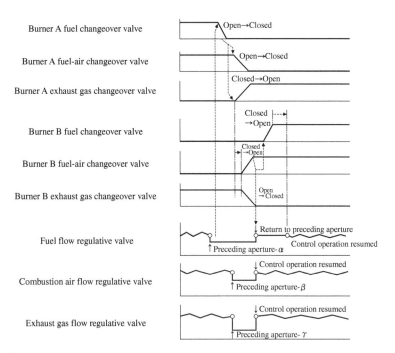

FIGURE 5.51 Concept of operating sequences of changeover valves and flow regulative valves. (From Hida, A., *Nippon Steel Corp. Tech. J.*, 363:49, 1997. With permission.)

5.3.2 SIGNAL PROCESSING METHOD

The basic signal processing method is in accordance with the study made by Yamatake-Honeywell Co., Ltd. It is described below. In a conventional continuous control method, feedback control is employed with measured signals. However, for regenerative combustion, modifications must be made to the signal processing method. The following three processing methods are available:

1. The controlled variables, such as fuel/airflow rate, temperature of preheated air, and so on, are fixed until the flow rate or pressure becomes stable from the moment of fuel cutoff.
2. Intermittent control is executed with a cycle excluding the fuel cutoff timing.
3. Measurement signals are compensated for with the effects of transient response and used for continuous control.

The first method is applied in an ignition sequence for an example and is the most typical control method under these circumstances. It provides stable performances with a high controllability on condition that tuning for the fixed period and bumping at the beginning of control are observed carefully. One drawback of this method is that responsiveness cannot be increased when the output is fixed for long periods depending on the changeover cycle or fuel cutoff duration. In the case of the second method above, the controlled process can be handled as a continuous operation with the changeover signal ignored. It is suitable for a moderate process, as it is not influenced by local variations and can handle variations with a long response time.

The last method also handles the controlled process as a continuous operation similar to the second method, except that the signals are made to appear continuous by the use of a model instead of ignoring the influence of disturbances. In regenerative combustion, a short batch process made up of combustion and cutoff is repeated. If the components before and after the cutoff action can be forecast to form a continuous process, it is not necessary to wait until the flow rate becomes stable. This method has one advantage, i.e., a conventional control method can be applied without modification.

Each of the three methods mentioned above is suitable for use with different types of facilities or combustion methods. Each type of furnace should use the appropriate optimal control method. The waveform and response of the operation terminal at changeover and the third investigated method are examined in the following.

The flow rate in the flow-down block at the time of cutoff is held at the preceding value immediately before cutoff because the response cannot be predicted. Correction is applied throughout the period from cutoff action to the convergence of the step response model.

The correction equation is expressed as shown below:

$f(t)$ = measurement signal
$g(t)$ = correction signal
$k(t)$ = model (first-order lag step response of gain 1)

Assuming $t = 0$, when the hold is released (cutoff valve: OPEN), $g(0)$ = hold value, and $f(0)$ = lower limit.

$$g(t) = f(t) + (g(0) - f(0))(1 - k(t))$$

When $k(0) = 0$ is established, the right side of the equation above is $g(0)$. This means that the hold value is the start level. Assuming the convergence time of the step response to be T, $k(T) = 1$ is established to result in $g(T) = f(T)$. This indicates a convergence to the measurement signal. The second term on the right side ($g(0) - f(0)$) is the function to converge to 0, which is added to the measurement signal. The error of the model is simply added eventually, converging to zero. The time constant of the step response model is the tuning parameter. A function of the first-order lag is usually incorporated in most controllers, and can easily be made with a difference equation. Figure 5.52 shows the COG flow data. The data appear to be somewhat overresponsive because the original data are controlled by hold processing only. The controller changed the output for no response in hold processing, as is clearly evident from the graph. This proved that controllability improved with an increase in the amount of information exceeding that in holding type processing when the flow rate is at a low level, thereby improving controllability.

Design Guidelines for High Performance Industrial Furnaces

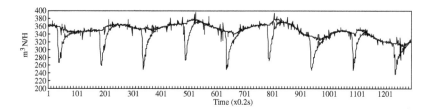

FIGURE 5.52 COG flow rate correction signal (at 100% combustion).

The following problems are anticipated with online processing of in-furnace pressure signals:

1. Correction of pressure drop at changeover
2. Control of peak pressure at ignition
3. Correction of burner-specific error

As for item 1, the immediately preceding value must be held, as it cannot be controlled with the flow rate. Item 2 above is also difficult to control as in the case of item 1. Since there is an obvious relationship between the peak level and convergence value, the controllability may be improved by reflecting the value from the relationship to the correction value. For item 3 above, some errors with a plurality of burners used can be ignored and others in the case of a pair of burners can be corrected by varying the combustion amount.

Accordingly, this type of problem can be excluded from this discussion. The waveform of the peak at ignition is of a typical lag-process impulse response. However, the impulse gain data cannot be determined using an online method and peak timing is also difficult to set. Correction using a modeling method is not practical. For the reason mentioned above, a method to forcibly attenuate the signal at the peak was examined.

The continuity of response cannot be guaranteed when the signal is attenuated as the response component is also attenuated. The attenuation does not occur in the same way as with the hold processing method. Changing the gain from 0 to 1 in proportion to the convergence of the peak component will result in the response component being increased as the peak component weakens. Compared with a case where hold processing only is applied, the controllability will increase on condition that the convergence time is short.

In the flow-down block at the time of cutoff, the immediately preceding value is held as the response cannot be predicted, and correction is executed from the moment of cutoff to the convergence of the step response model. The correction equation is expressed as shown below:

$f(t)$ = measurement signal
$g(t)$ = correction signal
$k(t)$ = attenuation model (monotonic increasing function from 0 to 1)

Assuming $t = 0$, when hold is released (cutoff valve: OPEN), $g(0)$ = hold value and $f(0)$ = lower limit.

$$g(t) = f(t) * k(t) + g(0)(1 - k(t))$$

When $k(0) = 0$ is established, the right side of the equation above is $g(0)$. This means that the hold value is the start level. Assuming the convergence time of step response to be T, $k(T) = 1$ is established to result in $g(T) = f(T)$. This indicates a convergence to the measurement signal.

The middle section in Figure 5.53 shows the result of correction using the monotonic increasing function ($y = Ax$). This appears quite practical, but the effect of attenuation on the control performance must be tested. Simulations will have to be conducted to verify the effect, and the optimum attenuation model will have to be selected.

The in-furnace pressure is controlled by changing the aperture of the damper and the rpm of the exhaust gas fan, as shown in Figure 5.54. The characteristics with respect to the degree of activation and the resultant furnace pressure can be defined on the basis of the responses to stepwise variations to these devices. Further, a change in the combustion amount changes the exhaust gas flow rate and in-furnace temperature, thus influencing the in-furnace pressure. The characteristics of the effect of disturbance on the control system involving the combustion amount and the in-furnace pressure can be defined from the responses to the stepwise variations. A simulation of a change in the furnace pressure in accordance with the changed

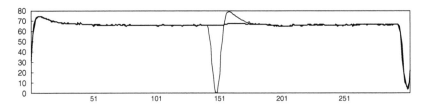

FIGURE 5.53 Controlled COG flow rate.

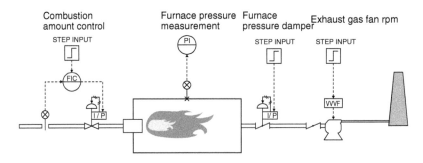

FIGURE 5.54 Schematic diagram for pressure measurement.

Design Guidelines for High Performance Industrial Furnaces 291

combustion amount was conducted by identifying the models with respect to the damper, the rpm of the exhaust gas fan (variable voltage, variable frequency, or VVVF), and the combustion amount.

The simulator is a Proportional Integral Derivative (PID) controller for feedback control. Two operation terminals for the furnace pressure damper and the rpm of the exhaust gas fan are involved. A subsystem to branch the control output is necessary. Several methods are available for this purpose. The split control method was employed, in which the low-rpm region of the exhaust gas fan is controlled by the damper and the rpm of the fan will assume control after the damper is fully opened. Thus, the damper and VVVF were modeled separately in this case.

The results of the simulation with the combustion amount of the simulator shown in Figure 5.55 reduced from 50 to 28% are shown in Figure 5.56. The simulator was set so that the output within 30% would be controlled by the damper and the output above this figure would be controlled by VVVF, while the damper was set to be fully opened at the 70% level. The figure clearly illustrates the alternate control

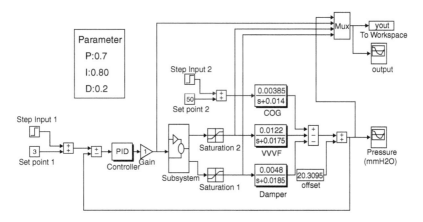

FIGURE 5.55 Simulation of changing combustion amount.

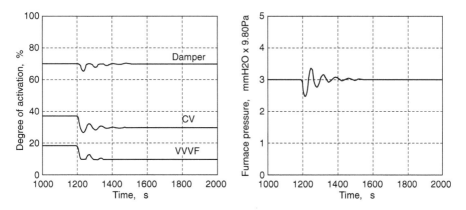

FIGURE 5.56 Results of simulation.

by the damper and VVVF. Tuning was made with reference to the parameters of the test furnace.

5.3.3 DISTURBANCE SUPPRESSION CONTROL OF DOOR OPEN AND CLOSE

A change in the furnace pressure upon charging steel material into a heating furnace is a typical example of a disturbance. However, the characteristics of such disturbances are not fully known. The operation of opening or closing the charging door of a furnace constitutes a typical pattern that is determined by such factors as the volume and heat capacity of the steel to be newly fed into the furnace. Because steel data can be partially obtained before charging, it can be considered a disturbance factor $D(k)$. With a certain amount of compensation added to the input, the variation in the quantity of state due to disturbance can be prevented in advance. For the purpose mentioned above, a Kalman filter is used as a parameter identification unit to control the variation in the quantity of state previously forecast on the basis of incomplete disturbance information. The filter can also be installed in the feedback control loop, as shown in Figure 5.57. Figure 5.58 shows a block diagram of the simulator using a Kalman filter.

Simulation was conducted by using a simulator for furnace pressure control equipped with feedforward control including an identification unit using this Kalman filter. The disturbance resulting from opening/closing the door can be expressed by a first-order lag model of step input, as in the case of the response by the control terminal. However, the time constant and the gain are different. In the current study, the empirical parameters were not available. Simulation was conducted by assuming several patterns for the parameters. As a result, it was learned that the Kalman filter can estimate the gain no matter what value it may take. Further, the simulation showed that a time constant with only a minimum degree of deviation with respect to the time constant in the basic response was sufficiently effective. The closer the time constant of disturbance is to the basic response, the more easily the disturbance can be controlled.

Yokogawa Electric Works Ltd and other manufacturers are promoting a change in the style of furnace operation from the so-called stockpile charging method in which slabs of conventional dimensions are controlled within the plant yard on a lot basis, to "schedule-free heating" method in which slabs of the desired dimensions

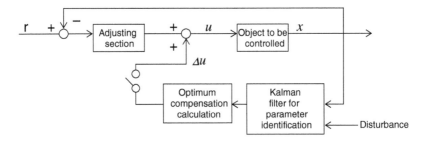

FIGURE 5.57 Compensation using Kalman filter to control disturbance.

Design Guidelines for High Performance Industrial Furnaces

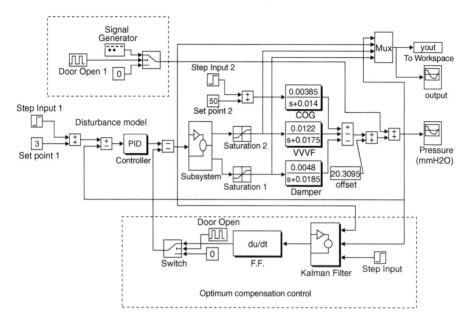

FIGURE 5.58 Flow of optimum compensation control.

are fed into a heating furnace on a desired timing basis. The method currently under investigation is described below.

The schedule-free heating method refers to a method of realizing the need to change operating conditions such as the target temperature, soaking level, etc. of a discharged steel workpiece at a given time by using a continuous heating furnace, as shown in Figure 5.59. The control system incorporated in this method, which processes the optimization task of minimizing the error between the set value and the actual temperature measurement, is known as the "online schedule-free heating optimum control system."

The online tracking function estimates the data that cannot be directly measured (temperature distribution in a heated steel workpiece and temperature distribution on the furnace wall) of all the data required for the control system of a continuous heating furnace. For example, the current temperature distribution data are estimated by giving measurable data (furnace temperature, air-to-fuel ratio, etc.) as well as previous temperature distribution data to the online model. Otherwise, if the temperature distribution data can be partially measured with a thermocouple or a radiation thermometer, the calculated estimate is corrected on the basis of the measured data.

The online optimum heat pattern calculation function is the core function of the entire control system. This function processes the optimization of a proper evaluation function at given control intervals by assuming the consequent furnace temperature in the future as a first-order-lag function of the set value at each moment in the control cycle and forecasting the temperature of a slab at the outlet by using the online model. The initial set value of the group of furnace temperature set values at given intervals obtained through the optimization process is selected as the optimum set value and passed to the interface.

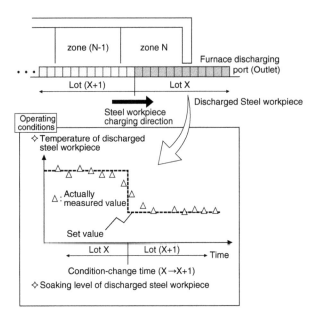

FIGURE 5.59 Schedule-free heating.

The interface function collects the information required for the online model of the online tracking function such as the temperature of the steel workpiece and furnace temperature obtained from a continuous heating furnace at fixed intervals. It also passes the optimum furnace temperature set values obtained by the online optimum heat pattern calculation function to the furnace temperature control function at fixed control intervals.

The furnace temperature control function determines the amount of fuel to be supplied in such a manner that the optimum furnace temperature set value is maintained. The operation monitor screen starts, stops, or displays the status of the functions of the digital control system that was actually used in this study.

In an online model, the imbalance portion of the heat balance of the in-furnace gas is calculated to determine the change in the enthalpy. The radiant heat flux is determined using the total absorptivity method (ϕ_{CG} method). The temperature distribution in a steel workpiece is calculated by dividing the workpiece in the thickness direction and using the difference calculation method. The distribution in the feeding direction is calculated for each workpiece independently. The radiative heat transfer and convective heat transfer from the gas to the surface of the workpiece that is in contact with the gas and a certain amount of heat loss caused at the bottom of the workpiece must be taken into consideration. For heat transfer through the furnace wall, the furnace wall and ceiling are considered a single entity in the calculation of heat transfer. As in the case of the steel workpiece, the temperature distribution on the wall is calculated by dividing the wall in the thickness direction and using the difference calculation method. The model calculation is made at fixed time intervals. With calculation for the movement time equivalent to the mesh width in

Design Guidelines for High Performance Industrial Furnaces

the movement direction from the movement velocity of the steel workpiece completed, the temperature of the steel workpiece is moved to the next mesh in the movement direction. It is assumed that the air-to-fuel ratio control and furnace pressure control have been fully carried out. The behavior over an extremely short period of time due to the performance of burners is ignored. Figure 5.60 shows the overall configuration of the model.

As an example of test results, Figure 5.61 shows the actual measurements changed in response to the altered operating condition (temperature set value of the discharged steel workpiece) with this control system implemented in the test furnace. This furnace has 4.0 width × 3.0 height × 8.0 m length, two heating zones, and 4.65 MW (400 × 10^4 kcal/h) combustion output. Although the model is not completely accurate, the control response was obtained as shown in the figure.

5.3.4 FUTURE TRENDS OF COMBUSTION CONTROL TECHNOLOGY USING HIGH TEMPERATURE AIR COMBUSTION

New concepts and methods in terms of both hardware and software for high performance stable combustion systems incorporating nonsteady dispersed flame characteristics of short-interval switchover regenerative-type combustion are expected to be introduced in the near future. Currently, the measurement of temperature depends on the use of thermocouples for temperature measuring method and the signal processing method. Determination of the positions and the number of measurement points is crucial. Fusion of measured temperatures and theoretical temperatures (analyzed temperatures) and their suitability with respect to control temperature (temperature specified for the purpose of control) all need to be taken into consideration. Further, the flame itself is expected to be measured. In addition, direct measurement of flow and reaction fields was considered to enable the micromeasurement and analysis including LIF, CARS, etc. of circulation flow and combustion

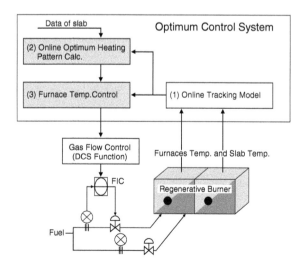

FIGURE 5.60 Basic configuration of optimum control system.

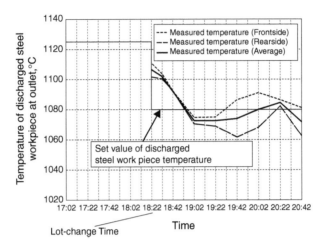

FIGURE 5.61 Temperature of discharged steel workpiece.

boundary flow in practical applications. Quantification of radical component measurements and practical application of the technique are also expected.

5.4 APPLICATION DESIGN OF HIGH PERFORMANCE FURNACES

5.4.1 Reheating Furnace

This specification is for a slab-reheating furnace that fully incorporates the results of the national High Performance Industrial Furnace Development Project. With this furnace, energy saving (over 30% reduction of CO_2) is realized. Further, the facility is reduced in size by over 20%, and low-NO_x combustion conformance with the environmental regulation level is seen.

The furnace incorporates a regenerative burner combustion system. With regenerative burners used to achieve the furnace performance outlined above, the combustion system of the furnace makes use of high temperature preheated air at a temperature higher than that necessary for autoignition (about 800°C). The furnace body and entire system is reduced in size. With the results of the High Performance Industrial Furnace Development Project incorporated, the dimensions of the furnace body have been decreased compared with conventional furnaces, to form a short, low heating furnace (a reduction of 20% in effective furnace length, 17% in furnace height, and 35% in furnace volume).

Further, since the diameters of the gas flue smokestack and of the combustion air piping can be reduced and a recuperator dispensed with, not only the furnace body itself, but also the entire facility can be designed to be smaller. The total construction cost of the facility, including the foundations and building can be considerably reduced from that required for a conventional furnace. The furnace features show improved performance. This includes maximized heat transfer; with flattened time-average temperature distribution and increased adiabatic flame temperature, the heat transfer effect with respect to steel materials is maximized. Also, with uniform heating, with the furnace temperature made uniform and the in-furnace combustion gas thoroughly stirred, uniform heating of steel materials and reduction of skid marks are realized with greater improvement compared with that achieved with the conventional technology. And finally, with critical heat recovered by the regenerative burners, heat loss from the furnace is minimized.

By employing control methods based on the characteristics of regenerative burners applied for heating control, furnace pressure control and oxygen control, both stable operation and high precision schedule-free heating are realized. Refractory materials are used to attain a high energy-saving effect and low heat inertia. A fiber lining structure is employed for the furnace ceiling and walls and heat insulation of the skid pipe is improved to realize a high energy-saving effect and schedule-free combustion (low heat inertia).

5.4.1.1 Specifications and Performance of Facility

1. Operating conditions
Object to be heated (slab):

a. Standard slab dimensions: 255 (thickness) × 960 (width) × 9016 (length) mm (Max. length: 9300 mm)
b. Type of steel material: standard carbon steel
c. Charging method: single-line charging or double-line charging

Heating temperature:

a. Charging temperature: 20°C
b. Discharging temperature: 1200°C (lowest average temperature in the direction of thickness)

Heating capacity: 285 t/h when a standard slab is heated continuously at the heating temperature indicated above.
Utilities:

a. COG
   1. Application: main burner fuel
   2. Low-calorific value: 4500 kcal/m_N^3 (18,840 kJ/m_N^3)
   3. Theoretical air amount (A_o): 4.65 m_N^3/m_N^3
   4. Theoretical exhaust gas amount (G_o): 4.98 m_N^3/m_N^3
   5. Rough estimate of consumption: 15,500 m_N^3/h
b. Cooling water
   1. Application: for cooling skid
   2. Supply pressure: 3.0 kg/m² gauge (29.5 Pa) or higher

3. Supply temperature: 32°C or less
4. Rough estimate of consumption: 320 m³/h

Note: The consumption values of the above utilities are calculated in accordance with the heat balance estimation. Refer to the attached material for detailed information on the results of the heat balance estimation.

2. *Heating furnace type*
Six-zone walking-beam continuous reheating furnace (Figures 5.62 through 5.64).

3. *Dimensions of furnace*

   a. *Effective length*: 30,000 mm. A heat transfer calculation is made to determine the furnace length on the assumption that the heat transfer to steel material in a high performance reheating furnace is about 1.1 times that of a conventional heating furnace, on the basis of the result of the High Performance Industrial Furnace Development Project. The reasons the furnace length can be reduced from that of a conventional reheating furnace are that the heat transfer rate to the steel material improves by 10%, and the average temperature can be lowered for the same lowest temperature of the discharged slab due to the effectiveness of uniform heating
   b. *In-furnace width*: 10,000 mm. The in-furnace width is determined with reference to the maximum steel length 9300 mm.
   c. *Furnace height*: upper region, 1800 mm; lower region, 2000 mm. It is known, theoretically, from the results of the High Performance Industrial Furnace Development Project, that heating capacity will not be affected even if the furnace height is lowered to the 1200 mm level. This is true for both upper and lower regions. However, heights of 1800 mm for the upper region and 2000 mm for the lower region are actually employed by taking the interference between the burner flame and skid/furnace bed scales, arrangement profile of burners in the upper and lower regions, and maintenance space into consideration.

4. *Combustion control zones*
See Table 5.18.

5. *Steel workpiece discharging interval.*
218 s at the operating rate of 285 t/h

6. *Unit fuel consumption*
The unit fuel consumption when the steel workpiece is heated continuously under the following conditions is 240×10^3 kcal/t (1.00 GJ/t), realizing an energy-saving effect of about 30% compared with a conventional heating furnace.

   a. Steel workpiece dimensions: $255 \times 960 \times 9016$ mm
   b. Type of steel: standard carbon steel
   c. Charging temperature: 20°C

Design Guidelines for High Performance Industrial Furnaces

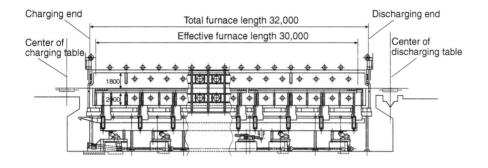

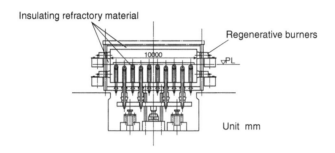

FIGURE 5.62 High performance heating furnace for slab.

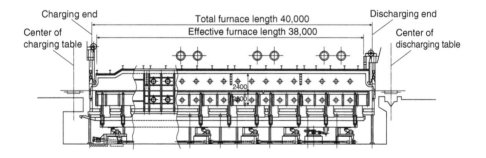

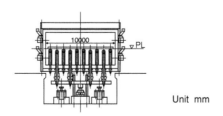

FIGURE 5.63 Conventional heating furnace for slab.

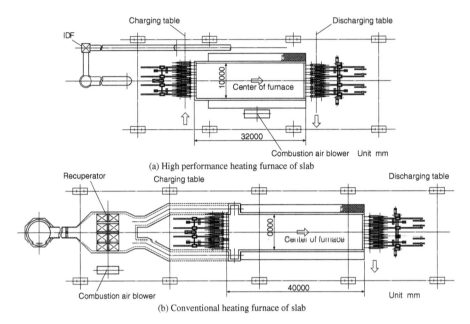

FIGURE 5.64 Overall plain view of high performance and conventional heating furnace for slab.

TABLE 5.18
Combustion Control Zones

Zone No.	Combustion Region	Burner Type	Number of Burners
1	Upper region	Side burner	12 burners (6 pairs)
	Lower region	Side burner	12 burners (6 pairs)
2	Upper region	Side burner	8 burners (4 pairs)
	Lower region	Side burner	8 burners (4 pairs)
3	Upper region	Side burner	8 burners (4 pairs)
	Lower region	Side burner	8 burners (4 pairs)

d. Discharging temperature: 1200°C (lowest average temperature in the direction of thickness)
e. Air ratio: 1.05
f. Amount of exhaust gas that passes through regenerative burners: 90% of combustion exhaust gas

7. *Discharging temperature deviation (skid mark)*

The skid mark that occurs when discharging the steel workpiece out of the furnace when the steel workpiece is heated continuously under the above conditions is 32°C, realizing a reduction in skid mark of about 50% from a conventional heating furnace.

Design Guidelines for High Performance Industrial Furnaces

8. *NO$_x$ emission level*

The average value of NO$_x$ emission over a duration of 1 h (converted to 11% O$_2$) is lower than the environmental regulation level.

5.4.1.2 Detailed Specifications of Facility

1. Burner

    a. Burner type: regenerative burner (regenerative changeover combustion burner)
    b. Capacity and number of burners (Table 5.19)
    c. Accessories
        1. Regenerative medium
        2. Combustion air changeover valve
        3. Exhaust gas changeover valve
        4. Fuel gas changeover valve
    d. Electric-ignition pilot burner
        1. Capacity and number of pilot burners: 23.3 kW/burner (20,000 kcal/h/burner), 56 pilot burners
        2. Accessories:
            Flame sensor
            Ignition plug

2. *Combustion air blower*

    a. Number of blowers: 1
    b. Air capacity: 1450 m$_N^3$/min
    c. Discharge pressure: 1100 mmH$_2$O (10.8 kPa)

TABLE 5.19
Capacity and Number of Burners

Zone No.	Combustion Region	Burner Capacity (kcal/h/burner) (MW)	Number of Burners (pairs)
1	Upper region	3,700 × 10^3 (4.30)	12 burners (6 pairs)
	Lower region	4,500 × 10^3 (5.23)	12 burners (6 pairs)
2	Upper region	2,200 × 10^3 (2.56)	8 burners (4 pairs)
	Lower region	2,600 × 10^3 (3.02)	8 burners (4 pairs)
3	Upper region	1,200 × 10^3 (1.40)	8 burners (4 pairs)
	Lower region	1,500 × 10^3 (1.74)	8 burners (4 pairs)
Total	Burner capacity	158,400 × 10^3 (184.2)	56 burners (28 pairs)
	Combustion amount	79,200 × 10^3 (92.1)	28 burners

3. Pilot air blower

   a. Number of pilot air blowers: 1
   b. Air capacity: 30 m_N^3/min
   c. Discharge pressure: 700 mmH$_2$O (6.86 kPa)

4. Exhaust gas induced fan

   a. Number of induced fans: 1
   b. Air capacity: 2200 m^3/min (at 150°C)
   c. Discharge pressure: 1100 mmH$_2$O (10.8 kPa)

5. Combustion air piping

Piping is necessary between the combustion air blower and the regenerative burners for the supply of combustion air to the burners. Air at room temperature flows through this piping.

6. Exhaust gas piping

Piping is also necessary for the low temperature exhaust gas from the regenerative burners to the exhaust stack via the exhaust gas induced fan. Combustion exhaust gas at 150 to 200°C flows through this piping.

7. Fuel piping

Piping from the COG intake point to the respective burners is also necessary.

8. Gas flue

A gas flue is necessary for high temperature exhaust gas discharged from the bottom section of the furnace-charging end.

5.4.1.3 Attachments

Comparison of performance between reheating furnaces is shown in Table 5.20.

   a. Specifications and performance (shown in Table 5.20)
   b. Heat-balance calculation table (shown in Table 5.21)
   c. Combustion facility capacity (shown in Table 5.22)
   d. Temperature rise curves and distribution of discharging temperature
      1. Temperature rise curves (Figure 5.65)
      2. Distribution of discharging temperature (Figure 5.66)

These specifications relate to the new installation of a reheating furnace using regenerative burners. The typical case of modification may include installation of a regenerative burner in a noncombustion zone in a currently operated reheating furnace and replacing the burner in the preheating zone with a regenerative burner for the purpose of saving energy and/or improving the heating capacity.

This section lists the major items to be considered prior to such modification work. The items listed here may be insufficient or exceed requirements, depending

TABLE 5.20
Comparison of Performance between Reheating Furnaces

	Item	High Performance Reheating Furnace	Conventional Reheating Furnace
1.	Heating capacity, ton/h	285	285
2.	Slab dimensions		
	Thickness, mm	255	255
	Width, mm	960	960
	Length, mm	9,016	9,016
3.	Charging temperature, °C	20	20
4.	Discharging temperature		
	Lowest average temperature in the direction of thickness, °C	1,200	1,200
	Cross-sectional average, °C	1,220	1,255
5.	Furnace dimensions		
	Effective furnace length, mm	30,000	38,000
	In-furnace width, mm	10,000	10,000
	Furnace height		
	Upper zone, mm	1,800	2,400
	Lower zone, mm	2,000	2,200
	Furnace bed load, $kg/m^2/h$	1,054	832
6.	Refractory material arrangement		
	Ceiling	Ceramic fiber	Amorphous refractory, fiber
	Walls	Ceramic fiber	Amorphous refractory, brick, etc.
	Skid	Plastic refractory + CF (50 mm) + (30 mm)	Plastic refractory + CF (50 mm) + (20 mm)
7.	Number of combustion control zones	6	6
8.	Furnace set temperature		
	Preheating zone, °C	1,300	1,300
	Heating zone, °C	1,380	1,380
	Soaking zone, °C	1,330	1,330
9.	Combustion air temperature (before burners), °C	1,150–1,230 (Furnace average: 1,175)	450
10.	Air ratio	1.05	1.2
11.	Unit fuel consumption		
	$\times 10^3$ kcal/ton	240	345
	GJ/ton	1.00	1.44
12.	Discharging temperature deviation (skid mark), °C	32	65
13.	NO_x value	Below the regulation level	Below the regulation level
14.	Heating time, h	1.8	2.3

TABLE 5.21
Heat Balance Calculation Table

	Item	High Performance Reheating Furnace		Conventional Reheating Furnace	
		× 10³ kcal/ton	MJ/ton	× 10³ kcal/ton	MJ/ton
1.	Combustion heat of fuel	240.0	1,004.8	345.0	1,444.4
2.	Sensible heat of fuel	0.4	1.7	0.5	2.1
3.	Scale formation heat	9.3	38.9	9.3	38.9
4.	Sensible heat of combustion air	1.6	6.7	2.7	11.3
5.	Take-in heat of charged slab	2.1	8.8	2.1	8.8
6.	Heat recovered by combustion air	100.7	421.6	58.9	246.6
	Total heat input	354.1	1,482.5	418.5	1,752.2
7.	Take-out heat of discharged slab	198.5	831.1	203.9	853.7
8.	Heat loss by exhaust gas	127.3	533.0	172.1	720.5
9.	Heat loss by cooling water	16.8	70.3	27.5	115.1
10.	Heat radiation through furnace body, etc.	11.5	48.1	15.0	62.8
	Total heat output	354.1	1,482.5	418.8	1,753.4
	Thermal efficiency, (7)/(1)%	82.7		59.1	
	Waste heat recovery rate, ((4) + (6))/(8)%	80.4		35.8	
	Energy-saving rate, %	30.4		Base	

on the application purpose of individual users. Consider these to be an example of a set of requirements.

Items to be considered for modification:

a. Optimum layout range of regenerative burners, optimum burner layout, capacity, and the number of burners
b. In-furnace gas flow in the zones with regenerative burners and the distribution of in-furnace temperature after modification
c. Temperature of exhaust gas at the outlet end of the furnace after modification
   → Identification of the necessity for the application of heat-resisting measures on the existing recuperator
d. NO_x value after modification
e. Maximum furnace temperature (or permissible furnace temperature) in the zone with regenerative burners
   Refractory material arrangement in the zone with regenerative burners
f. Change in furnace pressure after modification
   → Influence on the charging side opening
g. Discharging method of exhaust gas after it passes regenerative burners

TABLE 5.22
Combustion Facility Capacity

	Item	High Performance Reheating Furnace	Conventional Reheating Furnace
1.	Burner		
	Number of burners	56	60
	Facility (combustion) capacity		
	kcal/h	$79{,}200 \times 10^3$	$118{,}000 \times 10^3$
	MW	92.1	137.2
2.	Combustion air blower		
	Number of blowers	1	1
	Air capacity, m_N^3/min	1,450	2,400
	Discharge pressure, mmH_2O	1,100	1,300
	Discharge pressure, kPa	10.8	12.7
3.	Exhaust gas induced fan		
	Number of blowers	1	
	Air capacity m^3/min (at 150°C)	2,200	
	Discharge pressure, mmH_2O	1,100	
	Discharge pressure, kPa	10.8	
4.	Air heat exchanger	(Regenerative medium)	(Air recuperator)
	Air temperature, °C	1,150–1,230 (Furnace average: 1,175)	500
	Temperature efficiency %	92.0	61.5

h. Interference between the current facility and regenerative (air and exhaust gas) piping
i. Installation space for newly employed induced draft fan
j. Consumption of utilities after modification

5.4.2 BILLET REHEATING

This specification relates to the experimental design of a billet-reheating furnace that incorporates the results of the national High Performance Industrial Furnace Development Project. The furnace is a six-zone walking beam reheating furnace. The billets, delivered in fixed positions on the furnace side, are fed by a charge roller from the furnace side into the furnace. The billets fed into the furnace are forwarded one after another by the walking beam that maintains continuous rectangular movement. The billets heated and soaked inside the furnace are then transferred to a discharge roller to be continuously discharged through the side of the furnace as demanded by the rolling process.

This is a reheating furnace that incorporates the results of the national High Performance Industrial Furnace Development Project. With this furnace, energy saving (over 30% reduction of CO_2), facility reduction in size by over 20%, and low NO_x combustion conforming to the environmental regulation level are all realized.

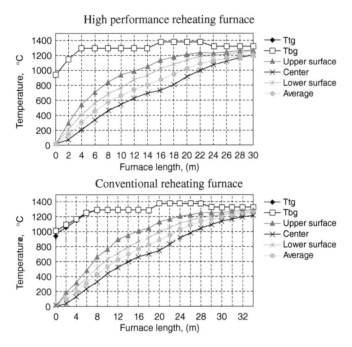

FIGURE 5.65 Temperature rise curves.

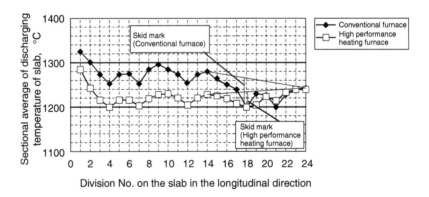

FIGURE 5.66 Distribution of discharging temperature.

This furnace has the following characteristics. The furnace incorporates a high velocity air jet regenerative burner combustion system. With high velocity air jet regenerative burners used to achieve the furnace performance mentioned above, the combustion system of the furnace utilizes high temperature preheated air at a temperature higher than that necessary for autoignition (about 800°C).

The heating furnace body is reduced in size. The dimensions of the furnace body have been decreased compared with conventional furnaces, to form a short, low heating furnace. Since a conventional billet-reheating furnace is designed for long, wide objects, axial flow heating burners are necessary to secure uniform temperature distribution in the direction of the furnace width. However, with this combustion system comprising high velocity air jet regenerative burners, the uniformity of temperature distribution in the direction of the width is realized through a sideways heating method. The use of the sideways heating method has made it possible for the furnace to be designed as a simple boxlike unit. Furthermore, there is no noselike section. Thus, the height of the furnace is reduced.

The furnace features the following improvements:

1. *Maximized heat transfer*: With flattened time-averaged temperature distribution and increased adiabatic flame temperature, the heat transfer effect with respect to steel materials is maximized.
2. *Uniform heating*: With the in-furnace combustion gas stirred thoroughly, uniform heating of steel materials is realized to the same degree as achieved with conventional technology.
3. *Minimized heat loss*: With critical heat recovered by the regenerative burners, heat loss is minimized.

By employing control methods based on the characteristics of regenerative burners applied for heating control, furnace pressure control, and oxygen control, stable operation is realized.

Refractory materials and a fiberlining structure are employed for the furnace ceiling and walls to realize high energy savings and low heat inertia. In addition, a ceramic fiber lining is employed for the in-furnace skid pipe to improve the thermal insulation effect.

The facility building and foundation are also reduced in size. With the employment of regenerative burners, the recuperator used in a conventional furnace can be dispensed. As the furnace length and height are reduced, the heights of the facility building and foundation can also be reduced. As a result, the installation space for the furnace and related equipment becomes smaller than that for a conventional furnace.

Product quality and yield are improved. As a result of improved in-furnace heat transfer capacity with the use of regenerative burners, the heating time required for steel workpieces is reduced. This leads to a reduction of scale formation on the surface of steel workpieces and also a reduction in decarburization.

5.4.3 HEAT TREATMENT FURNACE

A cross section of a semicontinuous heat treatment furnace (STC furnace), consisting of a charging table, heating room, and discharging table, is shown in Figure 5.67 as an example of a production heat treatment furnace. Materials for heat treatment on the charging table are transferred into the furnace by driving rollers, heat-treated and come to the end of the heat-treating process by being transferred to the discharging table. The furnace is filled with atmospheric gas to prevent oxidation, and carburization, decarburization, and radiant tubes are installed on both the upper and lower sides of the heat-treating materials as well as under them.

Agitating fans are installed in the ceiling of the heating room to improve temperature distribution.

5.4.3.1 Heat Balance and Evaluation Method of Furnace Performance

The ϕ_{CG} model, which has a term of surface area of gas, is generally used for estimation of overall heat absorptivity. With an indirect heating furnace, which uses radiant tubes (RT), a heat transfer model is used for calculation without consideration of gas radiation, as shown in Figure 5.68. The ϕ_{CH} model has no term of gas radiation, in contrast to the ϕ_{CG} model for direct-heating furnace.[1] Table 5.23 compares the ϕ_{CG} model, ϕ_{CH} model, and the model equation. As shown in the model equations, the ϕ_{CH} model does not include terms of surface area of gas A_G and emissivity of gas ε_g, and it is determined by the geometrical relation of positions of heating area and heat-receiving area.

The ϕ_{CH} can be calculated by Equation 5.3:

$$q = \sigma\phi_{CH}\left(T_H^4 - T_C^4\right) \quad (5.3)$$

where q is the heat flux (W/m²), T_C is the average temperature of surface of the coils, T_H is the average temperature of RT, and σ is the Stefan–Boltzmann constant.

The ϕ_{CH} and λ on plate-like material are factors that are concerned with elevating the temperature of heated materials in the RT type heat treatment furnace, as shown in Figure 5.69. The heat flux at the surface of materials being heat-treated can be calculated if ϕ_{CH} is defined and, further, the process of elevating temperature both of stable and of nonstable heating conditions can be calculated if the thermal conductivity λ is defined. The calculation model of ϕ_{CH} is shown in Equation 5.4 with the emissivity of surface of RT, ε_H, and of surface of coil, ε_C:

$$\frac{1}{\phi_{CH}} = \frac{1}{\overline{F_{CH}^*}} + \frac{A_C}{A_H}\left(\frac{1}{\varepsilon_H} - 1\right) + \left(\frac{1}{\varepsilon_C} - 1\right)$$

$$\overline{F_{CH}^*} = F_{CH}^* + F_{RH}^*\left[\frac{F_{RC}^*}{1 - F_{RR}^*}\right] \quad (5.4)$$

$$= F_{CH} + F_{RH}\left[\frac{F_{RC}}{1 - F_{RR}}\right]$$

$$= F_{CH}$$

Calculation of the shape factor F_{CH} of a heating surface projected from the side of heat-treating materials is easier if the data of angles related to the tube arrays that are listed in the heat transfer engineering handbook are used, because RT is in tube form.[2] The thermal conductivity of materials is listed according to material composition in the related heat transfer engineering handbooks. When the heat transfer is

Design Guidelines for High Performance Industrial Furnaces 309

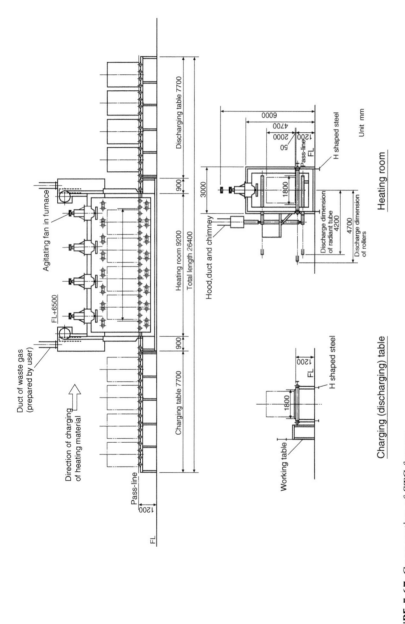

FIGURE 5.67 Cross section of STC furnace.

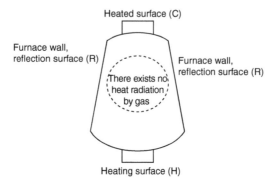

FIGURE 5.68 Heat transfer in furnace in case of no radiation combustion gas.

TABLE 5.23
Comparison and Model of ϕ_{CH} and ϕ_{CG}

	ϕ_{CH}	ϕ_{CG}
Furnace	Furnace that has no radiation gas in furnace, such as a radiant tube heat treatment furnace, electric heating furnace, etc.	Furnace that has radiation combustion gas in furnace, such as direct heating furnace
Model equations	$\dfrac{1}{\phi_{CH}} = \dfrac{1}{\overline{F}^*_{CH}} + \dfrac{A_C}{A_H}\left(\dfrac{1}{\varepsilon_H} - 1\right) + \left(\dfrac{1}{\varepsilon_C} - 1\right)$	$\dfrac{1}{\phi_{CG}} = \dfrac{1}{\overline{F}^*_{CG}} + \left(\dfrac{1}{\varepsilon_G} - 1\right)$
	$\overline{F}^*_{CH} = F^*_{CH} + F^*_{RH}\left[\dfrac{F^*_{RC}}{1 - F^*_{RR}}\right]$	$\overline{F}^*_{CG} = \varepsilon_g \left(\dfrac{A_G}{A_T}\right) \times$
	$= F_{CH} + F_{RH}\left[\dfrac{F_{RC}}{1 - F_{RR}}\right] \leftarrow \varepsilon_g = 0$	$\left[1 + \dfrac{A_R}{A_C}\dfrac{1}{1 + \left(\dfrac{A_T}{A_G} - \varepsilon_g\right)\dfrac{1}{F_{RC}}}\right]$
	$= F_{CH}$	
	A_C = area of heated surface	A_T = total surface area of furnace wall
	A_H = area of heating surface	A_G = surface area of gas body
	ε_C = emissivity of heated surface	A_R = area of reflection surface
	ε_H = emissivity of heating surface	ε_g = emissivity of gas
	F_{ij} = shape factor	
	F_{ij}^* = shape factor under consideration of gas radiation	
	$\overline{F_{ij}^*}$ = shape factor under consideration of both gas radiation and reflecting surface	

Design Guidelines for High Performance Industrial Furnaces

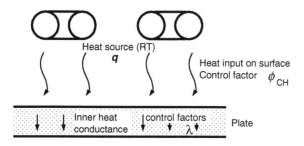

FIGURE 5.69 Control factors according to temperature elevation of heating materials.

in one dimension of the thickness direction of the plate, the principal equation (Equation 5.5) can be used for calculation of the temperature distribution in the plate:

$$\frac{dT}{dt} = a\left(\frac{d^2T}{dx^2}\right) = \frac{\lambda}{\rho c}\left(\frac{d^2T}{dx^2}\right) \tag{5.5}$$

where
- c = specific heat of material
- a = thermal conductivity = $\lambda/(\rho c)$
- ρ = specific weight
- λ = heat conductivity of material

It may not be easy to solve this differential equation analytically when the boundary condition is complex. Usually, the numerical calculation of secondary differential equation is performed after this secondary differential equation is converted to a finite-difference equation by Taylor expansion.

Assuming that the temperature of a certain plane i in the depth direction of the plate is T_i and the temperature of a plane that proceeds by the minute distance of Δx into the inner direction of heat flow is T_{i+1}, the minute distance of Δx to outer direction is T_{i-1}, as shown in Figure 5.70. The transient heat conduction expressed by Equation 5.5 can be converted to Equation 5.6.

$$T_i^{(j+1)} = T_i^j + a\frac{\Delta t}{(\Delta x)^2}\left(T_{i-1} - 2T_i + T_{i+1}\right) \tag{5.6}$$

According the boundary conditions at the ends of a plate, the varying temperature distribution in the plate can be obtained by numerically solving Equation 5.6.

For calculation of ϕ_{CH} and λ, coiled wire is the same as plate, if we know the ϕ_{CH} of the surface of the coil and its heat transfer coefficient on the process of raising the temperature of the coil. But even if coils partially touch, there are still many air clearances. Consequently, while the heat conduction in the direction of the circle of coiled wire can be determined only by the solid thermal conductivity of the coil, the heat transfer in the radial direction is decided by three factors. These are the

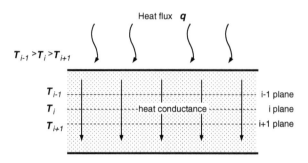

FIGURE 5.70 Temperature distribution in infinite length plate.

conduction of heat by contact of wires, the convection of heat in the air clearance between wires, and heat radiation between wires.

Usually some effective heat conduction models on the packed bed are reported in manuals of heat transfer engineering. These models can only be used for spherical solids and therefore there exists no model that can be applied for wire. We can calculate the real thermal conductivity of wires by means of a method of inverse calculation of the λ through the differential equation of heat conduction (cylindrical axis indication) by using measured data of heat distribution in coiled wires.

5.4.3.2 Furnace Scale-Up for Commercial Production

While the tests were conducted in an experimental furnace, the method of scale-up for commercial application by applying the rule of similarity is important to obtain high efficiency in a production furnace. In this section, the conditions to realize the rule of similarity between the test furnace and the high efficiency production heat treatment furnace are discussed. The conditions to realize the rule of similarity between the test furnace and the highly efficient production heat treatment furnace and the condition to realize the rule of similarity of heat transfer phenomena between the two furnaces compared are as follows:

1. *To realize the geometrical similarity*

The relation, $D_1/W_1 = D_2/W_2$, in Figure 5.71 must be realized. As the dimensions of the above-mentioned experimental furnace are D_1 (average value) = 1.37 m, W_1 = 2 m, the relation, $D_1/W_1 = 1.37/2 = 0.685$, can be obtained. If $D_2/W_2 = 0.685$ in another high performance furnace, then consistency of the geometrical similarity between the two furnaces can be realized.

2. *To derive the nondimensional control equations*

The relation of geometric positions of heat-treating material and RT alone determines the term of radiation, because the ϕ_{CH} model that is applied to radiation of RT furnace has no term for the gas surface area. Because the rule of geometric similarity between the experimental furnace and the high performance heat treatment furnace is realized by the relation of previously described items in this section, the two furnaces have the same value of ϕ_{CH}. The rule of similarity on heat transfer in the furnace can therefore be decided only by the term of convection.

Design Guidelines for High Performance Industrial Furnaces

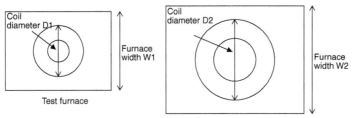

FIGURE 5.71 Test furnace and high performance heat treatment furnace.

The heat transfer coefficient in the furnace consists of six parameters, as shown in Equation 5.7:

$$\alpha = f(L, V, \rho, \mu, \lambda, C_p) \tag{5.7}$$

where
- L = key length
- ρ = density of liquid
- λ = heat conductivity
- V = average flow rate
- μ = viscosity of liquid
- C_p = average specific heat of liquid

The dimension analysis of Equation 5.7 by the Buckingham theorem provides that the parameters are 6 – 3 = 3. Consequently the well-known nondimensional control equation of convective heat transfer can be obtained as the following expression:

$$\mathrm{Nu} = f(\mathrm{Re}, \mathrm{Pr}) \tag{5.8}$$

The nondimensional control equation for two furnaces can be maintained equal by applying Equation 5.9:

$$\mathrm{Nu}_x = 0.33\,\mathrm{Re}^{1/2}\,\mathrm{Pr}^{1/3} \tag{5.9}$$

3. *To derive the nondimensional boundary conditions*

From Equation 5.8, Nusselt numbers Nu can be maintained equal by combining the values of Reynolds number Re and Prandtl number Pr between the two systems. In this, Pr of these systems can be maintained equal under the same kind of gas, because all the parameters are determined by the kind of atmospheric gas. If the values of the Re of the two systems are the same, then the conditions mentioned above to realize the rule of similarity on heat transfer in furnaces are satisfied.

From the following relation, the average diameter of coil in the test furnace $D = 1.37$ m and the average flow velocity around the coil in the test furnace $V = 3.36$ m/s, we can obtain the following relation:

$$\mathrm{Re}_1 = \frac{\rho V D}{\mu} = 30.8 \times 10^4$$

where $\rho = 1.205$ (kg/m³) and $\mu = 0.0180 \times 10^{-3}$ (Pa·s).

Correspondingly, the Re of the high performance heat treatment furnace is

$$Re_2 = 66944V \times 1.55 = 30.8 \times 10 \qquad (5.10)$$

If the average diameter of the heat-treating coil in the high performance furnace is 1.55 m, then

$$D_2/W_2 = 1.55/W_2 = 0.685 \qquad (5.11)$$

Consequently, the selection of fan capacity can be decided as $V = 2.97$ m/s from Equation 5.10 and a high performance heat treatment furnace can be designed that has the same characteristics of heat transfer as the test furnace by adopting the furnace dimension of $W_2 = 2.26$ m from Equation 5.11.

5.4.3.3 Test Design of Heat Treatment Furnace

This section discusses the design dimensions necessary for the actual RT-type heat treatment furnace and, in particular, the design dimensions for general use on both batch furnaces and continuous furnaces, as well as an outline of the heat treatment that is applied based on the results obtained by development of the high performance furnace. The heat treatment results in a 30% reduction in CO_2, and in a low NO_x level, which satisfy environmental regulation.

The features of minimum loss of waste gas and uniformity of RT temperature are realized by using the RT-type regenerative burner. Minimum loss of waste gas and uniformity of RT temperature are realized by using the RT-type regenerative burner, and as a result maximum heat transfer to heating materials is obtained, as well as optimum agitating fans in the furnace. Uniformity of convection heat transfer in the surrounding heating materials is realized by using agitating fans, which maintain uniform flow of the atmospheric gas in the heat treatment furnace. Uniformity of temperature of both furnace and heating materials can be realized by means of the above-mentioned features. Stable furnace operation and uniformity of RT temperature can be realized by adoption of the heating control system suitable for the RT-type regenerative burner.

The actual design method is outlined below:

1. The flow of design (Figure 5.72)
2. The study of user's specifications
   a. Heat source: kinds of fuels, calorific values, supply pressure, etc.
   b. Use materials for heat treatment: name, quality, and shape (maximum size, minimum size, dimensions, and weight of standard materials)
   c. Temperature: normal temperature, minimum temperature; objects to be kept within the required temperature range, number of measuring points, and measuring methods
   d. Treating capacity: the specification of capacity is required in terms of kg/h or t/h in the case of continuous furnaces and kg/batch or t/batch in case of batch furnaces; if not, alternative terms (for example, monthly treating amount and operation time) are required

Design Guidelines for High Performance Industrial Furnaces 315

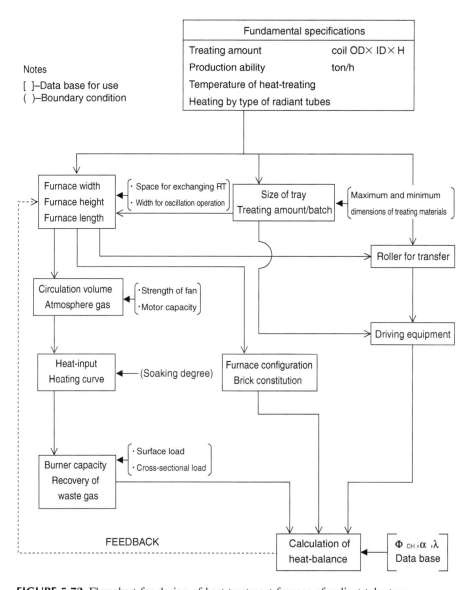

FIGURE 5.72 Flowchart for design of heat treatment furnace of radiant tube type.

Annual operation time = 12 months × 25 days × 8 hours × operating time 70% = 3360 h

Annual production rate = 600,000 pieces/year × single weight 50 kg/piece/annual operating time 3360 hours
= 892.86 kg/h

Capacity = 900 kg/h

e. Cooling water: which source of water (industrial water, well water, city water, others) is available and the temperature, capacity, and pressure of water supply must be confirmed; with industrial water, the water quality must be checked, because accidents due to corrosion may occur
f. Temperature curve: temperature curve must be carefully checked, because of occasional errors during inspections
g. Others: fluctuation of current voltage, an emergency current supply, atmosphere in furnace, control units of temperature, cranes, setting place, accessory equipment, etc. must be confirmed. Also specify other issues, if any.

Based on these specifics, a plan and the fundamental specifications can be drawn up.

3. Fundamental plan of design, which is based on the following considerations:
    a. Countermeasures for energy saving: installation of preheating zone (particularly in case of continuous furnace); heat recovery from trays and heating materials; reduction of the accumulated heat in walls and the radiation heat from walls by selection of insulated materials
    b. Countermeasures for labor saving: installation of automatic charging of materials, automatic oil feeding, etc.
    c. Legal regulation: confirmation of regulation values of NO_x, SO_x, and so on, which can vary according to location; noise regulation values: select burners and blowers with low noise specifications
    d. Furnace configuration: there exist various configurations according to the type of continuous furnace; for a semibatch heat treatment furnace for coiled wire, both heating and cooling are conducted in the same chamber
    e. Others: saving operating cost, countermeasures to product liability (PL) regulation, maintenance, etc.
4. Deciding fundamental specifications
    a. Deciding furnace width: adoption of the width of heating materials + about 640 mm
    b. Deciding furnace height:
        1. See Figure 5.73; clearance between furnace ceiling and fan decided as indicated by Figure 5.74
        2. Dimension determined by reference to change of fan axis length owing to heat and slackness of fan blades (about 200 mm)
        3. In general about 290 mm but in some cases the length must be greater because of the relation of the heat uniformity of the heating materials and the angles between RT and the heating materials
        4. The same as that of 3
        5. The same as the diameter of RT
    c. Deciding furnace length (in the case of continuous furnace):
        1. Tray size (from sizes of heating materials)
        2. Loading weight per tray

Design Guidelines for High Performance Industrial Furnaces

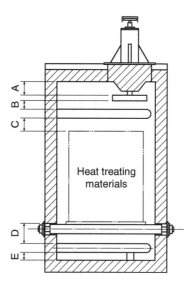

FIGURE 5.73 Dimension of furnace height.

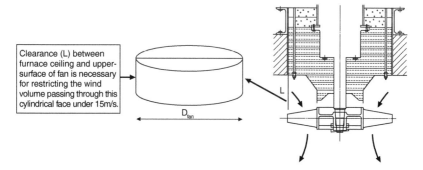

FIGURE 5.74 Clearance between furnace ceiling and fan.

3. Cycle time according to treating capacity (traveling speed)
4. Furnace length regarding the heating time, soaking time, and cooling time
   (Deciding furnace length for a batch furnace: length of heating materials + about 2 m)

d. Necessary heat capacity in furnace: in general, the heating materials in a heat treatment furnace are treated by the heat pattern as shown in Figure 5.75. The heat capacity necessary for conducting the heat pattern is calculated as follows
   1. Heating materials H_{CP}

$$H_{CP} = M(C_F - C_{room})$$

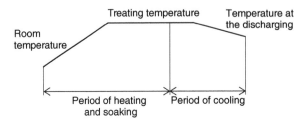

FIGURE 5.75 Heat pattern of heat treatment furnace.

C_F = heat content at heat-treating temperature
C_{room} = heat content at room temperature
M = mass of heating materials

2. Heating metal fittings in furnace H_{MF}

$$H_{MF} = M_{MF}(C_F - C_{exit})$$

C_F = heat content at heat-treating temperature
C_{exit} = heat content at temperature of discharging point
M_{MF} = mass of metal fittings

3. Heating furnace wall H_{WL}

$$H_{WL} = C_{INS}M_{INS}(T_F - T_{exit})$$

T_F = average temperature of wall at treating
T_{exit} = average temperature of discharging point
C_{INS} = specific heat of insulator
M_{INS} = mass of insulator

4. Emission heat capacity from furnace wall H_{emis}

$$H_{emis} = qA\Delta t$$

q = unit emission heat
A = surface area of furnace
Δt = time of heating and soaking

5. Heating atmosphere gas H_{circ}

$$H_{circ} = \sum_j \eta \Delta T_j \Delta t$$

Design Guidelines for High Performance Industrial Furnaces

η = unit consumption
$\Delta T j$ = average temperature rise of jth element
Δt = operating time

6. Heat loss by cooling water H_{loss}

$$H_{loss} = \eta c_{water} \Delta t_{water} \Delta t$$

η = unit consumption
c_{water} = specific heat of water
Δt_{water} = temperature difference of water
Δt = heating and soaking time

7. Other losses (6) × 10%
8. Total heat capacity $H_{total\ loss}$ = (6) + (7)

$$H_{total\ loss} = 1.1\ \eta c_{water} \Delta t_{water} \Delta t$$

9. Total combustion amount Q_{comb}

$$H_{total\ loss} = (viii)/\eta_c = 1.1 \frac{\eta}{\eta_c} c_{water} \Delta T_{water} \Delta t$$

η_c = efficiency of combustion

10. Unit fuel consumption $Q_{fuel\ consumption}$

$$Q_{fuel\ consumption} = H_{total\ loss}/M_{coils}$$

M_{coils} = weight of coils

  e. Positions of fans in furnace: the positions of fans can be determined by the rule of geometrical similarity with the test furnace.
  f. Capacity of fan: the ratio of region downward from the total volume of wind generated by the fan was confirmed to be 0.39 under the actual data of wind rate distribution, as shown in Figure 5.76. The capacity of the fan can be decided by the assumption that half of the total volume of wind generated by fan flows into the region. The capacity of the fan Qm³/min can be expressed by Equation 5.12.

$$Q = AV_{av} \cdot 60 \qquad (5.12)$$

where A[m²] is the shaded area in Figure 5.77, and V_{av}[m/s] is the average wind rate around the coil equal to that of the test furnace.

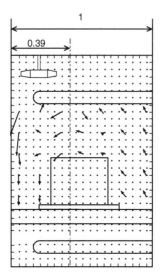

FIGURE 5.76 Distribution of wind rate generated by fan.

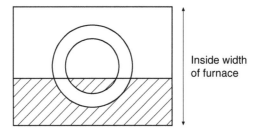

FIGURE 5.77 Region of "down blowing" in furnace.

g. Calculation of heating and cooling: calculation of heating and cooling can be conducted by using ϕ_{CH} of surface of coils and the heat conduction coefficient of inside coils. In this case, the calculation of discrete differential equation of heat conduction can be done on a personal computer. Details can be seen in the paper "The Method of Numerical Calculation of Non-stable Heat Conduction" (The Iron and Steel Institute of Japan).

5.4.4 Melting Furnace

5.4.4.1 Energy Savings and Exhaust Gas Regulation

It is required that the fuel used for aluminum-melting furnaces be cost-effective, and often oil such as heavy oil Type A is used. Aluminum-melting furnaces are batch-operated, and the temperature inside the furnace is raised from 700°C to approximately 1150°C, when the melting of aluminum starts. As a means of recovering waste heat, metal heat exchangers are used in most cases, but recently regenerative burners using balls or honeycombs have become available, making it possible to build higher performance aluminum-melting furnaces.

Table 5.24 shows the comparison of the energy efficiency of a conventional aluminum-melting furnace and a high performance furnace of 40 t/ch (tonnage per charge) each.

On reduction of NO_x (converted to 12%O_2), oil burning regenerative burners used for aluminum-melting furnaces must comply with NO_x values regulated not only internationally (180 ppm, converted to 12%O_2) but also domestically (around 130 ppm). Two types of burners can satisfy these regulations. One is a concentrated collision flame oil burner (F1). This type of burner reduces NO_x by lowering the flame temperature. It does this by jetting fuel from the burner center and by expelling in the furnace atmosphere from around the combustion air. This type of burner has good combustion efficiency, but the NO_x value is slightly higher. The second is the distributed collision flame oil burners (F2). Combustion air is jetted out of the burner center and fuel is jetted from around the center, which results in distributed flames, thereby lowering both the flame temperature and the NO_x. With this type of burner, the NO_x value is lower but the combustion efficiency in the low furnace temperature zone is somewhat inferior. The test results of a kerosene oil burner burning 2093 MJ/h are shown in Table 5.25.

On reduction of CO_2, the application of regenerative burners to an aluminum-melting furnace will lead to a reduction of energy by 30%. The heating source of an aluminum-melting furnace is primarily fossil fuel, and the following is the trial calculation assuming a reduction of CO_2 made feasible by this reduction of fuel consumption in Japan.

Assumptions for the calculation:

1. Aluminum demand in Japan: $\dot{M}_{Al}$ =3 million t/year
2. Fuel unit consumption of a conventional furnace: Q_{fuel} = 2930 MJ/t
3. Ratio of melting furnaces in Japan with regenerative burners: r = assumed as 50%
4. Combustion heat of fuel oil: H_f = 42.2 MJ/kg

TABLE 5.24
Comparison of Energy Efficiency

	Heat Exchanger	Average Temperature of Combustion Air	Furnace Efficiency	Waste Heat Recovery Rate	Energy Saving Rate
Conventional furnace	Metallic recuperator	200°C	40%	17%	30%
High performance furnace	Regenerative burner	720°C	60%	63%	

Both capacities are the same, 40 ton/ch (tons per a charge).

TABLE 5.25
Test Result of Kerosene Oil Burner

	NO_x Value (Converted to 12% O_2) ppm	Air Temperature, °C	Furnace Temperature, °C
Oil burner of concentrated collision flame type (F1)	40–90	800	1060
Oil burner of distributed collision flame type (F2)	20–40		

5. Volume of combustion gas generated: $V_c = 13.5$ m$_N^3$/kg of fuel oil
6. Percentage of CO_2 in the combustion gas: $F_{CO2} = 11\%$

According to the above assumptions, the reduction of fuel oil consumption is expressed as follows:

$$R_{fuel} = 0.3 r \, \dot{M}_{Al} Q_{fuel} / H_f = 31,200 \text{ t/year}$$

and the reduction of CO_2 gas is

$$R_{CO_2} = 1000 R_{fuel} V_c F_{CO_2} M_{CO_2} / 22.4$$

where M_{CO_2} = molecular weight of CO_2 = 44.

5.4.4.2 Size Reduction

In the case of an aluminum-melting furnace, the cold material charged inside rises as high as 2 to 3 m, and a large space is required inside. But once it is melted, the depth of the molten bath will be 40 to 60 cm, meaning the inner space required for the furnace is large for the former but small for the latter, a consequence of a batch furnace. Easier working conditions inside the furnace also have to be considered. A larger door will improve the working efficiency of charging the material, removing slag and cleaning the furnace interior, but a smaller door will not only hinder ease of working but will also be a source of trouble by being easily damaged.

In consideration of all these factors, it is necessary to achieve a reduction in the size of a furnace so that the initial construction costs and the amount of heat loss are reduced. One of the ways to realize this is to obtain an appropriate distance between burners and aluminum material. The test results are shown below:

Design Guidelines for High Performance Industrial Furnaces

Overview of the tests

Objective: tests for the comparison of melting time and fuel consumption by changing the distance between burners and material
Equipment: inner dimensions of the melting furnace (in mm):

length: 3200 × width: 1200 × variable height: (H = 1750 to 1150)

Burner Type: one pair of regenerative burners using oil as fuel (alternative 20 s operation)
Combustion Capacity: 1255 MJ/h (flame length = 1400 mm)
Position: flame directly downward from the furnace ceiling

Test results

The comparisons of the test results of melting time and fuel consumption in percentage are shown in Table 5.26.

Summary

Despite the fact that the temperature inside the furnace when the melting was completed was 1020 to 1030°C and the temperature during melting was almost the same, the results were best for the case of "Middle" (H = 1.45 m) regarding both melting time and fuel consumption. The following reasons may account for this.

High: Emissive heat transfer from the internal gas layer inside the furnace was the most effective but since the temperature inside the furnace was as low as approximately 1000°C, the expected efficiency might not have been achieved.
Middle: The heat transfer effect by convection of flame collision was very effective and the radiated heat from the ceiling and from the gas layer both worked as expected. This probably explains why this produced the best results overall.

TABLE 5.26
Test Results of Melting Time and Fuel Consumption[a]

Distance Between Burners and Material	Melting Time, %	Fuel Consumption, %
High, H = 1.75m	96	98
Middle, H = 1.45m	86	89
Low, H = 1.15m	100	100

[a] With 100% for the case of short distance.

Low: Possibly excessive direct flames on the surface of the cold aluminum and the molten bath led to imperfect combustion to the detriment of heat transfer by convection. Because the ceiling was low, the radiation effect from the ceiling was also low and not much could be expected either of the radiation heat from the gas layer, which probably caused the heat transfer to be the worst.

The test results stated above indicate that the size reduction should be arranged in such a way that the distance between burners and material equals the length of the flame.

5.4.4.3 Method of Improving the Heat Transfer Efficiency inside the Furnace

By improving the heat transfer efficiency and by reducing the furnace size, economic benefits in terms of initial costs and fuel cost can be anticipated. This is because of reduced heat loss from the furnace wall. Below are approaches to improve the heat transfer efficiency:

- Optimization of the distance between burners and material: The optimum design is to set the distance between burners and material equal to the length of the flame expected when the burners are at 100% combustion. The flame length depends on the burner capacity and the type of fuel used. However, the distance between burners and material increases as melting advances. This is the reason an average distance should be adopted. Care should be taken not to make the distance too short; otherwise imperfect combustion will take place.
- Adoption of burners with a flame of high momentum: The aluminum material will receive heat transferred by convection from the burner flames and radiation heat transferred from the furnace wall and the gas layer. In the case of aluminum-melting furnaces, the furnace temperature is in the midrange, and it is understood that heat transfer by convection is more effective than radiation heat transfer in terms of heat transfer rates, which is the reason it is necessary to choose high momentum burners whose jet flow velocity is larger (jet flow velocity of 100 to 120 m/s).
- Optimization of the gas flow inside the furnace: The gas inside the furnace flows out mostly through regenerative burners, and it is necessary to decide their location and angle so that the optimum heat exchange with the material is ensured. It should be designed to prevent the combustion gas shortcuts and flows to the burner port before it hits the material.
- Tilting of the burners themselves: This involves changing in a certain frequency the place where the flames hit the material by mechanically swinging the burners themselves. It was learned from tests that the flame shift-speed of 3.3 m/min on the surface of the material produced the best results and showed improvement by 5% both in terms of melting time and fuel consumption compared with the case without tilting. It was also

learned that any faster speed would cause turbulence in flame forming and, therefore, lower performance. Any slower speed would also produce an inferior effect.

- Agitation of the molten bath: A thermal layer phenomenon toward the deeper direction of the molten bath becomes stronger because some solid aluminum remains immersed in the molten bath at the furnace bottom and because the heat transfer rate of an aluminum molten bath is low. Moreover, the bath surface will be covered by dross that forms through the process of melting and drifts up, which makes it difficult for the radiated heat from the atmosphere and the furnace wall to enter the bath. Therefore, it is necessary to angle the jet flow from the high momentum burners and try to remove the floating dross cover. At the same time, a forced agitation system of the bath is necessary to break the thermal layer and increase the heat transfer of radiated heat.

5.4.4.4 A Design Example of High Performance Aluminum-Melting Furnace

An approach for a high performance aluminum-melting furnace is to design the furnace based on a conventional furnace, but with modifications to integrate the foregoing elements for high performance furnaces. Therefore, the design procedure of a high performance aluminum-melting furnace is shown in Figure 5.78. Figure 5.79 is an example of a high performance aluminum-melting furnace. Its major features are reduction of CO_2 by more than 30% as energy saving, reduction of furnace floor area by 16% as size reduction, low NO_x meeting environmental regulations, and reduction of metal loss as energy saving and waste reduction, as well as a melting furnace adopting the regenerative burner combustion system. To realize the above-mentioned high performance of the furnace, a combustion system that exploits the combustion characteristics of high temperature air is adopted.

The furnace volume is determined by the working conditions, such as the charge volume of the bulk cold material and the required casting volume. But thanks to the improvement of melting rate by the increased heat transfer, the reduction of furnace floor area is realized within the allowable limits of working conditions and the depth of the bath.

The maximum heat transfer effect of aluminum that has a low capability of radiation is obtainable by collision heating the center of the material heap for melting, using directional burners of high velocity jet flow (high momentum) coupled with a tilting movement.

Raising the entry angle of high velocity jet flow (high momentum) burners, and by their sheer velocity, the floating dross covering the surface of the bath is removed and the heat transfer by convection is improved. Also, by agitating the bath using a forced agitation system, the thermal layer is broken and the radiation heat transfer efficiency is improved. The following types of regenerative burners use oil as fuel and are adopted for aluminum-melting furnaces. They are chosen depending on the type of furnace.

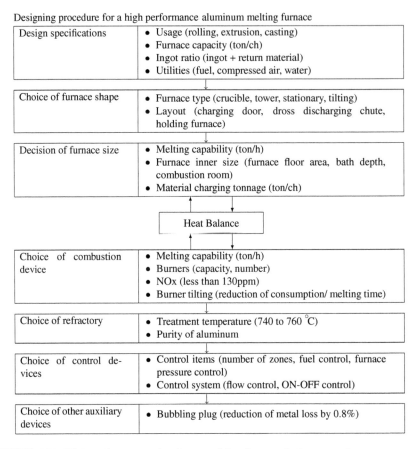

FIGURE 5.78 High performance aluminum melting furnace design procedure.

- Concentrated collision flame oil burners (F1) (NO$_x$ lower than 100 ppm, converted O$_2$ 12% mol). This type uses high velocity jet flow (high momentum), indispensable for melting, and in parallel two-stage combustion and combustion gas self-circulating system. This type of burner can easily meet the regulated NO$_x$ levels. It also has a high combustion performance.
- Distributed collision flame oil burners (F2) (NO$_x$ lower than 50 ppm): Because the fuel is blown from outside the combustion air, the flame temperature is lower; thereby NO$_x$ is lower than in the two-stage combustion type of burners. However, the combustion performance at the low temperature zone is somewhat questionable.

Finally, by blowing in treatment gas from porous plugs in the furnace floor, the separation of slag from the bath and drying of the slag are carried out to reduce metal loss incidental to the removal of the slag. By agitating the bath, thereby breaking the thermal layer, the metal loss attributable to the oxidation by the excessively high temperature of the bath surface is also reduced.

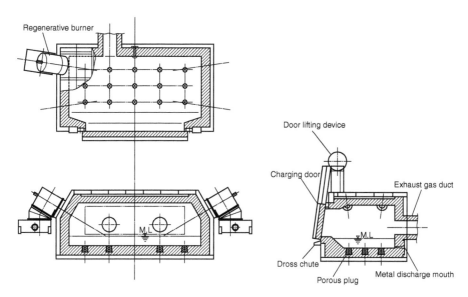

FIGURE 5.79 High performance aluminum-melting furnace.

5.5 FIELD TRIALS AND EXPERIENCES OBTAINED THROUGH FIELD TEST DEMONSTRATION PROJECT

The Third Conference of Parties on the Framework Convention on Climate Change (Kyoto Conference on the Prevention of Global Warming, COP3) was held in December, 1997, and the "Kyoto Protocol" was adopted. This protocol stipulated that Japan should achieve more than 6% reduction of greenhouse gases from the 1990 emission level within the period from 2008 to 2012. To meet the target, it was considered crucial to reduce the total amount of energy consumed by the industrial sector, which accounts for about 50% of total energy consumption in Japan. Various industries, in particular the iron and steel industry, which uses industrial furnaces, and the petrochemical industry, were expected to work toward achieving energy saving.

Under these circumstances, one part of the research and development of the High Performance Industrial Furnace Development project, which had been promoted from 1993 by NEDO, was completed in 1998. The project resulted in technological advances including new technologies applicable to energy saving purposes. MITI and NEDO initiated a joint research program called High Performance Industrial Furnace Field Test Demonstration Project for a period of 3 years from 1998 to 2000 as a step toward the practical application of the technology on the basis of the results of the development project mentioned above.[3] This program is now under way.

This section details facts related to the program and the data, although partial, obtained from the program to the end of 1998. In addition, information on the applications for the field test in fiscal year 1999 is included. The project aims to increase the use of High Performance Industrial Furnaces through the evaluation of the actual implementation of the system and verification of the system as a practical technology.

5.5.1 OUTLINE OF THE FIELD TEST PROJECT

The range of results available from the joint research activities under this project and the attributes of the data obtained may be defined briefly as follows. This High Performance Industrial Furnace Field Test Demonstration Project[4] dealt with various types of furnaces — heating furnaces, heat treatment furnaces, and melting furnaces for industrial use. The data for a pair of furnaces, an existing furnace and a furnace in use for production, modified to function as (or be replaced by) a high performance industrial furnace, were compared. The intent was to identify the effects and operating characteristics of the new system introduced as a high performance industrial furnace.

Detailed data on the normal operation of the furnaces in similar situations were to be collected in the first year of the joint research, followed by collection of long-term running data for the furnaces in the second, third, and fourth years.

Figure 5.80 shows the execution scheme of the current High Performance Industrial Furnace Field Test Demonstration Project. For the execution of joint research activities, we first accepted a wide range of applications for this joint research project, then examined the contents of the applications to determine which facilities would be suitable. Table 5.27 shows the basic schedule of this project. The budget of the project for fiscal 1998 to 1999 was 3.9 billion yen (approximately $32 million U.S.) per year (see Table 5.28). The results of case analysis of the joint research project on the basis of the types of furnaces, industries, fuel, and burners are discussed below.

5.5.2 APPLICATIONS FOR THE FIELD TEST IN FISCAL YEARS 1998 AND 1999

Figure 5.81 shows the number of furnaces adopted for field tests in fiscal 1998 and 1999. The types of furnaces under the field test are divided into the three types, namely, heating furnaces, heat treatment furnaces, and melting furnaces. In addition, heating furnaces are grouped under the following categories: continuous heating furnaces, batch-type heating furnaces, and ladles. Heat treatment furnaces are grouped under the following categories: continuous heat treatment furnaces, batch-type heat treatment furnaces, and gas treatment furnaces. Melting furnaces form an independent group. Partly due to the difficulty of sorting the data according to the groups, the seven furnace types mentioned above were designated for the analysis.

Analysis of the applications seemed to indirectly identify the respective industries' expectations toward the ease of introduction of and the great merits of high performance industrial furnaces. The number of heat treatment furnaces exceeded other types, with heating furnaces the second largest group. The total number of applications adopted in fiscal 1999 was slightly lower than in fiscal 1998. The share ratios of the respective furnace types in the 1998 and 1999 applications were almost the same.

Design Guidelines for High Performance Industrial Furnaces 329

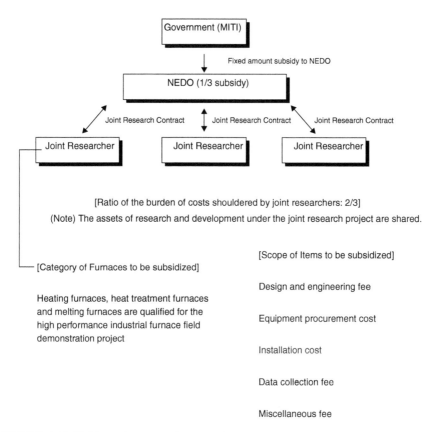

FIGURE 5.80 Field test demonstration execution scheme.

TABLE 5.27
Schedule of Field Test

	Schedule							
	1998	1999	2000	2001	2002	2003	2004	2005
High Performance Industrial Furnace Field Test Demonstration Project	Installation/Data collection			Long-term running data collection				

The case analysis for 1998 on the basis of industry types is shown in Figure 5.82. The introduction of high performance industrial furnaces in the iron and steel industries accounted for 50% (29 cases), followed by the metallic machine manufacture industry at 31% (18 cases). Breaking down the percentage for the iron and

TABLE 5.28
Budget of Field Test

	Budget (Equipment Expense + Management Expense)
1998 fiscal year	3.92 billion yen
1999 fiscal year	3.90 billion yen

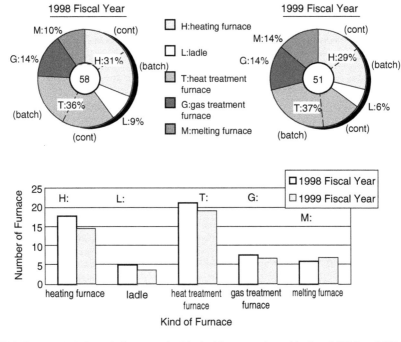

FIGURE 5.81 Relationship between the kind of furnace adopted in fiscal 1998 and 1999 and number of each furnace type.

steel industry further, the share consists of 42% blast furnace makers and 38% casting or forging steelmakers — the two groups account for 80% of the share of the iron and steel industry. For the breakdown of the percentage of the metallic machine manufacturing industry, the share consists of 38% of the aluminum industry and 33% of the heat-treating processing industry. To promote the application in a wider range of industries in the future, it will be necessary to conduct analyses in industries other than blast furnace makers in the iron and steel industry, the casting or forging steel industry, the aluminum industry, and the heat-treating processing industry.

A case analysis on the basis of the types of fuel is shown in Figure 5.83. Eight types of fuels were involved in fiscal 1998, with LNG (13A) outnumbering others and accounting for almost half of the total. Liquid fuels accounted for only a little

Design Guidelines for High Performance Industrial Furnaces

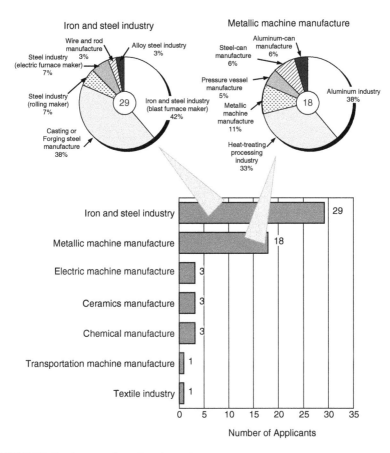

FIGURE 5.82 Product number of each product type.

less than 25%. Generally speaking, small to medium-sized businesses mostly use liquid fuels. The results of the field test conducted this time especially in relation to liquid fuels are very important.

A case analysis on the basis of burner types is shown in Figure 5.84. It shows the types of burners, on the basis of regenerative media types, adopted in the high performance industrial furnaces under this field test project. The burners are largely divided into two types: direct firing and radiant tube burners. Gas treatment furnaces are classified as an independent category because they employ a honeycomb as the regenerative medium, and the device itself cannot be clearly categorized as a type of burner. As a result, direct-firing burners accounted for 66% of all burners, while radiant tube burners and gas treatment furnaces, respectively, accounted for 21 and 13%.

As for the structures of regenerative media, ball-type media accounted for 48% and honeycomb-type media 36%. With the ratio for gas treatment furnaces included, the honeycomb media ratio increased to 50%. The structure of regenerative media is an important factor when determining the performance of high temperature air

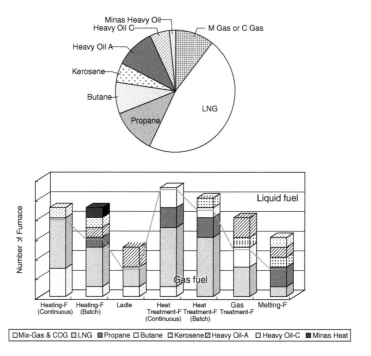

FIGURE 5.83 Relationship between kind of fuel and number of applicants.

Burner type and Media type			Number of F
Direct Firing	Ball		27
	Honeycomb	twin	9
		single	2
Gas Treatment Furnace			8
RT	Honeycomb		10
	Ball		1
	other		1
Total			58

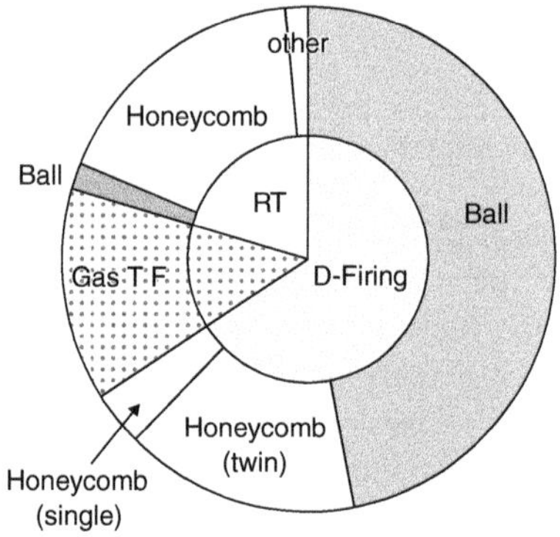

FIGURE 5.84 Relationship between kind of burner and number of applicants.

combustion. The results of this field test must be carefully examined to identify the functional advantages and practicality of the structures.

Combustion methods can be divided into a direct injection method in which fuel is directly injected into a furnace and an indirect injection method. The combustion method is selected in accordance with the shape of the furnace, the type of fuel, etc. The combustion method is the key factor influencing NO_x emissions, combustion quality, flame length, and in-furnace temperature distribution. The difference between the types of combustion methods is also discussed in the report of "High Performance Industrial Furnace Development Project." Joint research activities include the study and analysis of combustion methods.

5.5.3 Characteristic Aspects of the 1998 Field Test Project

The objectives, purposes, application methods, and effect of the introduction of high-performance industrial furnace technologies with respect to the seven types of furnaces were examined by taking typical cases in fiscal 1998 as examples.

With continuous heating furnaces, nine cases were considered. Among them, four were high performance industrial furnaces modified by employing regenerative burners only in the preheating zones or the heating zones corresponding to preheating zones. They were all for blast-furnace makers, aiming to save energy, to reduce NO_x emissions, and to reduce size. In one case, a two-furnace operation was changed to a one-furnace operation. In another case, a three-furnace operation was modified to a two-furnace operation to realize a successful reduction of the facility by reducing the heating time. Furthermore, with two electric furnace makers, the length of their furnaces was reduced by more than one third. (One of the makers modified the furnace by reducing the length by more than half and increasing the charging temperature.)

For batch heating furnaces, as a result of the field test of high performance industrial furnaces of this type, the energy-saving effect was seen to be the most outstanding. Most of these furnaces were batch-type heating furnaces for forging, in which waste heat recovery was rarely executed and heating and machining processes were repeated resulting in very large unit fuel consumption. With these furnaces transformed into high performance industrial furnaces, their energy consumption of 2.0 to 25.1 GJ/t (500 to 6000 Mcal/t) prior to the modification was reduced by 30 to 60%.

Ladle preheating (heat-retaining) units are designed for the purpose of saving energy and uniform refractory heating. The applications in fiscal 1998 included five ladle preheating units (including a ladle for heat-retaining purposes). The capacity ranged from 40 to 200 t/ch. The energy-saving effect attained by high performance industrial furnaces in these cases amounted to between 36 and 57%.

Continuous heat treatment furnaces consist of many furnace types, many materials, and many shapes of objects to be treated. Not only the workpieces to be treated but also other objects, such as trays, are often heated in these furnaces, with the heat capacity of such trays, etc. in some cases larger than the workpieces themselves. Therefore, the energy-saving effect varies enormously from one case to another. Nine cases of 11 used radiant tube burners. In particular, two cases aimed at drastic energy savings by changing from electric heating to gas-fired heating with a resulting reduction of energy consumption of more than 60%.

Batch heat treatment furnaces, as with continuous heat treatment furnaces, consist of many furnace types, materials, and shapes of objects to be treated. Many of those furnaces are used for processing automotive parts. Four cases of ten used radiant tube burners. In these types, too, four electric heating furnaces were changed to gas-fired furnaces with resulting achievement of an energy saving of 55 to 80%.

Gas treatment furnaces refer to direct combustion gas treatment furnaces (incinerators) used as deodorizing machines for exhaust gas generated at painting or printing plants. The purpose of the application for the test was to realize energy savings. Table 5.29 shows the kinds of exhaust gas treated in fiscal 1998. The treatment gas in the gas treatment furnaces themselves is combustible. The fuel required for the facility incorporating this high performance industrial furnace was used only to raise the temperature on facility start-up and the establishment of an ignition flame, and it became apparent that a minimum amount of fuel was necessary to operate the modified system.

Melting furnaces adopted in fiscal 1998 numbered six, which included five aluminum-melting furnaces and one glass-melting furnace. The purpose of modifying the melting furnaces into high performance industrial furnaces was to achieve energy savings and size reduction. Two aluminum-melting furnaces, which had not been equipped with a waste energy recovery function, recorded an energy savings of more than 60%. Other furnaces with different types of waste energy recovery systems realized energy savings of about 30%.

5.5.4 Effects of Modifications in the Field Tests

The effects achieved in the field tests are arranged on the basis of the data included in the respective reports submitted by 58 joint researchers. However, it is necessary to study the conditions under which the data were collected to compare with the data in the same table. Follow-up investigation and analysis of this aspect will be conducted in the future. An outline of the results to date is given.

The waste heat recovery rates were markedly improved from those of conventional furnaces. All types of furnaces achieved heat recovery rates higher than an average rate of 60% (Figure 5.85).

Waste Heat Recovery Rate:

$$\eta = \frac{V_A \left(C_{PAo} \cdot T_{Ao} - C_{PAi} \cdot T_{Ai} \right)}{V_E \left(C_{PEi} \cdot T_{Ei} - C_{PEa} \cdot T_{Ao} \right)} \times 100 \tag{5.13}$$

TABLE 5.29
Kinds of Treatment Gas

The beverage can outside painting exhaust gas
Function material film machine drying line exhaust gas
Aluminum-powder drying kiln exhaust gas
Clothes pigments drying exhaust gas
Calcium carbonate baking exhaust gas
Construction materials painting exhaust gas
Metallic drum can painting drying exhaust gas

Design Guidelines for High Performance Industrial Furnaces

where

V_A	= total airflow (m_N^3/h)
V_E	= total exhaust gas flow (m_N^3/h)
T_{Ai}	= inlet air temperature (°C)
T_{Ei}	= inlet exhaust gas temperature (°C)
T_{Ao}	= preheating (outlet) air temperature (°C)
C_{PAo}	= specific heat of preheating air (kJ/m³ K)
C_{PAi}	= specific heat of inlet air (kJ/m³ K)
C_{PEi}	= specific heat of inlet exhaust gas (kJ/m³ K)
C_{PEa}	= specific heat of exhaust gas at inlet air temperature (kJ/m³ K)

An energy saving of slightly more than 10% was achieved in the continuous heating furnace group. The value is reasonable with respect to the unit fuel consumption for this type of furnace, as it was already an energy efficient system.

An energy-saving ratio higher than 90% was achieved in the gas treatment furnace group. The object to be heated in the gas treatment furnace itself is combustible and fuel is required only for ignition. In other words, combustion of fuel seldom occurs throughout the operation except during ignition. For this reason, the energy-saving ratio of this group exceeded 90%.

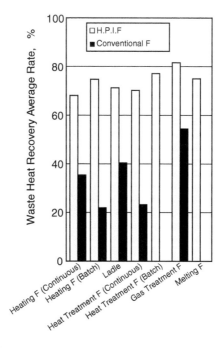

FIGURE 5.85 Waste heat recovery rate.

The energy-saving ratios of the other furnace groups increased virtually in proportion to the respective processing temperatures (Figure 5.86).

Rate of Change of Unit Energy Consumption before and after the Modifications:

$$\eta = 1 - \frac{U_{am}}{U_{bm}}$$

where

- η = energy saving ratio
- U_{am} = unit fuel consumption after modification
- U_{bm} = unit fuel consumption before modification

Because continuous heating furnaces consume a large amount of fuel, the amount of energy saved was substantial, although the energy saving ratio appeared to be low numerically. On the contrary, the amount of energy saved in batch-type heat treatment furnaces was small as both the size of each facility and fuel consumption were generally small (Figure 5.87), i.e.,

$$\Delta E = \eta E_{bm}$$

where

- ΔE = amount of energy saved
- E_{bm} = energy consumption before modification

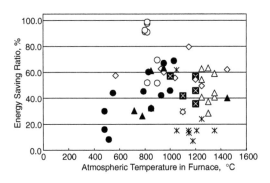

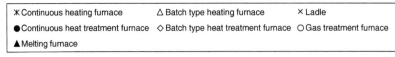

FIGURE 5.86 Relation between atmospheric temperature in furnace and energy-saving ratio.

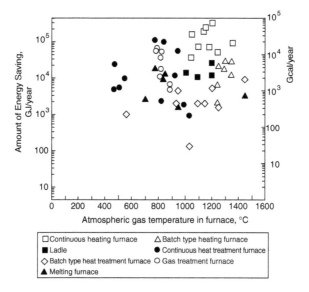

FIGURE 5.87 Relation between atmospheric temperature in furnace and amount of energy saving.

The legally permitted values of NO_x emissions vary with the type and capacity of furnaces. Table 5.30 shows the typical legal restriction values concerning the 58 furnaces included in the field test project. Many local governments have designated standard values in excess of these restriction values, which are problems for facility operators. Because facility operators see no point in lowering their NO_x emission levels below the current legal restriction values or standards, it is beyond the scope of the work of this field test project to define clearly the technological capacity of

TABLE 5.30
Legal Restriction Value of NO_x

Items	Conversion O_2, %	Legal Restriction Value, ppm	Notes
Heating F	11	100–180	In some localities within the
Heat treatment F	11	80–180	metropolitan area, strict values
Melting F	12	130–200	lower than $1/3$ of the legal restriction values are applied

NO_x ppm: $C = (21 - O_n)/(21 - O_s) \times C_s$

O_n = Conversion O_2 %

O_s = Measured O_2 %

C_s = Measured NO_x ppm

high performance industrial furnaces to lower NO_x emission levels (as to how low the level can be reduced). However, it must be noted that the NO_x emission level was reduced by application of these newly developed high performance industrial furnaces to the same extent or more than that reached with conventional technology. With the conventional furnaces, the NO_x level increased to a level of several hundred ppm when the combustion air was heated to a high temperature (800°C or higher), so that heating the combustion air to a high temperature could not be used in industrial technology (Figure 5.88). The main achievement of this field test was that the emission level was below 50 ppm (converted to 11% O_2) in the range higher than 1000°C.

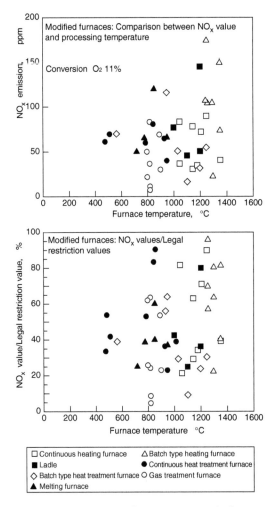

FIGURE 5.88 Relationship between atmosphere temperature in furnace and NO_x emission.

5.5.5 SUMMARY

This section dealt with the applications adopted for fiscal 1999 and the results of the High Performance Industrial Furnace Field Test Demonstration Project in fiscal 1998 in the form of intermediate results. Specifically, the results mainly classified the cases and typical applications with respect to their performance, which are based on the data obtained from the 58 joint research cases in fiscal 1998.

It was estimated, on the basis of the research conducted in 1998, that more than 30,000 furnaces throughout Japan could be modified into high performance industrial furnaces. The number of furnaces this field test project will be able to cover over the 3-year project period will account for 0.5% (about 160 cases) of the total number. This percentage is very small; however, the field test project will certainly serve as crucial information for the dissemination of this technology. Therefore, from now on, the authors plan to emphasize the amalgamation of information aimed at promoting the application of high performance industrial furnaces.

REFERENCES

1. S. Matsunaga, *Theory and Application of Heat Transfer in Furnace*, Tokyo Techno Center, Tokyo, 1979.
2. The Japan Society of Mechanical Engineers (JSME), *Data Book: Heat Transfer*, 4th ed., JSME, 1986.
3. NEDO ed., *Yearly Reports of High Performance Industrial Furnaces Development Project (1998, 1999, 2000)*, Tokyo, Japan. Available at URL: http://www.tech.nedo.go.jp/
4. H. Murakami, T. Saito, J. Hayashi, and A. Hida, *Proc. IJPGC Conf. FACT*, 23, 239–244, 1999.
5. J. W. Deardorff, *J. Fluid Mech.*, 42(2):453–480, 1971.
6. H. Taniguchi et al., *Proc. Int. Heat Transfer Conf.*, 1986, pp. 757–762.
7. M. Uede, K. Tanaka, M. Imada, and K. Murakami, *Proc. ASME IJPGC Conf.* 2000, paper no. 15082.
8. T. Ishii, C. Zhang, and S. Sugiyama, *Proc. IJPGC Conf., Vol.* 5, ASME EC, Denver, CO, Nov. 1997, pp. 267–278.
9. M. Morita and T. Tanigawa, *Chem. Eng. Articles*, 26:227–235, 2000.
10. A. Hida, H. Nozaki, and T. Morita, *Nippon Steel Corp. Tech. J.*, 363:49–54, 1997.

6 Potential Applications of High Temperature Air Combustion Technology to Other Systems

6.1 INTRODUCTION

The specific emphasis of high temperature air combustion so far has been on different kinds of furnaces and boilers for industrial applications. In all cases significant impact was achieved from the point of view of significant energy savings (up to about 60%), reduced pollutants to the environment including CO_2, and smaller (compact size) of the equipment. It is also clear that we have begun to think about the traditional definition of flame as one that gives heat and light. Indeed, for the high temperature air combustion case one can have a flame without visually observing it since the flame can be colorless. For much of the 20th century the technological advances on the combustion of fuels have not been at the level of those in the electronics industry; see Table 6.1.

This chapter provides our vision of the application of high temperature air combustion technology to other systems. High temperature air combustion technology, in principle, can be utilized in almost all kinds of combustion, power, and propulsion systems,[1-3] in addition to its demonstrated use in industrial furnaces, discussed in Chapters 1 through 5. Some development projects are already in progress to apply the HiTAC technology principles for use in coal burning boilers, heavy fuel oils, waste fuels and chemicals, and solid wastes. Some of the current development efforts using high temperature air combustion technology are given in Table 6.2.

Visions on potential applications of high temperature air combustion, including those already developed and currently being developed, are given in Figure 6.1. The near-term future developments and applications of this technology are summarized below:

1. Stationary gas turbine combustion using isothermal combustion. Included here is combined cycle with elevated steam temperature, and cogeneration with elevated steam temperature, micro gas turbines and independent electric power generation, and gas turbine/stationary boiler integrated system.

TABLE 6.1
Major Advances in the Electronics and Energy Sector during the Second Half of the 20th Century

Decade	Electronics Industry	Energy Industry
1950s	Vacuum tubes	High energy conversion
1960s	Transistors	Large scale units
1970s	Integrated circuits	Environmental pollution
1980s	Printed circuits and large-scale integration	High efficiency
1990s	Very large scale integration	High efficiency and low pollution
2000+	Super large scale integration, compact and high density systems	Energy and environment conservation and health effects from the burning of fuels

TABLE 6.2
Current Development Efforts on the Application of High Temperature Air Combustion to Other Technologies

Application	Issues with Original Fuel Type/Material	Final State of Fuel or Products
Waste materials (fuel)	Difficult or unwanted fuel	Conversion to gaseous fuel of uniform composition from which thermal energy can be recovered
Refuse derived fuel (RDF), waste fuels, such as biomass, municipal, hospital, farm and industrial wastes, and chemical wastes	Difficult to burn Incineration is not the solution	Conversion to gaseous fuel Combustion of gaseous fuels results in very high combustion efficiency
Volatile organic compounds and odors	A pollutant to the environment	Complete destruction of the VOCs and odors without forming hazardous pollutants

2. Fuel reforming steam and high temperature combustion air.
3. Power generation from wastes and low quality fuels.
4. Combustion of low heating value fuels.
5. Fuel cells.
6. Household thermal appliances. Included here are radiant tube furnaces providing hot air by a blower and water heaters.
7. Engines for power and propulsion (e.g., automobiles).
8. Ash vitrification.

Potential Applications of High Temperature Air Combustion Technology

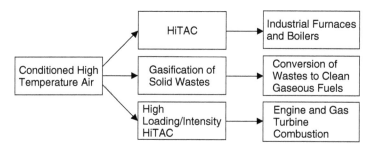

FIGURE 6.1 Applications of high temperature air combustion in furnaces and boilers, thermal destruction of solid wastes, and engines.

High temperature air combustion technology was initially developed for different kinds of heating processes. Now the HiTAC is being developed for use in the thermal destruction of solid wastes and turbulent high intensity HiTAC technology (high loading HiTAC) for use in gas turbine engines. The above processes are summarized in Figure 6.2.

High temperature air has been used to convert several kinds of municipal and industrial solid wastes into useful gaseous energy using high temperature air provided by the NFK manufactured Air Enthalpy Intensifier[4] (called AI) and steam. Some other potential applications include thermal destruction of hazardous chemical wastes and

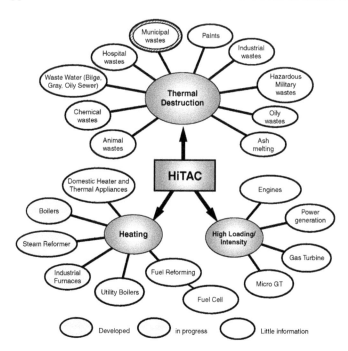

FIGURE 6.2 Potential applications of high temperature air combustion, including applications already developed or at present being developed.

military wastes, fuel reforming, waste water, ash vitrification, engines (such as stationary gas turbine and micro gas turbine, internal combustion), and compact-size combustion systems such as micro-combustors; see Figure 6.2. The chemical wastes include wastes generated by the petrochemical industries, paints, oils, and residues. The technology is particularly attractive in systems that require fuel lean combustion. It is expected that many new applications and innovations will evolve with further research and development focused on this innovative combustion technology.

We describe below the application of high temperature air combustion technology for the thermal destruction and energy utilization of negative-costs waste fuels.

6.2 COMBUSTION OF WASTES AND SOLID FUELS

Combustion of any moisture-containing organic materials occurs in the following four stages: drying, pyrolysis, gasification, and oxidation of volatiles. In the traditional case the combustion air used for solid waste fuels (municipal and industrial wastes and low grade coals) is characteristically preheated to about 300°C. The combustion reaction essentially involves two components: fuel waste and oxygen. A simplified combustion reaction can be represented as:

Hydrocarbon fuel (waste) + Oxidizer → Products + Heat + Pollution

All industrialized countries worldwide have been actively engaged in discussing the growing environmental problems, such as global warming, acid rain, and destruction of the ozone layer. Many of the pollutants are also known to cause adverse effects on human health. Indeed, some of the combustion-generated pollutants are carcinogenic and mutagenic. Attention must be given to environmental pollution in addition to conservation of natural resources. Recently, much attention has been given to recycling, and secondary or tertiary use of wastes and energy recovery from wastes. Emission of dioxins and furans from the thermal destruction of wastes is also an important issue from plastic-containing wastes. Principles of high temperature air combustion technology are also potentially attractive for use in the thermal destruction of plastic-containing wastes. The energy resource recovery technology applied to waste materials is one such constituent of the high temperature air technology. Technology used for transforming the organic portion of the materials present in wastes to attractive fuels also includes refuse-derived fuel (RDF).

The United States produces about 4 pounds per person per day of solid waste. In the United Kingdom the waste generated is about 1.5 kg per person per day. In Japan the amount of general waste disposed per person per day is approximately 1 kg. Wastes produced in Japan today are divided into two categories of industrial and general wastes in accordance with the Waste Disposal and Public Cleansing Act. General disposal of such large quantities of wastes is processed according to the following: 74.3% incineration, 11.3% conversion to resources, and 14.4% buried in landfills. The landfill option of the past for the disposal of wastes is becoming increasingly small everywhere in the world since it is difficult to secure land for the disposal of wastes. The future trend for solid waste disposal seems to be further reduction in the amount of waste produced, recycling, and cascade use of resources.

In addition, significant efforts continue to be made to convert wastes to fuels. In the following we describe how some of the important features of this technology have already been utilized for converting wastes to useful energy.

Local governments and businesses, which dispose of enormous quantities of industrial wastes, have started to examine practical methods to reduce wastes and to convert waste to energy by applying thermal recycling technology. These systems are already in operation in some areas. Power generation using waste as the fuel (one type of thermal recycling technology) is now considered to be a new energy-producing process. In Japan, the outline of "New Energy Introduction Plan" issues in 1994 set target levels for the waste utilizing power generation at 2 GW in 2000 and 4 GW in 2010. In the United States solid wastes alone have the potential to produce about 20 GW electricity with only 30% plant efficiency.

High temperature air, preheated to 1000°C and higher, has rarely been utilized in industrial practices for waste treatment and energy recovery. At such high temperatures most fossil fuels autoignite. If the heat recovered from the exhaust gas is used to preheat the combustion air, and the waste (organic portion of municipal, industrial, farm, and hospital solid wastes) is burned with this preheated air, then the enthalpy of the exhaust gases is said to be effectively used to increase the temperature in the combustion region. Recent studies at NKK Corporation, Tokyo Institute of Technology (TIT), and other institutions reveal many benefits of utilizing high temperature air for the thermal decomposition of several different kinds of wastes and coals using a regenerative high temperature air facility designed by NFK.[5] It is to be noted that, with some proper measures of heat recovery, prevention on the formation of dioxins and furans can be made by rapid cooling of the exhaust gases to low temperatures,[6] although the TIT group had not addressed this issue at the time of this writing.

Pyrolysis and gasification of solid fuels and wastes using elevated temperatures has been used for a long time. In former times, coal was gasified with the aid of gasifying material such as gas or steam using elevated temperatures. It should be noted that for pyrolysis the material is heated in the absence of oxygen. The TIT studies used high temperature air to gasify coal and solid wastes. They used a two-chamber arrangement with provision for feeding the waste from the side and hot air from the top in the top chamber. The two chambers were connected together with a throat, which also had ceramic balls. The syngas formed and the slag flowed over the ceramic balls into the bottom chamber. In this chamber the slag was collected and the syngas was removed from one side of the chamber. Any loss of ceramic balls into the bottom chamber had to be replenished from the top chamber via an opening for the solid fuel feed. Gases of low to medium calorific value have been produced from the gasifier using over 1000°C high temperature air. In these studies the high temperature air was obtained from a regenerator facility manufactured by NFK, Japan. The NKK and TIT demonstrations are device application of high temperature air to pyrolyze or gasify the solid fuels or wastes. If the wastes fuel or coal utilizes both high temperature and steam (or some other gasifying material), it is appropriate to consider the overall gasification and steam-reforming reactions. The high temperature air gasification and steam reforming with wastes may be represented by the following very simple reaction:

High Temperature Air Gasification

$$\text{Carbon (Hydrocarbon)} + O_2 + N_2 \rightarrow CO + CO_2 + H + H_2O + N_2 + [-\Delta Q_1] \cdots \text{Exothermic}$$

High Temperature Steam Reforming

$$\text{Carbon (Hydrocarbon)} + H_2O \rightarrow CO + H_2 + [+\Delta Q_2] \cdots \text{Endothermic}$$

where the − or + sign in the above equations, involved with ±Q, represents the usual meaning in thermodynamics, i.e., heat evolved from the system is given with a negative sign (exothermic reaction) and the heat supplied to the system is positive (endothermic). It should also be recognized that the above reactions are very simple representations. The exact mechanistic pathways for the reaction are much more complex than that given above.

In wastes containing moisture, high temperature air promotes reactions associated with pyrolysis, gasification, and steam reforming — all occurring simultaneously. Higher temperatures are advantageous for both pyrolysis and gasification and to form slag. It should be noted that the temperature of air should be higher than the melting point of ash for slag to form. However, much higher air temperatures can result in ash vaporization. In case the ash vaporization occurs, caution must be exercised to capture the submicron-sized particles from the syngas evolved. The properties of the gases evolved can be controlled somewhat with the aid of some gasifying agent. A schematic diagram of gasifier used for wastes and low grade coals is given schematically in Figure 6.3. The coal or waste is fed from the top into the gasifier. The waste rests on top of a grate so that the high temperature air can be fed from the bottom chamber. The TIT group used relatively large-size (about 2 in. diameter) ceramic balls in place of the grate and the high temperature air was supplied from the top of the gasifier. High temperature air was obtained from a regenerative air enthalpy intensifier manufactured by NFK. The syngas obtained had a heating value of approximately 1000 to 1500 kcal/m_N^3. No information is given on the removal of submicron-sized particles evolved from the gasification or pyrolysis process. Particle-containing gas stream can be cleaned for particulates using a suitably designed cyclone. The syngas produced is claimed to provide clean combustion in furnaces and combustors without any additional concerns for the waste and emission of undesirable pollutants.[5] A schematic diagram of the combined industrial gas turbine and furnace is shown in Figure 6.4. The gases evolved from the gasification unit can also be used in micro gas turbines for independent power generation;[3] see Figure 6.5. Many other designs for the gasifier can also be used in place of the slope-shaped gasifier. The relative merits or drawbacks of each system, or some other configuration, require further examination. The nonleachable slag produced from the gasification process can be used for building and construction material. In the TIT studies the low to medium gases produced from the pebble bed gasifier have been used in a boiler to produce steam. In principle, a visionary zero waste emission production facility, incorporating a waste to energy unit, can be

Potential Applications of High Temperature Air Combustion Technology 347

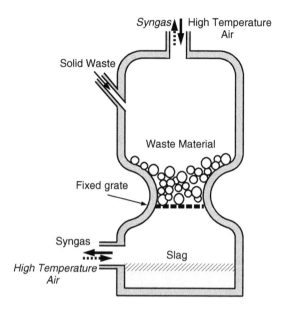

FIGURE 6.3 A schematic diagram of twin chamber "bed type" gasifier.

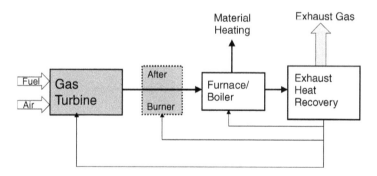

FIGURE 6.4 A schematic diagram of industrial gas turbine and furnace.

produced; see Figure 6.6. Of course, this is an ideal representation for a plant with the aim to stimulate and challenge engineers on the design of future facilities.

Much of what has been discussed so far involves the transformation of waste to syngas chemical energy in only one stage in the bed. A gasifying agent can be used to improve the calorific value of the gas produced. One such agent includes steam but another suitable gas can also be used. A schematic diagram of the steam-assisted gasification system is shown in Figure 6.7. The objective of the gasifying material is to further enhance the conversion of carbon in the waste (fuel) to syngas fuel. The syngas evolved can also be used for power generation, using, for example, high temperature air principles in micro gas turbines. Typical mole fraction of the steam in air in Figure 6.7 would be 10 to 20%, depending on the properties of the fuel.

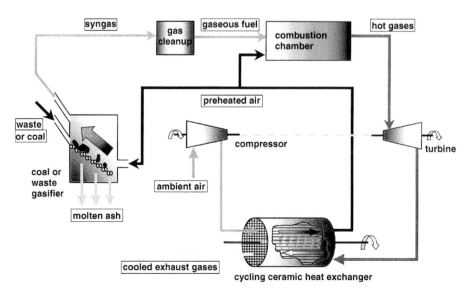

FIGURE 6.5 A schematic diagram of high temperature waste gasification and industrial gas turbine system.

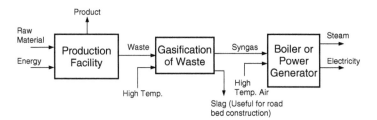

FIGURE 6.6 Prospects for future ultra-low waste generation from a production facility.

Part of the energy in the syngas may be used for high temperature air/steam preheating and the remainder may be used for power generation, steam generation, and industrial heating. With a proper design various undesirable constituents in the waste can be recovered, such as heavy metals (Fe, Cu, Al), inorganic compounds in the form of nonleachable slag, and sulfur and chlorine.[5] The proof of concept for the overall process shown in Figure 6.7 requires validation. The main features of this waste-to-energy conversion are as follows:

1. Conversion of solid wastes and low grade fuels (including coals and high-moisture-containing municipal and industrial wastes) to clean gaseous fuels having low to medium calorific value. The composition of the gaseous fuel can be adjusted via some control of the chemistry, e.g., gasification, pyrolysis, or reforming of the waste with steam and high temperature air, thus increasing the calorific value of the gases produced.

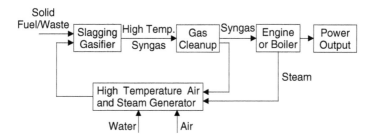

FIGURE 6.7 Integrated boiler/engine using syngas produced from solid waste using high temperature air and steam.

2. Large volume reduction of almost all types of wastes.
3. Permanent disposal method for solid wastes, including low grade coals.
4. Minimum amount of dioxin generation due to the reactor operation at high temperatures under reducing conditions.
5. Negligible amount of dioxin generation with the refined gas since almost all of the chlorine has been removed in the preprocessing stage. Any dioxin present will be decomposed in the high temperature air combustion process.
6. Rapid reduction of the exhaust gas temperature as the gases pass through a very large surface area honeycomb regenerator used to preheat the air. The regenerator acts as an efficient heat recovery system and prevents the formation of dioxins via rapid cooling of the exhaust gases to a temperature below the region where the dioxins are formed (see the discussion below on the formation of dioxins and furans). This is in contrast to the conventional system in which exhaust gases are cooled with a water spray so that the heat is lost to the exhaust and high levels of dioxins are formed.

Possible measures to reduce emissions of dioxins and related compounds from waste combustion furnaces are: (1) reducing the volume of such materials as they may generate dioxins and furans in the furnace, (2) preventing reproduction of dioxins and furans at low temperatures, (3) decomposition or elimination of dioxins and furans by catalysts or adsorption. Effective measures for the thermal destruction of wastes are to implement the three Ts (high Temperature, high Turbulence and long residence Time). This is consistent with the basic characteristics of high temperature air combustion. Fluidized-bed combustion can be suitable for burning refuse derived fuel (RDF). However, because of the low temperature operation of fluidized beds, proper measures must be taken to reduce the formation of dioxins and furans. A study conducted by IHI on simulated waste consisting of RDF and PVC laminated paper, using a fluidized-bed furnace and high temperature air, showed reduction of dioxin and related compounds by rapid cooling of the exhaust gases. This reduction is consistent with the following general information on the formation of dioxins.

6.2.1 FORMATION OF DIOXINS AND FURANS

The dioxins and furans are polychlorinated dibenzo-para-dioxins (PCDDs) and polychlorinated dibenzofurans (PCDFs). Dioxins and furans are chlorinated tricyclic aromatic compounds with similar chemical properties. Of some 210 dioxin compounds, only 17 compounds are of significant interest because of their toxicity. The most toxic compound is 2,3,7,8-tetrachlorodibenzo-para-dioxin (2,3,7,8-TCDD), and a few other dioxin compounds cause carcinogenic or co-carcinogenic effects on humans. Further, chlorine and fluorine in wastes is directly responsible for the formation and emission of dioxin and furan from incinerators. Chlorine in municipal solid waste can characteristically vary in the range of 0.3 to 0.7% by weight and the fluorine content can be 0.5 to 2% of the chlorine content. Factors influencing the formation of dioxin/furan are given below:

1. Oxygen: Thermal degradation experiments using polychlorinated biphenyls (PCB) have shown that the yield of dioxins increases with the increase in oxygen concentration. However, it is also reported that increased oxygen concentration decreases the yield of dioxins when dioxins are produced only from their precursors.
2. Temperature: The temperature band of 250 to 400°C is most critical for the dioxin and furan formation. Maximum levels of dioxins and furans are formed at a temperature of about 300°C. The levels of both dioxins and furans are about ten times higher at 300°C as compared with 100°C.
3. Water vapor: Dioxins and furans increase significantly (by some 76%) in wet samples as compared with dry samples.
4. Carbon: High carbon content in the fly ash results in higher formation of dioxins. Carbon acts as a sorbent for the precursor compounds for dioxin and furan formation.
5. Chlorine: High chlorine content in the fly ash favors the formation of PCDD/PCDF.
6. Hydrogen chloride/hydrochloric acid (HCl): Presence of HCl is not directly responsible for dioxin and furan formation, but it promotes the formation of chlorobenzenes and precursors of dioxins and furans. Excess oxygen leads to the formation of chlorine gas by the following chemical reaction:

$$2HCl + \frac{1}{2}O_2 \rightarrow H_2O + Cl_2$$

High levels of chlorine gas present simultaneously with fly ash favor the formation of PCDD/PCDF.

6.2.1.1 Refuse (or Waste) Derived Fuel

RDF or WDF (waste derived fuel) is a fuel produced from the combustible (organic) portion of the waste. The composition of RDF is better defined as compared with the solid wastes. The process of making RDF involves sorting the waste, followed by size reduction, pulverization, drying, particle size conditioning, and transformation to size and shape. In Japan, the term RDF is more frequently used than WDF. These two terms refer to the same thing, and there are no clear definitions or standards for them. It is expected that the relevant standards are likely to be established in the near future in Japan. However in the United States, the ASTM (American Society for Testing and Materials) classifies the RDF products in various categories;[7] see Table 6.3. The RDF-5 shown in this table is equivalent to the RDF generally referred to in Japan with a diameter of several tens of millimeters. The RDF can also be gasified to form medium to high calorific value gas using high temperature air/steam or oxygen enriched high temperature air/steam. The advantage here is clean gas with no particulates or fly ash.

6.2.1.2 Applied Technology for RDF

Formerly, municipal waste was simply incinerated for disposal. The energy was not actively utilized for power generation purposes or in heat utilization applications other than the incineration facility itself. Today, municipal waste combustion technologies are being further developed with recycled material with a goal approaching zero emission and using the energy evolved. The following section provides a comparison of characteristics between the RDF combustion technology and municipal waste incineration technology.

6.2.1.3 Changes in the Calorific Value of Municipal Wastes

Figure 6.8 shows the energy balance of RDF with the energy consumption required for the production of RDF (taken as the base level).[8,9] The manufacturing process

TABLE 6.3
Classification of Refuse Derived Fuel (RDF)

Classification	Description	Type
RDF-1	Municipal waste to be converted to fuel excluding bulky waste	—
RDF-2	Mixed municipal waste, 95% of which has been filtered through 6-in mesh. Iron components are removed or retained depending on the needs	Fluff RDF
RDF-3	Municipal waste, 95% of which has been filtered through 2-in mesh. Glass and inorganic substances are removed	
RDF-4	Combustible powdered waste, 95% of which has been filtered through a 2-mm (#10) mesh screen	Powder RDF
RDF-5	Combustible waste formed into pellets, cubes, or briquettes using compression method	Densified RDF
RDF-6	Combustible waste processed into liquid fuel	—
RDF-7	Combustible waste processed into gaseous fuel	—

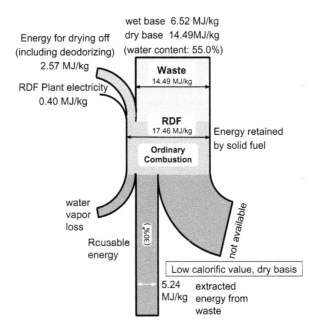

FIGURE 6.8 Approximate energy balance for an RDF manufacturing process.

of RDF includes a variety of steps and consumes a large amount of energy. To utilize RDF effectively as a fuel, it is very important to identify clearly the differences between the thermal energy obtained from RDF and the energy originally contained in the waste. According to the figure, the net amount of available energy is about 40%. Nagata and Ureshino[10] conducted an energy balance analysis of RDF power generation with respect to the life cycle of RDF. The analysis revealed that the drying process is the key to making RDF power generation reasonable with respect to life cycle assessment (LCA). In summary, it is advantageous to have low water content in industrial wastes from the viewpoint of energy balance and RDF production.

6.2.1.4 Problems with Waste Derived Fuel Production and Combustion

RDF technology was originally developed to cope with the problem of high levels of dioxins emitted from "batch type" small-size incinerators. The RDF manufacturing process needs to overcome the following problems to achieve the full practical use of RDF:

- Seasonal changes in the shape of the fuel and the amount of waste collected as well as local calorific value fluctuations of the wastes
- Treatment of odor and contaminated water at RDF manufacturing plants
- Energy saving considerations in the RDF manufacturing process (involving both the drying and pulverization processes)

Potential Applications of High Temperature Air Combustion Technology 353

- Optimization on the amount of lime as antiseptic and the relationship between the amounts of lime as antiseptic and as desalination agent
- Determination of the optimum shape of RDF (dimensions, density, strength, etc.)

In addition to the problems listed above, some combustion problems related to the manufacturing of RDF also remain to be solved. RDF contains lime to control the generation of toxic substances such as hydrogen chloride gas and dioxin. However, the quantitative database related to desalination, including the optimum desalination temperature and temperature dependency of desalination speed, etc. is not satisfactory. The problems related to RDF combustion process are as follows:

- Development of the control technology to suppress the formation of pollutants in a furnace (NO_x, SO_x, particulate, chlorine substances, toxic metals, etc.)
- Monitoring technology for pollutants (particulate, chlorine substances, toxic metals, etc.)
- RDF supplying method (stability of supply, feeding positions in the case of an air bubble fluidized-bed, etc.)
- Selection of the optimum combustion furnace (stoker, air bubble, or circulating fluidized-bed, rotary kiln, etc.)
- Combustion method (simple combustion, partial-combustion gasification, thermal cracking/combustion, thermal cracking/melt combustion, etc.)
- Measures for corrosion of heat exchanger tube

The problem of municipal waste is becoming more and more serious and the scope of the issue has expanded far beyond the range of RDF production and combustion technologies. Regarding RDF production technology, energy saving and environment protection measures must be developed. Combustion technology for RDF must be developed to achieve higher efficiency and more environmentally benign characteristics.

6.3 BURNING OF COALS AND LOWGRADE COALS

The traditional methods for the conversion of coal to energy are given in Reference 11. The developments of advanced methods for burning coal are under way at many institutions and laboratories, including International Flame Research Foundation (IFRF), Toyohashi University, Shizuoka University, University of Utah, Ohio State University, Pennsylvania State University, and MIT. An attractive feature of using high temperature air, in place of low or room temperature air, is to convert coal (including low grade coals and solid wastes) into clean gas of uniform gas composition and to eliminate the ash, particulates, and vapors that are formed during normal burning of the coal. The heating value of the gases evolved from coal gasification is around 2000 kcal/m_N^3. The slag is molten and can be vitrified. The issue here is either to increase the heating value of the gases evolved or to develop an efficient method for burning low calorific value gas. If these gases are burned using high temperature air, then the combustion efficiency is very high with minimum pollution. Stable flame can be obtained with HiTAC using very low heating value fuels. High temperature air can be effectively used for clean conversion of coal to gaseous fuels and its subsequent conversion to thermal energy. The HiTAC technology can also be developed to increase the calorific value of the gases evolved.

6.4 VOLATILE ORGANIC COMPOUNDS

High temperature air combustion technology can be effectively used for the destruction of volatile organic compounds (VOCs) and use the subsequent thermal energy evolved. Combustion systems have been used for the thermal destruction of odors associated with smelly vapors for a long time and are highly efficient and reliable for the removal of odors. However, in most cases auxiliary fuel is required in the combustion system since the concentration of hydrocarbon vapors in the odors (fuel) is not high enough to sustain combustion. The direct combustion process developed in the United States during the 1950s was widely used for the destruction of VOCs in the domestic market. The catalytic combustion process featuring a low combustion temperature of 300 to 350°C (as opposed to combustion temperature in excess of 750°C) was developed around 1970 for direct combustion process. Regenerative heat combustion process was developed in the United States in 1975 as a version of a direct combustion system with high heat recovery rate for greater energy savings. The market for low to medium waste gas treatment was expanded due to the stringent regulations on VOC emissions (implemented around 1990). Thus, the regenerative heat combustion process was employed in the market, which expanded rapidly in Europe and the United States. Recently, Japan has also made much progress for the destruction of VOCs using high temperature air combustion principles. Practical applications involve introduction or technical development and various types of equipment based on the three-tower system. Development efforts to reduce dioxin, based on the high temperature air combustion technology principles in exhaust, are in progress.

6.5 ASH MELTING

High temperature air combustion technology also has the potential for use in the ash melting and treatment process. The ash melting of bottom and fly ash lies in the range of 1300 to 1400°C . It will be quite natural to apply the high temperature air combustion technology to this process. The reburning process in a stoker furnace resembles, in principle, the high temperature air combustion process. Therefore, its application to ash melting and handling is conceivable. It should be noted that all of the above factors might not be applicable to some furnaces, depending on the type of regenerative media or available regenerative process. Therefore, one must design and develop a suitable type of the regenerative media as well as the process (for example, installation of preduster, automatic replacement system) from the beginning to develop this application and technology.

6.6 COMPACT BOILERS

The basic structure of a boiler consists of a radiant component and a convective component. The convective section of the boiler is much larger than the radiative section. Most of the heat transfer occurs in the radiative section of the boiler. Therefore, any attempt to reduce or eliminate the convective section of the boiler will result in a compact size boiler. The compact boilers have small or no convective section and take advantage of the features associated with the high temperature air combustion processes (very high heat flux plus feedback of the exhaust thermal energy into the combustor). The radiative heat flux with the high temperature air combustion is much higher (in excess of 500%) than the conventional boiler. The use of compact boilers in urban areas is particularly attractive from the point of view of space and low emission levels, including NO_x.

6.7 GAS TURBINE COMBUSTION, MICRO GAS TURBINES, AND INDEPENDENT POWER PRODUCTION

The characteristic performance goals for a gas turbine include time/range, complexity/weight and size, cost, reliability/durability, practicality, and emissions/signature. Combustion under lean premixed conditions is preferred for achieving low NO_x emission and high efficiency. However, under very fuel-lean conditions it is not possible to stabilize a flame. Flame stabilization becomes an issue when the combustor size is small. Flame quenching becomes an issue for smaller size because of the large ratio of surface area to volume of the combustor. This issue becomes even more important when using gases of very low heating value. Application of high temperature combustion air easily provides stable flame with even low calorific value gas (or liquid) fuel provided the temperature of the combustion air is high (above the autoignition temperature of fuel).[3] Since the autoignition temperature for most hydrocarbon fuels is much below 1000°C the fuel injected into the gas turbine combustion chamber will ignite and burn quickly. It is to be noted that the pattern factor associated with high temperature air combustion is very good because of the wider flame volume and low to negligible temperature fluctuations in the flame. The ignition delay can be tailored to satisfy the needs of the combustor. The high temperature air combustion technology, therefore, has good potential for application to stationary gas turbines, in particular micro gas turbines for independent power production (IPP). The emission of CO and hydrocarbons is negligible with most hydrocarbon fuels under high temperature air combustion conditions. Soot, particulates, and ignition delay can be an issue with high carbon ratio fuels so that one needs to develop means of burning residual fuel oils in stationary gas turbines. For gaseous and light liquid fuels soot and particulates are not a major issue. Some data are available in the literature on the use of heavy fuel oils with high temperature combustion air.[12] Emission of NO_x is also low since the thermal field uniformity in the combustion chamber is very high, which also results in a good pattern factor. The thermal field uniformity under high temperature air combustion conditions is far superior to any other known combustion method. An interesting gas turbine concept at present being developed at Nagoya University (Japan) is called the chemical gas turbine.[13] This concept combines the conventional gas turbine combustion technology and is based on converting the fuel to low molecular weight gases (and hence called the chemical gas turbine) prior to combustion using high temperature air. The combustion occurs under fuel-lean conditions. Limited studies on chemical gas turbine combustion have shown good concept for chemical gas turbine, but data are lacking under real engine-operating conditions. Further combustion challenges and potential for gas turbine combustion are given in Reference 14.

A schematic diagram of gas turbine using high temperature air combustion is essentially that shown in Figures 6.4 and 6.5. In this gas turbine configuration most of the heat from the exhaust gases is recovered and fed back into the furnace, boiler or the gas turbine. In the proposed system the exhaust heat is recovered, using, for example, a high efficiency heat exchanger and transported back to gas turbine combustion chamber. A boiler or furnace may be utilized in conjunction with the gas turbine to enhance overall system efficiency. The relative proportions of thermal energy transported back to the furnace, afterburner, and gas turbine require careful tailoring for the specific system under consideration.

6.8 PAINTS, OILY WASTES, AND HEAVY FUEL OILS

High temperature combustion technology offers significant potential for the thermal destruction of waste paints, oily wastes, and other viscous hydrocarbons. The technology can also be developed for the clean and highly efficient combustion of heavy fuel oils. At the time of writing this book only very limited data are available on heavy fuel oils using high temperature combustion air.[12] Much of the work on heavy fuel oil using high temperature air combustion has been conducted in Taiwan and Japan. Negligible to no data or experiences are available on waste paints using high temperature air combustion technology, although we expect this topic to see progress in the very near future. However, it can be expected that HiTAC should offer good benefits for further development of this technology for waste paints, oily wastes, and other viscous hydrocarbons.

6.9 FUEL CELLS

Steam reforming of hydrocarbons is attractive from the point of view of forming increased amounts of hydrogen production at low cost. High temperature air combustion technology has been demonstrated to be instrumental for providing higher levels of hydrogen using steam reforming. High levels of hydrogen production at the laboratory scale have been shown using characteristic features of high temperature air combustion. The challenge here is to produce very large quantities of pure hydrogen at low cost. It is expected that development efforts to produce hydrogen at low cost will increase in the future. Another challenge would be to separate out CO from the hydrogen produced. In the example given below, carbon (or hydrocarbon) is converted to useful chemical energy (CO and H_2) with the steam reforming process.[15] The results of equilibrium calculation show that increased levels of CO

Potential Applications of High Temperature Air Combustion Technology

and hydrogen are produced at high temperatures using high temperature steam. Of course, a tradeoff exists between temperature and pressure for a given fuel.

Example

Calculate the equilibrium conversion for the reaction:

$$C_{(s)} + H_2O_{(g)} \leftrightarrow CO_{(g)} + H_{2(g)}$$

at 1, 10, and 34 atm for each of these temperature: 800, 1000, and 1500 K.

Solution

The following constants are given:

$$K_{800} = 0.04406, \ K_{1000} = 2.617, \ K_{1500} = 608.1$$

from the following table. Also given are free energy (ΔG_T^0 kcal), and heat of reaction (ΔH_T^0 kcal).

T, K	ΔG_T^0, kcal	$\log_{10} K$	K	ΔH_T^0, kcal
298.16	21.827	−15.998	1.005×10^{-16}	31.382
400	18.510	−10.113	7.709×10^{-11}	31.723
500	15.176	−6.633	2.328×10^{-7}	31.978
600	11.796	−4.296	5.058×10^{-5}	32.164
800	4.966	−1.356	4.406×10^{-2}	32.371
1000	−1.912	+0.418	2.617×10^{-0}	32.445
1500	−19.107	+2.784	$6.081 \times 10^{+2}$	32.295

Basis: 1 mole of steam feed
Pressure: P atm
Converted moles of steam = x = fractional conversion

At equilibrium:

$$C + H_2O \leftrightarrow CO + H_2$$

Steam: $1 - x$ mol, CO: x mol, H_2: x, Total: $1 + x$ mol

Activities of gaseous components are considered equal to partial pressure, and the activity of solid graphite is unity. Thus,

$$a_{CO} = \frac{xP}{1+x}, \quad a_{H_2} = \frac{xP}{1+x}, \quad a_{H_2O} = \frac{(1-x)P}{1+x}$$

and the equilibrium constant:

$$K = \frac{a_{CO} \cdot a_{H_2}}{a_{H_2O}} = \frac{\left(P\dfrac{x}{1+x}\right)^2 \cdot P \cdot \dfrac{1-x}{1+x}}{P(1-x)} = P\frac{x^2}{1-x^2} \Rightarrow x = \sqrt{\frac{K}{K+P}}$$

Substituting the values of K and total pressure, give the following values of fractional steam conversion:

P (atm)	800K	1000K	1500K
	$K = 0.04406$	$K = 2.617$	$K = 608.1$
1	0.2054	0.8506	0.9992
10	0.0662	0.4554	0.9919
34	0.0360	0.2673	0.9732

These results illustrate well the trade-off between temperature and pressure effects, since increased temperature permits a higher operating pressure while attaining the same conversion. It demonstrates the importance of operating the reactor vessel at elevated pressure to decrease the size and cost. In this system the adverse effect of pressure can be compensated for by operation at a higher temperature. Similar considerations can be done for other hydrogen fuels, including liquid fuel (kerosene, approximate formula $C_{11}H_{21}$) and waste fuels. However, the calculations would require the gasification kinetics for the fuel examined. Data on some fuels can be obtained in advanced books on combustion.

6.10 HIGH TEMPERATURE AIR COMBUSTION USING PURE OXYGEN

Combustion temperature can be raised using pure oxygen in place of low oxygen concentration in air as the oxidant. Practical applications on the use of oxygen are limited to situations where very high temperatures are required, such as melting furnaces for glass and metal, because of the high cost for producing pure oxygen. However, with increased efforts for improvements in the production process for oxygen, significant cost reduction can be expected. One such attractive technology for O_2 production is membrane technology and vacuum pressure swing absorption (VPSA). Other membrane technologies are becoming attractive for producing oxygen at relatively low cost. Use of pure oxygen is more popular in the United States than in Japan because of the relatively low costs. The general trend in the United States has been to develop systems using oxygen-enriched combustion air rather than low oxygen concentration air.

Oxygen-enriched oxidant in itself has many benefits in the combustion of liquid fuels.[16] The use of oxygen-enriched atomization air, using, for example, an air assist atomizer, results in significant benefits with only a small quantity of oxygen. This increase of oxygen essentially makes a negligible change to the overall equivalence ratio of the fuel/air mixture. In spite of the benefits, however, the combination of oxygen-enriched oxidant and high temperature air results in harmful effects, such as intense oxidation on the inside of piping walls and very high temperatures. In addition, thermal dissociation of the gas also occurs at very high temperatures. These issues prevent using pure oxygen as the oxidant in the high temperature air combustion technology for other potential applications.

6.11 SUMMARY

Potential applications of HiTAC are envisioned to be much greater than originally sought at the onset of the program. Specifically in the application area of thermal destruction and energy utilization of solid wastes and low grade coals, promising outcomes have emerged from the fundamental and related laboratory-scale studies. During the past few years, the high temperature air combustion technology has been demonstrated, for most part, to effectively transform solid waste to clean energy as discussed in this chapter. Many of the applications cited here have been focused on the potential development of this technology for specific applications. However, to develop these applications fully, further interdisciplinary engineering work must be done to promote the proof of concept for a specific application prior to providing demonstration via bench-scale and pilot-scale experiments. In recent years, there has been a growing social demand for energy saving, reduction of pollutants emission, and prevention of global warming and the recurrence of the energy problem. In this aspect the HiTAC project has demonstrated excellent results on the enhanced performance of many kinds of industrial furnaces and boilers within the High Performance Industrial Furnace Development Project. In the future, it is expected that further international efforts will continue to develop other technological applications beyond those given in this chapter.

REFERENCES

1. A. K. Gupta. *Proc. of the Forum on High Performance Industrial Furnaces and Boiler* paper S2-5, pp. 66–1 to 66–18, Science Hall, Tokyo, March 8–9, 1999.
2. A. K. Gupta. *J. Energy Resource Technol., ASME,* 118:187–192, September 1996.
3. A. K. Gupta. *Flameless Combustion Workshop: Application to Gas Turbine Engines.* ALSTOM Power Technology, Baden, Switzerland, April 2, 2001.
4. T. Kiga, R. Hanaoka, M. Nakamura, H. Kosaka, T. Iwahashi, K. Yoshikawa, M. Sakai, K. Muramatsu, and S. Mochida. *37th AIAA Aerospace Sciences Meeting, Reno NV,* number 0730 in 99. AIAA, January 11-14, 1999.
5. K. Yoshikawa. *Proc. 2nd International Seminar on High Temperature Combustion in Industrial Furnaces,* pp. 1–22, Stockholm, Sweden, January 17–18, 2000. Jernkontoret- KTH.
6. A. K. Gupta. "*Incineration of Plastics*," *Polymers and the Environment*. Academic Press, New York, 2002.
7. M. Sigaki. *Waste Incineration Technology.* 2nd edition, Ohm Company, Ltd., 1998.

8. The Ministry of International Trade and Industry, Edition of The Minister's Secretariat. The statistic of petroleum consume structure in 1994. The Institute of Statistic on Trade and Industry, 1996.
9. T. Kagitani. *J. Japan Soc. Waste Manage. Expert*, 4(352):7, 1996.
10. K. Nagata and M. Ureshino. *J. Japan Soc. Waste Manage. Expert*, 4(282):7, 1996.
11. C. Y. Wen and E. S. Lee. *Coal Conversion Technology*. Addison-Wesley Publishing Company, Reading, MA, 1979, 330.
12. R. C. Chang and W. C. Chang. Research of high temperature air combustion fired heavy oil. *Proc. 2nd International Seminar on High Temperature Air Combustion*, Stockholm, Sweden, January 17–18, 2000.
13. T. Yamamoto et al. Prediction of NO_x emissions from high-temperature gas turbines: numerical simulation for low-NO_x combustion — a review. *Submitted to ASME J. Engineering for Gas Turbines and Power*, July 2001.
14. A. K. Gupta and D. G. Lilley. Combustion and environmental challenges for gas turbines in the 1990s. *J. Propulsion Power*, 10(2):137–147, March–April 1994.
15. E. Lois. National Technical University, Athens, Greece, Private Communications, 2001.
16. B. Habibzadeh and A. K. Gupta. Control of combustion in spray combustion. *Proc. International Symposium on Air Breathing Engines (ISOABE)*, Bangalore, September 6–10, 2001.
17. T. Hasegawa, S. Mochida, and A. K. Gupta. *J. Propulsion Power*, September 2001.
18. N. Konishi et al. *J. Propulsion Power*, December 2001.

Appendix A
Results of Investigations on the Current State of Japanese Industrial Furnaces

A.1 INTRODUCTION

To clarify the effect on energy saving and prevention of global warming by popularizing the achievements of the Development of High Performance Industrial Furnace Project, it became necessary to investigate the status of a number of industrial furnaces in Japan with regard to their energy consumption and heat efficiency. Consequently, the Energy Conservation Center conducted a questionnaire, interviews, and a statistical analysis of the investigations by request of the Japan Industrial Furnace Manufacturers Association.[1] The state of industrial furnaces and the effect of the popularization of the results of development are reviewed in the following sections.

A.2 ITEMS AND METHODS OF INVESTIGATION

The elements of the investigation comprise (1) a questionnaire, (2) an interview, (3) statistical analysis, and (4) an investigation of present conditions. In the statistical analysis, the number of industrial furnaces in Japan, energy consumption, and heat efficiency were assumed based on the results of the investigation. In the investigation of the present conditions, we extracted data from all the available related materials from the authorities concerned with the related industrial organizations, etc.

The elements of the investigation are shown in Table A.1, and the scope of the investigation is shown in Table A.2. The methods of investigation consist of questionnaire and interview inquiries with users and suppliers, as shown in Table A.3. As shown in Table A.2, the scale of facilities of the investigation objects is confined to combustion furnaces over 582 kW (500 Mcal/h) and electrical furnaces over 50 kVA. The food, textile, and paper pulp industries were excluded from the investigation, because almost all of them use heat sources of relatively small capacity, such as steam heating.

Interviews were conducted to improve the accuracy of the data, especially regarding heating furnaces, heat treatment furnaces, and aluminum-melting furnaces. That these were the objects of the development of high performance furnaces could

TABLE A.1
Items of Investigation

1. Kind of industrial furnace
2. Time of installation
3. Type of furnace (batch, continuous)
4. Type of heating
5. Heating capacity
6. Heat treating material
7. Annual treating amount
8. Average heating temperature
9. Kind of main heat source
10. Annual heat consumption by heat source
11. Unit heat consumption by heat source
12. Air ratio
13. Air preheater presence
14. Air temperature after preheating
15. Annual operation time
16. Heat efficiency
17. Number of industrial furnaces of the same specification

have caused incorrect calculation results on heat efficiency or energy unit consumption, if the calculations were conducted based only on the data obtained by the questionnaires.

A.3 RESULTS OF INVESTIGATION

A.3.1 RESULTS OF THE QUESTIONNAIRE WITH USERS

We sent the questionnaire to 8800 works. The questions focused on the current status of industrial furnace installation. We received replies from 2021 works, of which 513 use industrial furnaces with capacities over 500 Mcal/h or 50 kVA. The number of installed industrial furnaces by types of industries and the annual energy consumption is shown in Figure A.1. The number of installed industrial furnaces by types of furnaces and annual energy consumption is shown in Figure A.2. The unit consumption of the main industrial furnaces are arranged by types of industries in relation to the heating capacities (ton/h); typical examples are shown in Figure A.3.

A.3.2 RESULTS OF INTERVIEW WITH USERS

Results of the on-site interviews concerning the relationship among fuel unit consumption, heat efficiency, and heating capacities (ton/h) are shown in Figures A.4 through A.8. In most cases, both the fuel unit consumption and the heat efficiency show greater improvement, the higher the heating capacity. The values

TABLE A.2
Scope of Investigation

	Contents		Details			Notes
1.	*Type of industry*: 10 industries, listed in the right column, in mining and manufacturing industries according to Japan Standard Industry Classification	(1) (2) (3) (4) (5) (6)	Chemical industry Petroleum and coal products manufacturing industry Ceramics and soil products manufacturing industry Iron and steel industry Non-ferrous and metals manufacturing industry Metals products manufacturing industry	(7) (8) (9) (10)	General machinery instrument manufacturing industry Electric instrument manufacturing industry Instrument for transportation manufacturing industry Precision instrument manufacturing industry	Iron and steel industries: including casting and forging industries
2.	*Type of furnace*: 19 types of furnaces, listed in the right column, based on Japan Standard Products Classification	(1) (2) (3) (4) (5) (6) (7) (8) (9) (10)	Blast furnace Steel melting furnaces Arc furnaces Steel induction furnace Steel vacuum melting furnaces Nonferrous and metals melting furnace Nonferrous and metals induction furnace Nonferrous and metals vacuum melting furnace Metals soaking furnace Metals heating furnace	(11) (12) (13) (14) (15) (16) (17) (18) (19)	Metals heat treatment furnace Surface heat treatment furnace Surface treatment furnace Atmosphere gas denaturing furnace Metals sintering, roasting furnace Ceramics baking furnace Ceramics melting furnace Chemical industry furnace Drying furnaces	Excluding incinerators
3.	*Facilities scale*	(1) (2)	Combustion furnaces: facilities over 582 kW (2.1 GJ/h) Electric furnaces: facilities over 50 kVA			

of small furnaces with low heating capacities showed fluctuation, because these small furnaces treat many kinds of materials. They operate in different conditions and are therefore not stable. A summary of the above-mentioned analysis results is shown in Table A.4. The heat efficiencies of continuous furnaces are 52 to 55% for large scale, 37 to 45% for middle scale, and 27 to 38% for small scale. In contrast, the heat efficiencies of batch furnaces are at most 26 to 42% for large scale and 10 to 34% for middle and small scale, values that are generally much lower than expected.

TABLE A.3
Method of Investigation

1. User inquiries
   1.1 Questionnaire inquiries
      a. Selection of objectives for questionnaire:
         Works with over 30 employees in the ten industries listed in Table A.2 were selected based on the statistics on industrial works by the Management and Coordination Agency. At selection of the objectives, we used random sampling method suitable for statistical analysis.
      b. Inquiries table
         (1) The same format was used for both users and suppliers.
         (2) The format is mainly table type and selection type.
   1.2 Interview inquiries
      a. The objectives of inquiries were listed among heating furnaces, heat treatment furnaces, and aluminum-melting furnaces, which are the target of the development of a high performance furnace.
      b. Number of inquiries:
         (1) Heating furnaces: 6 types (continuous, batch, large/middle/small) × 3 cases
         (2) Heat treatment furnaces: 6 types (continuous, batch, large/middle/small) × 3 cases
         (3) Aluminum-melting furnaces: 3 types (large/middle/small) × 3 cases
2. Supplier inquiries
   2.1 Questionnaire inquiries
      a. The objectives are the members of Japan Industrial Furnace Manufacturers Association
      b. Questionnaire inquiries on the delivery results of industrial furnaces for the last 20 years
   2.2 Interview inquiries
      a. Selection based on the results of the questionnaire inquiries
3. Analysis of current state
      a. Investigation of the related statistics of Ministry of International Trade and Industry, Agency of Natural Resources and Energy, Environment Agency, etc.
      b. Investigation of data of the related industrial organizations: The Japan Iron and Steel Federation, the Japan Institute of Gas, the Japan Petroleum Federation, the Japan Institute of Electric Heating, the Japan Forging Industry Federation, etc.

A.3.3 RESULTS OF ESTIMATE OF NUMBER OF INSTALLED INDUSTRIAL FURNACES AND ENERGY CONSUMPTION

The number of operational industrial furnaces and energy consumption were each estimated statistically using the results of questionnaires based on the number of groups within various categories of industries. The results are shown in Table A.5. From the table, according to industrial companies with more than 30 employees, the number of combustion furnaces consuming over 582 kW (500 Mcal/h) and electric furnaces consuming over 50 kVA is estimated to be about 39,300 in total. The annual total heat consumption is 2901.6×10^9 MJ/year (693×10^9 Mcal/year) or 75×10^9 l/year of converted crude oil.

Appendix A

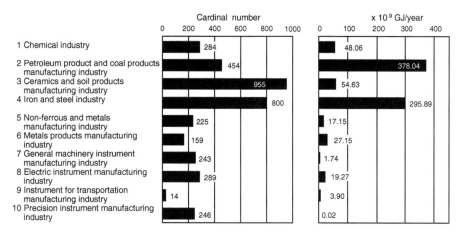

FIGURE A.1 Number of industrial furnaces and annual energy consumption by types of industry.

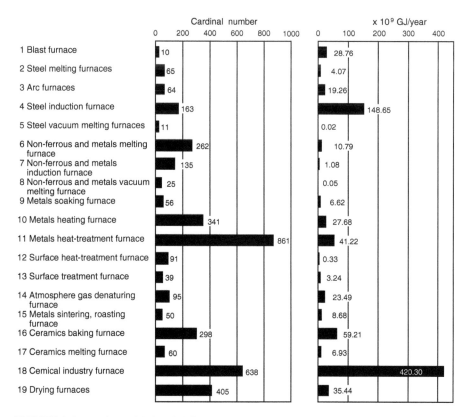

FIGURE A.2 Number of industrial furnaces and annual energy consumption by types of furnace.

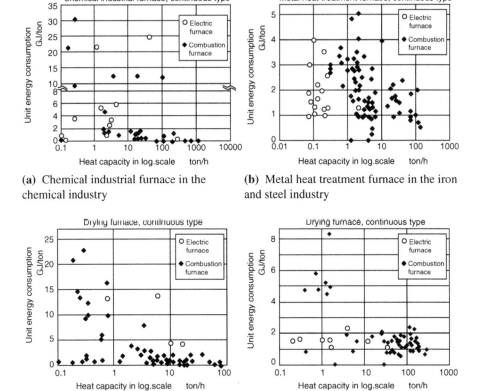

FIGURE A.3 Heating capacity and unit energy consumption of various furnaces.

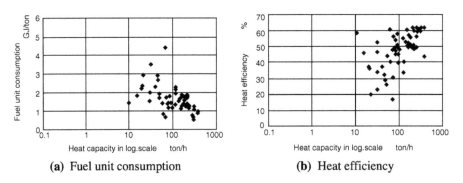

FIGURE A.4 Heating capacity of metals continuous heating furnace.

Appendix A

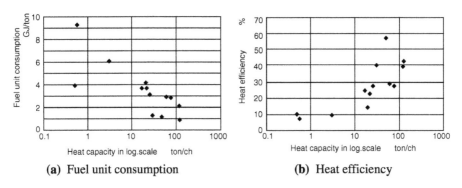

FIGURE A.5 Heating capacity of metals batch heating furnace.

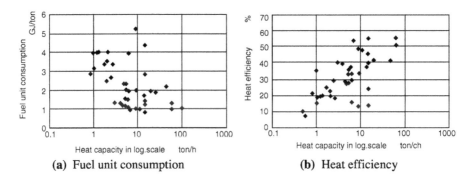

FIGURE A.6 Heating capacity of metals continuous heat treatment furnace.

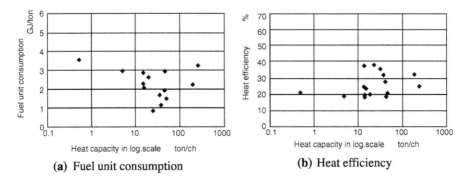

FIGURE A.7 Heating capacity of metals batch heat treatment furnace.

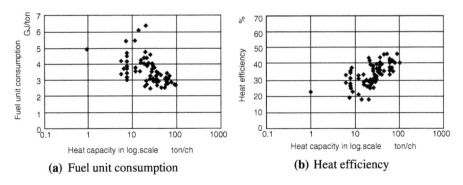

FIGURE A.8 Heating capacity of aluminum-melting furnace.

TABLE A.4
Heating Capacity, Fuels Unit Consumption, and Heat Efficiency According to Type of Furnace

	Type of Furnace	Classification		Heating Capacity	Fuel Unit Consumption, GJ/ton	Heat Efficiency, %
1.	Heating furnace	Continuous	Large	>100 ton/h	1.31	55
	(for metals use)	furnace	Middle	40–100	1.70	45
			Small	<40	2.18	38
		Batch furnace	Large	>80 ton/ch	1.49	42
			Middle	10–80	2.65	33
			Small	<10	6.28	10
2.	Heat treatment furnace	Continuous	Large	>20 ton/h	1.23	52
	(for metals use)	furnace	Middle	5–20	1.80	37
			Small	<5	2.17	27
		Batch furnace	Large	>25 ton/ch	1.67	26
			Middle	10–25	2.09	24
			Small	<10	2.97	18
3.	Aluminum-melting	Batch furnace	Large	>50 ton/ch	2.97	40
	furnace		Middle	15–50	3.47	34
	(for nonferrous use)		Small	<15	4.71	27

A.4 EVALUATION BASED ON RESULTS OF INVESTIGATION

A.4.1 Evaluation of Estimated Number of Industrial Furnaces

1. Facilities emitting soot that are controlled by the law of prevention of atmosphere pollution mainly consist of furnaces with capacity over 50 l/h

TABLE A.5
Annual Energy Consumption, Assumed Value

	Type of Furnace	Assumed Number of Furnaces	Assumed Consumption (10^3 GJ/year)	Converted Crude Oil (10^6 l/year)
1	Blast furnace	44	73.61	1,901
2	Steel melting furnace	853	21.95	567
3	Arc furnace	484	87.19	2,252
4	Steel induction furnace	1,843	21.67	560
5	Steel vacuum melting furnace	330	0.57	15
6	Nonferrous and metals melting furnace	2,360	72.59	1,875
7	Nonferrous and metals induction furnace	1,101	6.31	163
8	Nonferrous and metals vacuum-melting furnace	130	0.27	7
9	Metals soaking furnace	348	20.53	530
10	Metals heating furnace	3,809	118.01	3,048
11	Metals heat treatment furnace	9,159	198.02	5,114
12	Surface heat treatment furnace	1,981	5.79	149
13	Surface treatment furnace	474	68.70	1,774
14	Atmosphere gas denaturing furnace	1,562	273.73	7,069
15	Metals sintering roasting furnace	557	33.85	874
16	Ceramics baking furnace	6,512	658.85	17,015
17	Ceramics melting furnace	650	65.36	1,688
18	Chemical industry furnace (excluding incinerator)	2,527	1,012.17	26,141
19	Drying furnace	4,591	162.59	4,199
	Total	**39,316**	**2,901.75**	**74,941**

of converted crude oil or over 200 kVA according to the reports by the Agency of Natural Resources and Energy, even if the estimated capacity differs according to the kind of facility. The number of installed industrial furnaces (cardinal number), excluding boilers and waste incinerators, has decreased from the peak value in 1982 and was about 30,000 in 1992.

2. The number of industrial furnaces and facilities using fire and heat and the energy consumption of the facilities, investigated by the Agency of Natural Resources and Energy in 1975, show that the number of facilities, excluding boilers and wastes incinerators, was about 33,300.
3. Table A.6 shows a comparison of the following three investigations:
    a. The number of facilities emitting soot, as cited under the law of preventing atmosphere pollution in 1992
    b. Investigation by the Agency of Natural Resources and Energy in 1975
    c. The number of facilities assumed statistically according to this investigation in 1995

    The number of facilities with capacity using over 50 l/h converted crude oil or over 200 kVA based on this investigation is about 35,300. About

TABLE A.6
Comparison of Main Statistics on the Number of Industrial Furnaces

	Fiscal Year			
	(a) 1992	(b) 1975	(c) 1995	(c) – (b)
Data source	Facilities emitting soot as notified under the law of preventing atmosphere pollution	Investigation by Agency of Natural Resources and Energy	Assumed values based on this questionnaire inquiry	Amount increased during the two decades (from 1975–95)
Name of facilities				
Heating furnace		2,475	4,046	1571
Heat treatment furnace	8,503	8,078 5,603	15,757 11,611	7,679 6,008
Drying furnace	7,606	5,845	4,590	–1,255
Metals melting furnace	4,623	4,772	4,326	–446
Ceramics baking furnace	3,755	5,215	6,512	1,297
Petroleum heating furnace	1,619	1,764	2,526	762
Others	3,891	7,561	5,542	–2,507
Total of industrial furnaces	29,997	33,235	39,253	5,979

10% of the electrical furnaces under 200 kVA are included in this investigation because the capacities of the electrical furnaces in this investigation are confined to those over 50 kVA. Therefore, the values obtained during this investigation are similar to results published by the Agency of Natural Resources and Energy in 1975.

4. The comparison between data (b) and (c), shown in Table A.6, reveals that a large number of heating furnaces and heat treatment furnaces have been manufactured and introduced over the past 20 years. This is also confirmed by the return of questionnaires from the furnace manufacturers conducted in this investigation.

The number of industrial furnaces in Japan estimated from the questionnaires is about 39,300 when the minimum furnace capacities are limited to those over 50 l/h of crude oil equivalent or 50 kVA (or 35,300 when the minimum furnace capacities are limited to 200 kVA).

A.4.2 Evaluation of the Presumed Values of Energy Consumption of Industrial Furnaces

1. As Table A.7 shows, the total energy consumption in the energy flow in Japan in 1994 was 376×10^9 l in converted crude oil, of which industrial use was 50.0% (188×10^9 l), public use 25.8% (97×10^9 l), and transportation use 24.1% (91×10^9 l). Next, the change of energy consumption in the field of industry is shown in Table A.8. According to the table, the net energy consumption in 1994 was 172×10^9 l. The input fuel in the industrial field was 175×10^9 l, which is slightly less than that of the above-mentioned value, 188×10^9 l.[2,3]
2. Table A.9 shows the comparison of energy consumption of industrial furnaces in Japan based on several sets of statistics. In Table A.9, the current energy consumption of direct heating furnaces is 78.6×10^9 l in converted crude oil. The value is about 4.7% larger than the 74.9×10^9 l of this investigation. But this investigation does not include all heating facilities (excluding foods, textile, paper pulp and other industries); consequently, the value of assumed energy consumption, 75×10^9 l in converted crude oil, for industrial furnaces would be acceptable.

A.4.3 Consideration of the Results of Interviews — Efficiency of Industrial Furnaces

Table A.10 shows the average heat efficiency by capacity, types of furnaces, and their total average values obtained from the results of the interviews shown in Table A.4. From Table A.10, the following conclusions can be drawn:

TABLE A.7
Final Energy Consumption in Japan in 1994

Final energy consumption 376 (100%)	Industry 188 (50.0)	Iron and steel industry	44
		Chemistry industry	48
		Ceramics and soil industry	14
		Paper and pulp industry	11
		Other manufacturing industry	43
		Others	20
		Nonenergy	9
	Social use 97 (25.8)	Domestic	52
		Business	45
	Transportation 91 (24.1)	Automobile	80
		Others	11

TABLE A.8
Change of Details of Energy Consumption in Industries
(converted crude oil × 10⁹ l)

				Consumption					
Fiscal Year	Number of Works	Input	Generation, Collection, or Production	Total[a]	For Raw Material	For Boiler	For Direct Heating	Others	Output
1990	50,157	178	60	223 (168)	91	45	80	7	36
1991	50,702	182	59	228 (173)	93	46	82	7	31
1992	49,855	177	55	225 (174)	92	46	79	8	34
1993	48,281	176	55	223 (172)	91	46	78	8	34
1994	46,502	175	55	222 (172)	88	46	79	9	35

[a] The number in parentheses is the total net consumption, which excludes the amount converted from other fuels within the works.

TABLE A.9
Comparison of Some Statistical Values of Energy Consumption in Industrial Furnaces

	Content of Statistics	Energy Consumption (converted crude oil liters)
1.	Statistic of consumption structure of petroleum, etc. (MITI) (1994)	78.6 × 10⁹ for direct heating
2.	Investigation materials by Agency of Natural Resources and Energy (1975)	88.7 × 10⁹
3.	Assumed values based on questionnaire inquiries (1995)	74.9 × 10⁹

1. The larger the capacities of furnaces, the higher heat efficiencies become in every kind and type of furnace.
2. The heat efficiencies of continuous furnaces are higher than those of batch furnaces.
3. Among these, the heat efficiencies of the continuous metals heating furnaces are high in general and those of the batch metals heat treatment furnaces are low.
4. The average heat efficiencies are 26% in small-scale furnaces, 35% in middle-scale, 43% in large-scale, and 35% on average.

Appendix A

TABLE A.10
Average Heat Efficiency (%) by Kind, Type and Scale of Industrial Furnaces

Kind	Type	Scale			
		Small	Middle	Large	Average
Metals heating furnace	Continuous	38	45	55	46
	Batch	10	33	42	28
Metals heat treatment furnace	Continuous	27	37	52	39
	Batch	18	24	26	23
Aluminum-melting furnace	Batch	27	34	40	34
Average	Continuous	33	41	54	43
	Batch	18	30	36	28
Average of continuous type and batch type		26	35	43	36
Total average			35		

A.5 EFFECT OF ENERGY SAVING BY DEVELOPMENT OF HIGH PERFORMANCE INDUSTRIAL FURNACES

A.5.1 Assumptions of Calculations

The energy saving effect by the development of high performance industrial furnaces was evaluated. The number of installed industrial furnaces, their energy consumption, and the assumed values statistically obtained by this investigation are used.

1. Furnaces

The subject furnaces are the following 11 types of furnaces within the 19 types listed in Table A.2.

- Melting furnaces (steel-melting furnaces, nonferrous and metals melting furnaces, ceramics melting furnaces)
- Heating furnaces (metals soaking furnaces, metals heating furnaces)
- Heat treatment furnaces (metal heat treatment furnaces, surface heat treatment furnaces, surface treatment furnaces)
- Ceramics baking furnaces
- Chemical industry furnaces
- Drying furnaces

The heat efficiencies of these furnaces are presumed to be as low as 35% on averages excluding cement sintering furnaces and petroleum heating furnaces, which are discussed in Section A.6. Figure A.9 shows the relationship between the heating capacities and the heat efficiencies obtained from the questionnaires for continuous metal heating furnaces and heat treatment furnaces. According to the results of the questionnaires, the recuperator, which is waste gas recovery equipment, is generally installed in continuous metals heating furnaces. On the other hand, most metals

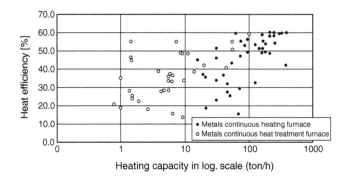

FIGURE A.9 Heating capacity and heat efficiency of metals continuous heating furnaces.

batch heating furnaces, metals heat treatment furnaces (continuous, batch), and aluminum-melting furnaces (batch) do not have recuperators.

As shown in Figure A.9, in the case of the middle and large continuous metals heating furnaces over a capacity of 10 ton/h, there exist some heating furnaces with heat efficiencies over 60% owing to the heat recovery systems. However, in the case of small-scale heating furnaces under 10 ton/h, several furnaces over 40 to 50% of heat efficiency exist, even if the average value is comparatively low. And, further, for batch heating furnaces, there are furnaces with heat efficiency over 40%, such as batch metal heating furnaces, batch metal heat treatment furnaces, and aluminum-melting furnaces.

The existing furnaces with recuperators have efficiencies of 40 to 60% in some quantities in every type of furnace. By introducing the result of the high performance industrial furnace development, the efficiency of heat recovery can be much improved compared with existing recuperators. Consequently, improvement of heat efficiency by about 50% in energy-saving ratio, that is, from the above-mentioned heat efficiency of 35% to around 53% on average, would be possible. The target value of the energy saving ratio by the development of high performance industrial furnace is over 30%. However, the energy-saving ratio rises to 34% with improvement of heat efficiency from 35 to 53%, as shown in Table A.11.

2. Cement kiln

In the above-mentioned 11 types of industrial furnaces, the NSP (new suspension preheater), which shows very high heat efficiency of 98%, was introduced in the 1970s and installed in 1994 including the SP (kiln with suspension preheater). Because greater improvement in heat efficiency is not expected, the cement-sintering furnace is excluded in the group of ceramics baking furnaces. The number of cement-sintering furnaces in 1994 was 81 furnaces and their energy consumption was 9.47 × 10^9 l (converted crude oil).

3. Petroleum heating furnace

The combustion amount and the heat efficiency of the petroleum heating furnaces are shown in Figure A.10; the heat efficiencies of large-scale petroleum heating furnaces with the air preheaters (APHs) show a high value of 90%. The heat

TABLE A.11
Energy Consumption of Industrial Furnaces and the Amount of Energy Saving by Introducing High Performance Industrial Furnaces

Type of Facilities	Ratio of Application, %	Number of Application (Cardinal number)	Energy Consumption (converted crude oil), 10^9 l/year	Ratio of Improvement of Heat Efficiency, %	Ratio of Energy Saving, %	Amount of Energy Saving, 10^9 l/year
1. Melting furnace		3,717	4.13			
2. Heating furnace		4,157	3.58			
3. Heat treatment furnace		11,614	7.03			
4. Baking furnace (cement)	100	6,512 − 81 = 6,431	17.01 − 9.47 = 7.54	50	34.0	15.24
5. Drying furnace		4,591	4.20	(Average 35–53)		
6. Chemical industry furnace (petroleum)		2,527 − 832 = 1,695	26.14 − 7.74 = 18.40			
Subtotal	—	32,351	44.88	—	—	15.24
7. Petroleum heating furnace	50	832/2 = 416	7.74/2 = 3.87	15 (78–90)	13.3	0.52
8. Industrial boiler	100	32,350	38.02	6 (85–90)	5.6	2.11
Subtotal	—	32,776	41.89	—	—	2.63
Grand total	—	65,117	86.77	—	—	17.87

Notes: (1) Heat efficiency = (Sensible heat of heating material /combustion heat of fuel) × 100(%). (2) Ratio of energy saving = (1 − heat efficiency before improvement/heat efficiency after improvement) × 100(%).

efficiency of middle- and small-scale petroleum heating furnaces without the air preheaters (APHs) is 78% on average. Because the efficiency of a furnace with this development would be better than that of the APH, about 90% heat efficiency could be achieved by introducing new technologies of high performance industrial furnaces to middle- and small-scale petroleum heating furnaces.

Because the energy consumption is at the same level for large-, middle-, and small-scale furnaces, we estimate, under the objective of 50% of the energy consumption, that the heat efficiency could be improved about 15% from the value of 78 to 90%.

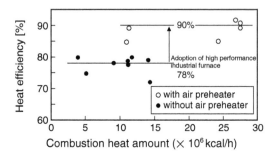

FIGURE A.10 Fuel consumption and heat efficiency in petroleum heating furnaces.

According to the most recent data (in 1993) of the Petroleum Federation on petroleum heating furnaces, the number of furnaces, 832, and the energy consumption, 7.74×10^9 l/year (converted crude oil), can be used for calculations.

4. *Industrial boiler*

The results of the development can also be applied to industrial boilers. Although the number of and energy consumption of industrial boilers is not confirmed in this investigation, the data on the number of boilers, 32,350, and the energy consumption, $38,028 \times 10^6$ l/year, from the data of the Agency of Natural Resources and Energy in 1975, are used for the calculation from the viewpoint of using the minimum values.

Industrial boilers are generally water tube type, where 80% heat efficiency can be obtained even in small-scale furnaces and 90% in large-scale furnaces owing to heat recovery of waste heat by means of an economizer or air preheater. Consequently, if the current heat efficiency of industrial boilers is 85% on average, the value of 90% heat efficiency could be achieved, as with petroleum heating furnaces.

A.5.2 Results of the Calculation

The summary of the assumptions and the results of the calculations are shown in Table A.11. The following points have been clarified:

1. The number of facilities for objects: N

$$N = \underset{\text{industrial furnaces}}{32,767} + \underset{\text{industrial boilers}}{32,350} = 65,117 \simeq 65,000$$

2. The energy consumption for objects: E (l/year)

$$E = \underset{\text{industrial furnaces}}{48.75 \times 10^9} + \underset{\text{industrial boilers}}{38.02 \times 10^9} = 86.77 \times 10^9 \simeq 87 \times 10^9$$

The value for the energy consumption corresponds to 23% of the final heat consumption, 46% of that in total industries, and 65% of that in industrial furnaces (excluding industrial boilers).
3. As shown in Table A.11, by introducing the results of the development of the high performance industrial furnaces the heat efficiency can be improved overall by 50% (from the current average of 35 to 53%), that of the petroleum heating furnaces by 15% (from the value of 78 to 90%), and that of industrial boilers by 6% (from the value of 85 to 90%); further, as a result energy savings of 17.9×10^9 l/year can be achieved. The value corresponds to 4.8% of the grand total energy consumption — $= 17.9 \times 10^9/376 \times 10^9 \times 100(\%)$.

A.6 SUMMARY

Japan Industrial Furnace Manufacturers Association is promoting activities of Development of a High Performance Industrial Furnace Project, which is a national project that has investigated the number of installed industrial furnaces, energy consumption, heat efficiency, etc., to clarify the effect of energy saving by introducing the results of the development. The results obtained are as follows:

1. The results of questionnaires for users of industrial furnaces showed that the number of facilities of industrial furnaces (excluding boilers and incinerators) is about 39,300 furnaces over 50 l/h of converted crude oil, or 50 kVA, and 35,300 furnaces over 50 l/h converted crude oil, or 200 kVA. Compared with the official data such as the number of facilities reported under the law of preventing air pollution or the investigation data of the Agency of Natural Resources and Energy, these values are reasonable.
2. The results of the questionnaires of users of industrial furnaces indicate that energy consumption (the consumed amount) of industrial furnaces (excluding boilers and incinerators) is annually 75×10^9 l (converted crude oil). Compared with the official data such as the statistical data of petroleum consumed structure (MITI) or the investigational data of the Agency of Natural Resources and Energy, these values are reasonable.
3. The results of the user interviews indicate that the heat efficiencies of metals-heating furnaces, metal heat treatment furnaces, and aluminum-melting furnaces average 26% for small-scale industrial furnaces, 35% for middle-scale, 43% for large-scale, and 35% on total average. Comparing continuous and batch furnaces, the heat efficiencies of continuous furnaces are higher than those of batch furnaces by 43 and 28%, respectively.
4. The energy-saving effects of introduction of the results of the development of high performance industrial furnaces are estimated as follows:
    a. The number of industrial furnaces: 65,100 furnaces (including 32,000 industrial boilers)

b. The energy consumption: 87×10^9 l/year (converted crude oil)
   c. The energy-saving amount: 17.9×10^9 l/year, when the improvement of general industry furnaces is 50% (the current value of 35 to 53%), that of petroleum heating furnaces 15% (78 to 90%), and that of industrial boilers 6% (85 to 90%)
   d. The energy-saving effect to the grand total energy consumption in Japan: 4.8%
5. According to the results of calculations based on combustion calculation of CO_2 reduction effect by introducing the results of development of high performance industrial furnaces, the total amount of CO_2 emissions in Japan is 320 million tons of converted carbon, in which industries annually emit CO_2 in converted carbon weight of 49.1 million tons by using gas and oil. Introduction of the results of the development of the high performance industrial furnace project can reduce emission of 14.7 million tons in converted carbon weight of CO_2, which corresponds to 4.6%, or about 5%, of total emissions in Japan.

REFERENCES

1. The Japan Industrial Furnace Manufacturers Association and Energy Conversion Center ed., *Current State Investigation from Several Japanese Industrial Furnaces*, JIFMA, Tokyo, Japan, 1996.
2. Plan & Research Section, Director-General's Secretariat, *Total Energy Statistics* 1995 ed, Trade and Industry Research Co., Tokyo, Japan, 1995, pp. 14 and 369.
3. Research and Statistics Department, Minister's Secretariat, *1994 Oil etc. Comsumption Structure Statistics Tables*, Trade and Industry Statistics Association, Tokyo, Japan, 1996, p. 128.

Appendix B
Constants and Conversion Factors

B.1 UNIVERSAL CONSTANTS AND CONVERSION FACTORS

Universal constants

Universal ideal gas constant	$\bar{R}$ = 8.313	kJ/kmol·K
	= 1.9872	cal/mol·K
	= 1.9872	Btu/lbmol·°R
	= 1545.33	ft·lbf/lbmol·°R
Boltzmann's constant	k = 1.38054 × 10^{-23}	J/K
Planck's constant	h = 6.626 × 10^{-34}	J·s
Speed of light in vacuum	c = 2.998 × 10^{8}	m/s
Avogadro's number	N = 6.022 × 10^{23}	molecules/mol
Stefan–Boltzmann constant	σ_{calc} = 5.6685 × 10^{-8}	W/m²K⁴
	= 0.1714 × 10^{-8}	h·ft²°R
	σ_{expr} = 5.729 × 10^{-8}	W/m²K⁴
	= 0.173 × 10^{-8}	h·ft²°R
Atmospheric pressure	P_{atm} = 0.101325	MPa
	= 1.01325	bar
	= 1.01325 × 10^{5}	N/m²
Ice point at 1 atm	T_{ice} = 0.00	°C
	= 273.15	K
Gravitational acceleration	g = 9.80665	m/s²
	= 32.17	ft/s²
Natural logarithm	$\log_{10} x$ = 0.4343 ln x	
Faraday's constant	F = 9.6500 × 10^{4}	As/mol
Electron charge	e_0 = 1.062 × 10^{-19}	As
Important mathematical numbers	e = 2.71828	
	π = 3.141592	
	1° = 0.01745	rad

Conversion Factors

Acceleration	1 m/s² = 4.252 × 10^{7}	ft/h²
Area	1 in² = 0.645 × 10^{-3}	m²
	1 ft² = 0.0929 m²	
	1 yd² = 0.8361	m²
	1 mi² = 2.590 × 10^{6}	m²

Quantity	Conversion	Unit
	1 acre = 4047.0	m²
Calorie or mechanical equivalence of heat	1 cal = 4.1868	J
Density	1 lbm/ft³ = 16.01846	kg/m³
	1 slug/ft³ = 515.379	kg/m³
Energy	1 Btu = 1055.0	J
	1 kW·h = 3600.0	kJ
	1 ft·lbf = 1.35582	J
	1 kcal = 4.1868	kJ
	1 kg_f m = 9.80665	J
Force	1 lbf = 4.448	N
	1 kgf = 9.80665	N
	1 dyne = 10^{-5}	N
Heat flux	1 Btu/h·ft² = 3.154	W/m²
	1 kcal/m²·h = 1.163	W/m²
Heat transfer coefficient	1 Btu/h·ft²·°F = 5.6786	W/m²K
	1 kcal/m²h°C = 1.163	W/m²K
Heat transfer rate	Same as power	
Kinetic viscosity (v), thermal diffusivity (α), mass diffusivity (D)	1 stokes = 1.0×10^{-4}	m²/s
	1 ft²/s = 0.0929	m²/s
	1 ft²/h = 2.581×10^{-5}	m²/s
Heat content, latent heat, specific energy, specific enthalpy, etc.	1 kcal/kg = 4.1868	kJ/kg
	1 Btu/lbm = 2.326	kJ/kg
Length	1 in = 2.54×10^{-2}	m
	1 ft = 0.3048	m
	1 yd = 0.9144	m
	1 mile = 1.609×10^{3}	m
Mass	1 lbm = 0.4536	kg
	1 U.S.ton = 907.19	kg
	1 ton = 1000.0	kg
	1 oz = 28.35×10^{-3}	kg
Mass transfer coefficient	1 ft/h = 8.467×10^{-5}	m/s
Power	1 PS = 735.4988	W
	1 kcal/h = 1.163	W
	1 Btu/h = 0.293	W
	1 hp = 745.7	W
	1 PS = 0.98632	hp
Pressure, stress	1 Pa = 1	N/m²
	1 psi = 6895	Pa
	1 atm = 1.013×10^{5}	Pa
	1 bar = 1.000×10^{5}	Pa
	1 torr = mmHg = 133.32	Pa
	1 mmAq = 1 mmH$_2$O = 9.80665	Pa
	1 ftH$_2$O = 2989	Pa
Specific heat, specific entropy	1 Btu/lbm·°F = 4.188	kJ/kgK
	1 kcal/kg·°C = 4.182	kJ/kgK
Speed	1 mi/h = 0.447	m/s
	1 km/h = 0.278	m/s
Temperature	1°C = 1.0	K

Appendix B

	1°F =	(5/9)	K
	T°C =	T K − 273.15	
	T°C =	5/9(T°F − 32)	
	T°F =	T°R − 459.67	
Thermal conductivity	1 Btu/h·ft°F =	1.7307	W/m·K
	1 kcal/m·h°C =	1.163	W/m·K
	1 cal/cm·s°C =	418.68	W/m·K
Thermal resistance	1°F/Btu·h =	1.896	K/W
	1 ft²·°F/Btu·h =	0.1761	m²K/W
	1 m²h°C/kcal =	0.8598	m²K/W
Viscosity (μ)	1 kg/s·m = 1 N·s/m² =	1.0	Pa·s
	1 lbm/ft·h =	4.1338×10^{-4}	Pa·s
	1 lbf·h/ft² =	1.7235×10^5	Pa·s
Volume	1 liter(l) =	0.001	m³
	1 in³ =	16.39	cm³
	1 ft³ =	0.02832	m³
	1 yd³ =	0.7646	m³
	1 gal(U.S.) =	3.785	l
	1 gal(imperial) =	4.546	l
	1 pint =	0.5683	l
	1 fluid ounce =	0.029	l
Volumetric heat generation rate	1 Btu/h·ft³ =	10.3497	W/m³
	1 kcal/m³·h =	1.163	W/m³

B.2 NONDIMENSIONAL PARAMETERS

Reynolds number $\quad$ Re = uL/v
$\quad\quad v$ = kinetic viscosity = $\mu/\rho \quad$ m²/s
$\quad\quad \rho$ = density $\quad$ kg/m³
$\quad\quad \mu$ = molecular viscosity $\quad$ Pa·s

Prandtl number $\quad$ Pr = v/a or $(gc_p\mu)/\lambda$
$\quad\quad a$ = thermal diffusivity = $\lambda/\rho c_p \quad$ m²/s
$\quad\quad \lambda$ = heat conductivity $\quad$ W/mK

Peclet number $\quad$ Pe = Re Pr = uL/a

Nusselt number $\quad$ Nu = hL/λ
$\quad\quad h$ = heat transfer coefficient $\quad$ W/m²K

Grashof number $\quad$ Gr = $(g\beta L^3 \Delta T)/v^2$
$\quad\quad \beta$ = volume coefficient of expansion
$\quad\quad g$ = gravity
$\quad\quad L$ = characteristic length
$\quad\quad \Delta T$ = temperature difference

Graetz number $\quad$ Gz = Re Pr (d/L)
$\quad\quad d$ = inner diameter of tube $\quad$ m
$\quad\quad L$ = tube length $\quad$ m

Mach number $\quad$ M = u/u_{sound}
$\quad\quad u_{sound}$ = sound velocity

Knudsen number	$Kn = \lambda_p/L$	
	λ_p	= mean free path
	L	= characteristic length
Damköhler number	$Da = \omega/(v\rho)$	
	First Damköhler number, denotes ratio of characteristics times of flow and chemical reaction.	
	ω	= amount of productives by reaction a unit time a volume.
	v	= flow velocity
Karlovitz number	$K = g_u \delta/u$	
	g_u	= gradient of velocity
	δ	= thickness of preheating zone of flame front
Schmidt number	$Sc = v/D$	
	D	= molar diffusivity
Lewis number	$Le = D/a$	

B.3 NOMENCLATURE

The symbols appearing in this book are summarized in Table B.1; subscripts and superscripts are listed in Table B.2.

TABLE B.1
Meanings of Symbols

Symbol	Meanings
a	furnace wall area per unit length in Chapter 2, velocity gradient in Chapter 5, and thermal diffusivity
a_λ	monochromatic absorptivity of spectrum λ.
$a_{g,n}$	the fractional amount of energy in the spectrum for the gray gas of absorption coefficient
A	nondimensional band intensity in Equation 2.22
A_C	area of heated surface
A_H	area of heating surface
$A_{S,O}, A_{s,P}$	surface Damköhler numbers defined in Equation 2.40
A_{nh}	nonhomogeneous total band absorptance
A_w	furnace wall area, $m^2 = La$
b_m	the sum of reciprocal of the heat capacity
B	line broadening parameter in Equation 2.22
B_S	the frequency factor of the surface reaction
c_p	specific heat
C	half-line width in Equation 2.20
C_0, C_1, C_2, C_r	constant
$C\dot{m}$	heat capacity of object transferred by unit time
d_λ	monochromatic transmissivity of spectrum λ.
D	line interval, diffusivity
e	nondimensional line intensity in Equation 2.20

TABLE B.1 (CONTINUED)
Meanings of Symbols

Symbol	Meanings
E	energy
E_S	activation energy of the specimen
E_x	exergy
f_v	soot volumetric fraction
f_s	nondimensional combustion rate
F_{ij}	shape factor
g	concentration fluctuation intensity
G	Gibbs free energy
Gr	Grashof number
h	convection heat transfer
h_i	enthalpy
h_{rad}	heat transfer coefficient
H	enthalpy
H_0	calorific value of fuel
H_a	sensible heat of preheated air
I	spectral intensity of radiation
I_{soot}	total radiance from soot only
k	reaction rate constant, turbulence energy
k_s^v, k_a^v	spectral absorption coefficients of soot and gas
K	gross heat transfer coefficient of heat exchanger
K	complement number of material transfer factor in Equation 2.34
K_m	overall coefficient of heat transfer coefficient
L	length of heat exchanger in Chapter 3, optical length of nonscattering media in Sections 2.5.3 and 3.3, zone length in Chapter 5
L_m	length of heat exchanger
m_i	molecular weight of ith species
M	mass weight
P_e	pressure parameter in Equation 2.22
Pr	Prandtl number
Q	heat flow rate
Q_f	heat of formation of fuel
Q_{rad}	radiative heat transfer
r_T	temperature ratio
r_λ	monochromatic reflectivity of spectrum λ.
R	gas recycling rate in Sections 1.2.1 and 2.2, gas constant
R_D	radial position
Re	Reynolds number
t	time
T	temperature
T_{at}	atmospheric temperature
T_{cold}	initial temperature of fuel and air
T_{fmax}	maximum temperature in the furnace or flame
T_{fw}	wall temperature of furnace
T_{out}	flue gas temperature

TABLE B.1 (CONTINUED)
Meanings of Symbols

Symbol	Meanings
u	flow velocity
v_0	flow rate of fuel without air preheating
v_a	flow rate of fuel
V	velocity of uniform flow upstream
W	bandwidth parameter in Equation 2.22, molecular weight
X	positions in coordinate
Y	positions in coordinate in Section 2.2, mass rate in Section 2.5
Y	mass rate
ΔG_T^0	standard free energy
ΔH_T^0	standard heat of reaction
ΔT	temperature difference
$\Delta V, \Delta v$	velocity difference
Δt	division of time
Δx	division of x coordinate
Ψ^3	the pentagamma function
α	ratio of the recirculated heat to the total heat contained in the recirculated gas flow
α_k, β_k	the integrals of the soot volumetric fraction over the path length $(L - S_k)$ and $(S_{k+1} - 0)$
β	mass transfer number in Equation 2.34
β	mass transfer number
δ_{ij}	Kronecker's delta
$\dot{m}$	mass flow rate
η	efficiency
λ	thermal conductivity
μ	coefficient of viscosity
ν	wave length
υ	radiation wave number
ω_i	mass fraction
$\overline{F_{ij}^*}$	shape factor under consideration with radiation and reflection of surface
ϕ	equivalence ratio, concentration fluctuation
ϕ_{CG}	overall heat transfer coefficient (total absorptivity)
ϕ_{CH}	overall heat transfer coefficient calculated by Equation 5.3
ϕ_i	the production rate
ρ	density
σ	Stefan–Boltzmann constant
τ	time constant
τ_s^v, τ_g^v	spectral soot and gas emissivity
S	the mean formation rate
ε	dissipation rate of turbulence energy
ε_C	emissivity of heated surface
ε_H	emissivity of heating surface
ε_λ	monochromatic emissivity
ξ	heat exchange coefficient

TABLE B.1 (CONTINUED)
Meanings of Symbols

Symbol	Meanings
	Special Dimensions
m_N^3	gas volume at normal state (0°C, 1 atm)
ton/h	tonnage processing rate per hour
ton/ch	tonnage rate per charge

TABLE B.2
Meanings of Subscripts and Superscripts

Symbol	Meanings
0	ambient condition or input
a	air
ad	adiabatic
al f	adiabatic limit of flame
at	atmospheric
av	available
cold	initial state
C	carbon
ev	excess value
exh	exhaust
f	fuel
fw	furnace wall
g	gas
i	initial
i, j	index of species
in	input
in1	exit of heat exchanger
in2	after mixer
j	index which refers to stagnation flow
loss	loss
m	heating materials
O	oxygen
P	product
rec	recirculation
s	soot
w	wall

Subject Index

ϕ_{CG}, 51, 243, 245, 266, 284, 295, 308, 310, 385
ϕ_{CH}, 308, 310, 312, 313, 320, 385

A

absorption
   band, 19, 101, 102
   band spectrum, 21
   body, 19, 21
   coefficient, 51, 99, 101–103, 147, 148, 192, 194, 195, 383, 384
   selective, 19
absorptivity, 19, 21, 22, 190–192, 204, 295, 308, 383
   total, 243, 385
acetaldehyde, 212
acetone, 42, 43
acid rain, 344
acrylonitrile, 212
activation energy, 71, 122, 128, 163, 172, 183, 383
Adams-Bashforth scheme, 203
additional enthalpy combustion, 3, 49, 57, 59, 60
adiabatic condition, 174, 271
adiabatic flame temperature, 24, 51, 57, 69–71, 73, 78, 97, 230–232, 234, 307
adjacent gas, 18
after burning, 244
air
   dilution, 66
   excess ratio, 89
air assist atomizer, 358
air ratio, 213
air-to-fuel ratio control, 295
alternative operation, 323
aluminum titanate, 61, 106
aluminum-melting furnace, 334
American Society for Testing and Materials, *see* ASTM
annealing apparatus, 212
aromatic species form, 95, 96
Arrhenius
   expression, 173
   model, 173
   plot, 69, 70, 123, 128
   rate expression, 190
   reaction rate, 182
   type, 71

   type expressions, 182
   type formula, 172
ash
   deposition, 157
   fly, 141, 354
   melting, 354
   vaporization, 346
   vitrification, 342
ASTM, 133, 351
asymptotic analysis method, 166
atom concentration, 198, 199
autoignition, 8, 114, 297, 306
   limit, 82
   temperature, 10, 13, 15, 59, 114, 171, 244, 355
axial-flow-heating burners, 307

B

band model, 146, 196
band width parameter, 385
batch furnace, 314, 363
bath surface, 325, 327
bayonet angle, 106
bearing dehydrating apparatus, 213
BFG, 280
biomass, 116
   power generation, 215
blackbody, 19, 22, 190, 191
   emissive power, 22, 190, 191
   emitter, 147
   function, 147
   radiation intensity, 191
   spectral intensity, 147
blast
   furnace, 212
blue-green flame, 115, 116
bluish green, 4, 32
boundary layer, 18, 20, 62
Bowman scheme, 75
burned-gas
   recirculating effect, 46
burned-gas circulation, 238
burner
   pitch, 225, 248
   rim, 178, 188
   type, 240, 299, 301, 323, 332

burnout
  level, 141, 150, 152–154
  phenomena, 117

C

calorific value, 7, 8, 29, 120, 180, 231, 233, 234, 245, 278, 280, 281, 284, 298, 315, 345, 347–349, 351, 352, 354, 355, 384
carbon content, 13, 350
carbon monoxide, see CO
Carnot cycle, 26
CCD camera, 32
cement kiln, 374
centrifugal force, 109
ceramic
  ball, 4, 15, 61, 345, 346
  fiber, 61, 303, 307
  honeycomb, 9, 15, 61, 158
  layer, 32
ceramics
  baking furnace, 374
CFD analysis result, 263
chain terminating reaction, 176
characteristic length, 18, 19, 382
charging method, 297
chemical
  waste, 343
  wastes, 342, 344
chemical composition, 34
chemical equilibrium, 177, 181, 182
  temperature, 180
chemical gas turbine, 356
chemical reaction, 64, 70, 92, 159, 172, 175, 179, 382
  characteristic time, 185
  control, 73
  formation of chlorine gas, 350
  overall, 69, 70
  rate, 71
  rate model, 176
  scheme, 91, 98
chemical species, 84, 172, 178
CHEMKIN, 89, 91, 97
chlorinated compound, 350
city gas, 39, 45, 213, 214, 250
clean combustion, 346
clean energy, 359
CNG, 214
CO, 66, 68, 92, 97, 121, 127, 133, 134, 136, 137, 166, 355–357
CO flame, 161, 162, 166–168

coal
  burning boilers, 341
  combustion, 145
    system, 158
  flame, 136, 138, 145
  gasification, 353
  gun, 132, 133, 135, 149, 150, 152, 157
  particles, 125, 140
  velocity injection, 135
coal combustion
  high temperature air, 157
  modeling of, 121
coal flame, 157
coefficient
  absorption, 99, 204, 384
  convection heat transfer, 20, 51
  expansion, 382
  frequency, 172, 183
  gross heat transfer, 384
  heat conduction, 320
  heat exchange, 50, 59, 386
  heat loss, 54
  heat transfer, 50, 200, 243, 266, 271, 312, 313, 380, 382, 384
  mass transfer, 381
  of viscosity, 163
  overall heat transfer, 384, 385
  thermal absorption, 51
  viscosity, 163
COG, 280
cold model, 34
combined cycle, 212
  power generation, 212, 213
combined cycle power, 213
combustible limit, 3, 8, 10
combustible range, 105
combusting volume, 40
combustion
  domain, 10
  exhaust emission, 2
  exhaust gas, 299, 302
  limit, see combustible limit
  mechanism, 115, 157, 158
  noise, 59
  product recirculation, 49, 50
  reaction, 15
  spray, 105, 115
  stability, 34, 54, 59
  state, 47, 86
combustion flame, distributed, 41, 46
combustion heat model, simplified, 202
combustion limit, 30
combustion model, 172, 176–178, 181, 182, 189, 196, 200, 204
  simplified heat, 202

combustion reaction, 12, 16, 68, 86, 87, 92, 173, 204
  mechanism, 198
  model, 177
  zone, 84
commercial
  application, 312
  furnace, 238
conservation
  energy, 4, 6, 9, 200
  equation, 200
  of resources, 2
continuous casting, 212
continuous combustion, 10
convection
  effect, 104
  heat transfer, 20
convection effect, 104
convective heat transfer, 190, 295, 313
convective section
  the boiler, 355
conventional
  combustion technology, 356
  control method, 288
convergence time, 289, 290
COP3, 1, 327
cordierite, 106
counterflow
  configuration, 89
  diffusion flame, 68, 178
crude oil, 369–371, 374–377

D

Damköhler number, 86, 185, 244, 382
  first, 382
  surface, 163, 166, 383
density
  change, 103, 200, 203
  fluctuation, 103
deodorizing machine, 334
desalination, 353
desulfurization, 124
diesel engines, 61
diffusion
  coefficient, 91
  combustion, 89
  flame, 16, 33, 43, 60, 70, 79, 89, 91, 178
  rate, 64, 125, 126
  turbulent flame, 244
diluted air, 30–32, 71, 111, 115, 116
  combustion, 33
  with burned gas, 12
  with inert gas, 10, 16

diluted combustion, 263
  burner, 240
dilution
  effect, 75
  gas composition, 34
  process, 36, 38
  rate, 111, 114, 116
  rate by air, 111
  ratio, 72, 78
  with burned product, 20
dioxin, 30, 344, 345, 349, 350, 352–354
discrete transfer method, 203
dispersion combustion, 227, 228
dissipation rate, 175, 200, 385
dominant wavelength, 190
double film model, 121
droplet
  fuel, 109
drying
  furnace, 363
  process, 352

EBU, 175
eddy-break-up model, 175, see EBU
eddy-dissipation model, see EBU, 175
efficiency, 13, 26, 37, 71, 90, 136, 212–215, 218, 247, 249, 277, 305, 312, 319, 321, 324, 342, 345, 353, 355, 374, 375, 377
  combustion, 86, 322, 342, 353
  conversion, 4
  energy, 321
  heat, 361, 362, 372, 374–377
  heat exchanger, 356
  heat transfer, 324
  heating, 226, 230, 243, 245, 247–249
  isothermal combustion, 26
  of furnace, 4
  of heating, 17
  of isenthalpic combustion, 26
  ordinary combustion, 25
  radiation heat transfer, 326
  system, 356
  temperature, 227
  thermal, 59
  thermodynamic, 25, 26
  waste heat recovery, 13
  working, 323
electric
  ceramic fan, 213
  furnace, 363
  heating, 159, 310, 334, 364
  power generation, 341
  power generator, 131

radiant heater, 213
electromagnetic wave, 19
electron charge, 379
elementary reaction, 72, 172, 178, 181, 183
Elsasser model, 101, 194
emission
    CO, 34, 37
    index, 80, 82, 89
    level, 34, 37, 59, 80, 82, 87, 88, 299, 327, 337, 338, 355
    spectrum, 43, 45
    spontaneous spectrum, 43
emissive power, 22, 190
    monochromatic, 191
emissivity, 385
    heated surface, 385
    heating surface, 385
    monochromatic, 386
    path length, 192
empirical constant, 173, 175, 182, 185, 189
endothermic reaction, 92, 159, 166, 176
energy
    consumption, 1, 5, 215, 221, 328, 334, 335, 337, 351, 361, 364, 369, 371, 373, 375–377
    conversion, 3, 4, 214, 342
    equation, 203
    problem, 359
    recovery waste-, 334
    saving, 2, 4, 5, 9, 23, 37, 61, 157, 213, 221, 233, 236, 306, 315, 325, 328, 333–336, 341, 354, 359, 361, 373–375, 377
energy saving, 214, 217, 218, 334, 336, 352, 375
enthalpy balance, 174
envelope flame, 62
envelope-type flame, 64
Environment Agency, 211, 364
environment conservation, 342
environment protection measure, 353
environmental pollution, 116, 342, 344
equation
    energy conservation, 202
    mass conservation, 202
    momentum conservation, 202
equilibrium
    chemical, 177, 179
    constant, 181, 358
    state, 68
    temperature, 180
equivalence ratio, 10, 11, 23, 24, 45, 61, 71, 72, 78, 80, 91, 97, 98, 179, 180, 183, 185, 190, 358, 385
equivalent heat-transfer coefficient, 271
ethylene oxide, 212

evaluation function, 294
evaporating process, 111
exchange area, 20, 21
excitation state, 45
exergy
    change, 23
    loss, 23
exothermic reaction, 92, 159, 186, 346
external energy, 3
extinction
    flame, 62, 69, 71
    limit, 64

F

Favre average, 200
feedback control, 284, 288, 291, 292
finite-difference equation, 311
first-half reaction, 184, 186
first-order reaction, 121
flame
    sheet model, 177
    adiabatic temperature, 73
    chemistry, 91
    color, 4
    extinction, 61, 62, 64, 66, 69, 71
    flat, 71–73, 91
    fluctuation, 85
    front, 72, 73, 75, 78, 87, 161, 162, 172, 174
    thickness, 382
    holder, 10, 176, 178
    intensity, 31, 62, 64
    length, 136, 323, 324, 332
    lifted, 186
    model, 91
        flat, 78
    pattern, 244
    picture, 116
    propagation mechanism, 72
    sheet, 69
    spectral emission, 30
    stability, 3, 4, 85, 188
    stabilizing, 10
    strain rate, 64, 65, 68
    structure, 31
    thin, 172
    visible length, 136
flame length, 323
flameless oxidation, 6, 29, 81
flamelet regime, 86
flammability
    limit, 59
flat flame, 72, 73
flow pattern, 34, 38, 139

Subject Index

fluctuation intensity, 175, 383
flue gas, 30, 37, 50, 53, 79, 134, 140, 142, 145, 149, 150, 153, 155–157, 174, 202, 384
fluid-dynamic simulation, 202, 203
fluidized beds, 349
fluorescence signals, 43
fly ash, 350, 351
forging steel industry, 331
formation mechanism, 9
four-step model
    Jones', 186
    reaction, 183
    reaction mechanism, 183
    reaction model, 184
    Srivatsa's, 186
free energy, 357, 383, 385
free jet theory, 139
frequency factor, 122, 128, 163, 383
front stagnation region, 158
fuel
    composition, 266
    dispersion, 244
    droplet, 114, 115
    gas, 3
    liquid, 3, 12, 105–107, 109, 157
    pyrolysis, 244
    solid, 3
    waste, 344
fuel cell, 211, 215, 342
fuel reforming steam, 342
fuel-rich mixtures, 181
full reaction mechanism, 177–179
full reaction model, 178
furan, 344, 345, 349, 350
furnace
    AC electric, 212
    aluminum-melting, 325, 326, 334, 374
    basic reheating, 266
    batch, 315
    billet-reheating, 306
    blast, 212, 278, 363
    cement sintering, 374
    ceramic melting, 373
    ceramics baking, 363, 374
    commercial, 238
    conventional, 321, 325
    conventional industrial, 3
    DC electric, 212
    direct-heating, 308
    drying, 373, 375
    gas treatment, 332
    heat treatment, 4, 316
    heating, 221, 293
    high-cycle regenerative, 9
    high-performance reheating, 303
    industrial, 362, 364, 374
    melting, 4, 321
    metal heat-treatment, 373
    metal-melting, 373
    petroleum heating, 373, 374, 376
    pressure, 132, 218, 243, 245, 284, 289–293, 295, 297, 305, 307
    pressure control, 245
    pusher-type, 229
    recuperator-type, 221
    regenerating, 221
    regeneration, 39
    reheating, 4, 265, 266, 268, 278, 282, 298, 303–306
    reheating continuous, 266
    reheating furnace, 298
    shape, 51, 219
    sintering, 212
    slab heating, 219
    slab-reheating, 296
    temperature, 15, 17, 39–43, 45, 47, 50, 103, 110, 111, 113, 114, 116, 208, 219, 228, 229, 231, 235, 237, 243, 250, 251, 263, 266, 271, 279, 290, 294, 305, 322, 324, 332
    type, 330, 334
    volume, 136, 325
    wall, 17–19, 21, 99, 100, 103, 110, 133, 190, 196, 231, 243, 245, 267, 271, 294, 295, 310, 319, 324, 325, 383, 386

G

gas recirculation, 54, 55, 59, 60, 245
gas recycling rate, 11, 384
gas-burning hot air heater, 213
gasification, 344–348, 353, 358
Gaussian distribution
    clipped, 174
geometrical similarity, 313, 320
Gibb's energy, 23
glass melting furnace, 103
global climate, 1
global environment, 1
global warming, 5, 327, 344, 359, 361
governing equation, 72, 175, 200, 202, 203
Grashof number, 202, 382, 383
gravitational acceleration, 379
gray
    analysis, 196
    approximation, 99, 103
    body, 102
    gas, 102, 103, 192, 383

green color flame, 4, 32
GRI-Mech, 91
Guasare coal, 133
gun position, 149, 150, 152, 157

H

health hazard, 87
heat
    accumulator, 20
    capacity, 15, 17, 50, 59, 292, 318, 319, 334, 383
    exchanger, 5, 8, 13, 15, 50, 52–54, 59, 61, 108, 157, 227, 266, 269, 305, 321, 353, 384, 386
    extraction, 132, 232
    feedback, 3
    flow rate, 50
    flux, 156
    pattern, 215, 218, 245, 264, 294, 318
    recirculating combustion, 6, 7, 9, 14, 15
    recirculation, 3, 4, 7, 8, 24, 25, 49, 54
    reservoir, 61, 106, 108, 227
    resistance, 47
    resistance alloy, 13
    subtraction, 10, 24
    waste, 3
heat balance, 273
    analysis, 53
    model, 221, 264, 266, 272
heat conduction, 19, 23, 312
    model, 312
    unsteady, 320
heat conductivity, 313, 382
heat efficiency, 374
heat flux, 6, 18, 19, 29, 104, 145, 156, 195, 196, 200, 207, 225, 250, 251, 308, 355, 380
    distribution, 116
    profile, 145
    radiant, 195, 295
    radiation, 6
    radiative, 194, 355
    total, 196
heat radiation, 266, 267, 271, 280, 304, 312
heat recirculation, 3, 25, 60
heat recovery rate, 264, 269, 278, 280, 282, 321, 335, 354
heat release, 16, 89, 171, 204
    rate, 16, 18, 24, 32, 204
    zone, 16
heat transfer
    coefficient, convection, 20
    convection, 20
    emissive, 324
    radiation, 20
    through furnace wall, 271
heat treatment furnace, 313, 329, 330
heating
    electric, 310, 334
    furnace, 294, 333
    gas-fired, 334
    method, 293, 307
heating exchanger, regenerative, 266
heavy fuel oil, 131, 341, 355, 356
heavy metals, 348
HFO, 138
high performance model, 266
high temperature air, 5, 10, 30, 32, 38, 39, 61, 70, 85, 86, 110, 126, 158, 160, 161, 168, 240, 341, 343–349, 353, 354, 356, 359
    generator, 106
    oxygen-enriched, 351
high temperature air combustion, *see* HiTAC
high temperature spot, 236
high-cycle regenerator, 9, 37
high-momentum jet, 185
high-speed video camera, 84
high-temperature combustion, 237, 263
high-temperature preheated air, 176, 244, 297, 306
HiTAC, 1, 3–6, 8–11, 13, 15–22, 25, 29, 30, 38, 48, 59, 71, 79, 85–87, 89, 103, 104, 116, 131, 145, 156–158, 166, 171–174, 176, 177, 182, 183, 186, 189, 190, 195–198, 202–204, 211, 229, 230, 234–240, 244, 245, 248, 256, 261–263, 332, 341, 343, 344, 349, 354–356, 359
    technology, 2, 228
hollow-cone spray, 106
hot direct rolling, 212
hot slab, 212
hot spot, 29
hybrid car, 211
hydraulic power generation, 215
hydrogen chloride, 350, 353
hydrogen-air mixture, 24

I

ideal engine, 24
IFRF, 131, 133, 353
IFRF furnace, 156
ignition
    delay, 72, 355
    phenomena, 178

stand-off distance, 153
time, 92, 97, 207
imperfect combustion, 80, 324
in-flame measurement, 135
in-furnace
    pressure, 219
    temperature, 230, 232, 234
industrial
    analysis, 123
    boiler, 376, 378
    chemical analysis, 120
    combustion unit, 3
    device, 71
    furnace, 2, 4–6, 9, 18, 83, 103, 171, 172, 211, 213, 215, 219, 238, 240, 243, 296–298, 306, 328, 333, 337, 338, 341, 359, 361, 362, 364, 369–371, 375, 377, 378
    production, 2
    sector, 2
inert gas, 16
inorganic
    compound, 184, 348
    species, 186
intense oxidation, 359
intermediate chemical species, 183
intermediate-partition wall, 250
International Flame Research Foundation, see IFRF, 353
inverter control, 213
ion current, 256, 260, 261
iron oxide, 21
isenthalpic combustion, 24–26
isothermal combustion, 24, 26

J

jet-stirred reactor, 86
JIS standards, 119

K

Kalman filter, 292, 293
Karlovitz number, 382
kerosene oil burner, 322
kinetic energy, 200
kinetic mechanisms, 198
Kyoto Protocol, 1, 327

L

laminar flame, 64, 116, 183
large eddy simulation, 203

latent heat, 211, 213, 233, 380
LDV probe, water-cooled, 134
lean combustion, 24, 88, 344
lean fuel combustion, 228
Lewis number, 382
LIF measurement, 42
lifted flame, 112, 113, 115, 116, 157, 176, 178, 182, 244
light fuel oil flames, 135
light scattering technique, 34
line spectrum, 194
line tracking function, 294
liquefied petroleum gas, see LPG
liquid
    droplet, 106, 107
    film, 106, 213
    fuel, 109, 331, 355, 358
LNG, 240, 280
low NOx combustion, 121
low oxygen, 31, 32, 89, 171, 244, 263
    combustion, 177
    region, 152
LPG, 29, 39, 43, 45, 105, 109, 111, 214, 240, 280
LPG gas, 34
luminous flame, 19, 21, 42, 45, 99, 105, 190, 192, 244

M

maceral composition, 120
mass
    balance, 136
    diffusivity, 380
    flow rate, 50, 163, 168, 385
    fraction, 173
    rate, 163, 165, 166, 168, 385
    recirculation, 50
mass balance, macroscopic, 50
mass flow, coal, 133
mass flow rate, 168
mass ratio, 10
mass transfer number, 385
mechanisms, gas emission-absorption, 147
melting furnace, 321, 322, 328, 329, 334, 358, 361, 363, 364, 373, 374, 377
melting rate, 325
membrane technology, 358
metal heating furnace, 374
methane
    air flame, 16
    air mixtures, 178, 183
    combustion, 75
    flames, 91
METI, 211
micro gas turbines, 211, 341, 355

mineral composition, 120
Ministry of International Trade and Industry, *see* METI
MixGas, 280
mixing layer, 87
mixtures, fuel-rich, 180
model
    chemical-equilibrium, 178
    heat balance, 221
moisture, 134, 344, 346, 348
molecular weight, 72, 120, 163, 322, 356, 384, 385
molten bath, 323–325
molten glass, 103
monochromatic
    emissive power, 190, 191
    emissivity, 191
monochromator, 84
Monte Carlo method, 204
multistage combustion, 227

N

naphtha cracking reaction, 212
narrow-band model, 194
Navier-Stokes equation, 175
nitrogen-containing compound, 87
nitric oxide (NO), 1, 9, 15, 16, 66, 68, 69, 71, 73, 75, 78–82, 86–90, 198–202
nitric dioxide (NO_2), 66, 68, 86, 87
noise
    combustion, 59
    regulation values, 316
non-premixed combustion, 8, 9, 15, 200
nonequilibrium
    intermediate, 178
    process, 23
nongray
    approximation, 99
    characteristics, 102
    condition, 99, 102, 103
    gas, 100–102, 104, 146, 203
    model, 196
    radiation, 100, 195, 196
nonluminous
    combustion, 21, 196
    combustion gas, 190
    flame, 99, 103, 105, 196
nonuniform grid, 200
NOx, 2, 5, 6, 34, 36, 37, 39, 41–43, 45–48, 59, 60, 65, 66, 68–71, 116, 133–136, 138, 149, 150, 155–157, 201, 207, 227, 228, 235–238, 240

nuclear power generation, 215
number, mass transfer, 162, 385
numerical
    analysis, 102
    calculation, 68, 320
    code, 202
    direct simulation, 171
    experiment, 208
    method, 203, 208
    model, 198
    prediction, 89
    simulation, 5, 61, 171, 176–179, 182, 188–190, 194, 196, 200
    solution, 100
    technique, 178
Nusselt
    number, 314, 382
    second-order rule, 126

O

O_2, 136
OH radical concentration, 199
oily waste, 356
on-line tracking function, 294
one-step global reaction, 172, 173, 176–178, 182, 184, 186, 188, 189
opposing jet flow, 207
optical
    band-pass width, 84
    emission, 189
    emissions, 84
    fiber, 45
    length, 384
    measurement, 48, 127
    measuring method, 39
    thickness, 104
optimization process, 294
oxidation process, 75
oxidation-terminating reaction, 186
oxygen
    concentration, 5, 10, 13, 20, 30–33, 36, 45, 47, 48, 60, 66, 85, 86, 89, 152, 159, 161, 165, 166, 182, 186, 188, 200, 236, 350, 358
    control, 219
    level, 132
    low concentration, 6
    mole fraction, 199
    reaction order, 199
oxygen-free thermal decomposition, 110

P

PAH, 30, 43, 91–93, 244
particle surfaces, 105, 124
particles, radiate, 148
path length, 148, 192, 385
pattern factor, 355
PCB, 350
PCDD, 350
PCDF, 350
permanent disposal method, 349
petrochemical furnaces, 103
petroleum heating furnace, 377, 378
phase-interaction model, 171
photo detector, 219
PID controller, 291
plastic-containing wastes, 344
pollutant species, 86
pollution reduction, 9
polycyclic aromatic hydrocarbon, 30
polynuclear aromatic hydrocarbon, 244
polypropylene, 212
porous plugs, 327
post flame, 152
post-ignition period, 92, 98
potential core, 140, 260, 261
potential theory, 62
power generation
    biomass, 215
    combined cycle, 212
    hydraulic, 215
    nuclear, 215
    solar, 215
    thermal, 215
    waste, 215
    wind, 215
practical simulation, 194
Prandtl number, 202, 314, 382, 384
precursor compound, 350
preheated air, 1, 4, 9–13, 15, 29, 30, 32–34, 37, 39, 41–43, 61, 64, 69–71, 79, 82, 84, 90, 99, 103, 105, 108, 115, 136, 176, 182, 185, 188, 208, 227–229, 231, 233, 234, 237, 243, 266, 267, 269, 288, 345, 384
preparation period, 12
pressure effects, 358
process simulator, 265
prompt NO mechanism, 69
propane ignition, 91
propane-air diffusion flame, 195
pull-back ratio, 208
pulverized coal combustion, 135
pure oxygen, 358, 359
pyrolysis process, 346

Q

quantum mechanics, 194
quasi equilibrium, 23
quasi-steady-state, 198
quick-mixing-type burner, 237

R

radiance calculations, 148
radiant
    energy, 21, 190, 191, 203
    heat flux, 195, 295
    intensity, 21
    tube, 308
    tube burner, 334
radiation
    analysis, 103
    background, 135
    effect, 99, 101, 104, 324
    energy, 4, 21, 22, 102, 103, 191, 194
    far-infrared, 19
    gas, 308, 310
    heat flux, 6
    heat transfer, 19–22, 99, 100, 103, 105, 202, 325
    heat transfer rate, 53
    intensity, 196
    thermometer, 294
    transport, 196
    wall, 200
    wavelength, 99, 105
radiation effect, 99
radiation-heat-transfer model
    detailed, 202
radiative
    energy, 104, 192
    fluxe, 133
    heat flux, 194, 355
    heat transfer, 203
    heat transfer model, 200
radical emission, 115, 116
radiometer, narrow angle, 134
radiometer probe, ellipsoidal, 134
rate coefficient, 92, 98
rate-determining factor, 125
ray tracing method, 204
RDF, 342, 344, 349–353
    burning, 349
    combustion, 353
    combustion technology, 351
    combustion technology for, 353
    composition of, 350
    energy balance of, 351

plant, 352
power generation, 352
production, 352, 353
production of, 351
shape of, 353
technology, 352
reacted model, 177, 200
reaction
   broadened zone, 12
   chemistry, 29
   constant, 199
   distributed, 82, 86
   phenomena, 124, 125
   propane combustion, 91
   scheme, 72, 172, 176, 183
   time, 78, 244
reaction model, 174, 177, 178, 182, 186
   chemically-controlled, 173
   one-step, 186
reaction rate, 64, 69, 70, 92, 97, 172, 175–177, 181
   Arrhenius, 182
   combustion, 175
   constant, 127, 384
   time-averaged, 175
reaction zone, broadened, 16
reactor vessel, 358
READ, 203, 204
reburning, 198, 199, 244, 354
recirculating
   combustion, 8, 26
   flue gas, 54
   gas, 53, 60
   turbulent flow, 171
recirculation ratio, 54, 59
recirculation zone, 81, 139, 148, 152
recuperator, 305, 374
recycling rate, 10
reduced mechanism, 183
reduced-reaction mechanism, 181
refractory, 13, 20
refuse derived fuel, see RDF
regenerative
   burner, 4, 48, 208, 209, 297, 302, 304, 305, 307, 308, 321–323, 326, 333
   combustion process, 204
   furnace, 208, 264
   heating-exchanger simulator, 268
   media, 158, 332, 354
regenerator, 4, 9, 14, 15, 202
reheating furnace, 266, 279, 302
residence time, 68, 86, 235, 349
respective mole fraction, 19
reverse reaction, 176, 179, 181, 182, 184, 185, 189
Reynolds number, 202, 209, 314, 382, 384

Reynolds-average, 203
rich combustion, 88
rich hydrocarbon flame, 98
rolling process, 219, 306
RT-type regenerative burner, 314

S

scale-forming heat, 266
schedule-free combustion, 297
Schmidt number, 382
semi-finished steel, 264, 266–268, 274
sensible heat, 4, 9, 17, 61, 231, 233, 234, 266, 267, 269, 277, 280, 304, 375, 384
shutdown condition, 219
signal processing method, 287
single film model, 121
sintering furnace, 373
skid
   mark, 229, 297, 299, 303
   pipe, 297, 307
   support, 229
slab thickness, 247, 283
slow heat release, 13
solar power generation, 213, 215
solar power, 214
solid carbon, 157–159, 161, 163, 165, 166, 168
solid waste, 341, 343–345, 348–351, 353, 359
soot, 29, 146–148, 196, 244, 355, 369, 384, 386
   cloud, 192
   emission, 98, 99
   emissivity, 148
   particle, 145, 192
   particles, 148
   precursor, 43
   spectral, 148
   total radiance, 384
   volumetric fraction, 148, 383, 385
source term, 198–200, 204
spectral soot, 147, 385
spherical
   lens, 45
   solid, 312
spontaneous emission spectra, 46
spray combustion, 105, 111
spraying
   angle, 107, 109
   method, 109, 110
   pressure, 109, 115
stability
   combustion, 59
   flame, 30, 31
   limit, 30, 32, 182
stabilizing flames, 176

Subject Index

stable combustion, 3, 31, 236, 284, 287, 296
stable furnace operation, 315
stagnation
    flow, 158, 162, 163, 168, 386
    streamline, 66, 68
stationary gas turbine combustion, 341
steady combustion, 239
steady operation, 84
steel slab, 208, 219, 225
steel-making process, 215
Stefan-Boltzmann constant, 51, 148, 204, 308, 379, 385
step reaction model, 181
step response model, 289, 290
stoichiometric
    oxygen, 175
    ratio, 15, 136, 174
strain rate, 61, 62, 65, 68
    at extinction, 64
    critical, 64
    flame, 64–66, 68, 71
stretch rate, 178, 179
suction pyrometer, 132, 134, 207
sulfur oxide, 2
surface
    reaction, 121, 161, 163, 166, 383
    regression, 159
    roughness, 21, 103
surface temperature, 159
switched regenerator, 34
switching
    cycle, 36, 37
    device, 31
    high-cycle-, operation, 202
    high-speed, 108
    operation, 209
    time, 5, 37
syngas chemical energy, 347
syngas fuel, 347
synthetic air condition, 117

T

Taylor expansion, 311
temperature
    adiabatic equilibrium, 178
    contour, 138
    fluctuation, 13, 17, 33, 82, 88, 179, 355
    uniformity, 209, 256
    variation, 203, 244, 249
temperature distribution, uniform, 171
temperature limit, furnace materials, 54
temperature-rise curves, 304

thermal
    conductivities, 91
    conductivity, 4, 268, 308, 311, 312, 381, 385
    decomposition, 345
    degradation, 350
    destruction, 343, 344, 349, 354, 356, 359
    diffusivity, 380, 382, 383
    efficiency, 3, 4, 9, 13, 23, 54, 57, 59, 60, 247, 264, 266, 304
    expansion, 163, 168, 203
    insulation, 215, 217, 307
    layer, 325–327
    mechanism, 89, 200, 201
    power generation, 215
    radiation, 104
    recycling technology, 345
    resistance, 381
thermocouple, 32–34, 42, 68, 81, 131, 134, 294
thermofluid
    dynamics, 172
    simulations, 177
thin film, 43, 46
thin flame, 172
Third Conference of Parties, *see* COP3
three-dimensional radiation, 100
tilting movement, 326
time constant, 81, 289, 293, 385
time scale, 86
time-averaged temperature distribution, 82
town gas, 116
toxic substance, 353
transmissivity, monochromatic, 383
transparent flame, 244
transport
    air, 135, 149, 150, 152–155, 157
    equation, 198
    parameter, 91
    rate of reactant by diffusion, 69
transportation, 214, 363, 371
turbulence model, 171, 175, 209
turbulent
    combustion, 17, 83, 87, 173, 200
    flame, 13, 84, 87, 179
    model, 203
    motions, 179
    time scale, 86
twin fluid spraying method, 109
two-color thermometer, 159
two-line method, 84
two-stage combustion, 327

U

ultra-lean mixture, 2, 3, 8

unburned hydrocarbons, 138
uniform temperature distribution, 86, 201, 307
uniform-heating condition, 249
unit fuel consumption, 245, 247, 299, 303, 320, 333, 335, 336
unit heat consumption, 362
universal gas constant, 172
universal ideal gas constant, 379
unstable flames, 154
unsteady state condition, 208

V

vacuum tubes, 342
velocity
    contour, 139
    gradient, 62, 64, 69, 70, 158 161, 163, 166–168, 383
visible flame, 135, 136, 156
volatile
    bituminous, 133
    matter, 117, 120–124, 126–128
    organic compound, 342, 354

W

wake flame, 62, 64
walking beam, 229, 242, 298, 306
wall
    radiation, 145
    side jet boundary, 137
waste
    combustion furnace, 349
    disposal, 344

energy conversion, 348
energy recovery function, 334
    gas, 215, 217, 218, 228, 314, 373
    heat, 3, 4, 9, 212, 217, 227, 304, 335, 376
    heat boiler, 217
    heat recovery, 321, 333
    heat recovery rate, 335
    heat-recovery, 245
    incineration technology, 351
    incinerators, 369
    paint, 356
    paper pulp, 213
    power generation, 215
    treatment, 345
    water, 344
waste derived fuel (WDF), 350, 351
waste heat, 13, 37
wave number, 101, 102, 147, 194, 385
wavelength, 21, 22, 148, 190, 195, 385
wavelength conversion, 20, 21
weighting factor, 192
well-stirred reactor, 24, 73, 85, 91
wet-type coke quenching, 212
wind
    power generation, 215

Z

Zel'dovich
    NO formation model, 201, 202
    thermal mechanism, 69
Zel'dovich mechanism, 66, 68, 70, 87, 88
    extended, 75, 87, 198
zero emission, 351

Author Index

A

Arai, N., 159

B

Balakrishnan, A., 194, 195
Béer, J. M., 192
Becker, H. A., 61
Bermudez, G., 93
Blint, R. J., 68
Bowman, C. T., 72, 89, 91, 178, 199
Boyle, J., 93
Braud, Y., 134

C

Chan, S. H., 146
Chang, R. C., 355, 356
Chang, W. C., 355, 356
Chedaille, J., 134
Cheng, P., 200
Coffee, T. P., 183
Communal, F., 93
Cornforth, J. R., 4

D

De Soete, G. G., 199
Dearden, L. M., 15
Deardroff, J. W., 203
Dixon-Lewis, G., 72
Drake, M. C., 68
Dupont, V., 199

E

Ebisui, K., 79, 90
Edwards, D. K., 192, 194, 195
Essenhigh, R. H., 125, 126

F

Flamme, M., 9
Fujimori, T., 182
Fujita, O., 45
Fujiwara, T., 121
Fukutani, S., 72, 90

G

Gupta, A. K., 341, 345, 346, 355, 356, 358

H

Haas, J. H. P., 155
Habibzadeh, B., 358
Hanaoka, R., 343
Hanson, R. K., 199
Hardesty, D. R., 131
Hasegawa, T., 6, 9, 11, 61, 71, 90, 91, 131
Hayasaka, H., 203
Hayashi, J., 328
Hida, A., 287, 328
Hidaka, Y., 93
Hirano, T., 45
Hjertager, H., 175
Hoshino, T., 6, 9
Hottel, H. C., 192, 225

I

Ihara, T., 45
Ikeda, Y., 45
Imada, M., 225
Ishii, T., 200
Ito, K., 45
Iwahashi, T., 343

J

Johnson, T. R., 192
Jones, W. P., 183
Ju, Y., 88, 89

399

K

Kagitani, T., 351
Kamp, W. L. van de, 155
Kang, S. W., 124
Kansa, E. J., 123
Kasahara, M., 90
Katsuki, M., 71, 79, 90
Kee, R. J., 72, 89, 91
Kiga, T., 343
Kishimoto, K., 90
Kobayashi, H., 123
Kojima, J., 45
Kosaka, H., 343
Kreutz, T. G., 61
Kunioshi, N., 72, 90

L

Lallemant, N., 146
Law, C. K., 61
Lee, E. S., 353
Lilley, D. G., 356
Liñán, A, 166
Lindstedt, R. P., 183
Lloyd, S. A., 3, 7
Locquet, J. J., 146
Lois, E., 357

M

Magnussen, B. F., 175
Makino, A., 159–161
Matsumoto, M., 9
Matsunaga, S., 308
Matsuoka, Y, 211
Mihara, Y., 159
Miller, J. A., 72, 89, 91
Mochida, S., 61, 343
Modest, M. F., 194
Morita, M., 131
Morita, T., 287
Murakami, H., 328
Murakami, K., 225
Muramatsu, K., 343

N

Nagata, K., 352
Nakamura, M., 343
Nakasima, T., 45
Naruse, I., 123, 127, 128
Niioka, T., 8, 88, 89, 131

Niska, E. J., 123
Nozaki, H., 287
Nusselt, W., 126

O

Ohtake, K., 121
Oike, H., 90
Okazaki, K., 123

P

Penner, S. S., 194
Perlee, H. E., 123
Pfefferle, L. D., 93
Pitt, G. J., 123

R

Rahbar, S., 6, 61
Richter, H., 91
Riechelmann, D., 182
Rupley, F. M., 72, 91

S

Saito, T., 328
Sakai, M., 343
Salimian, S., 199
Sarofim, A. F., 192, 225
Sato, J., 61, 182
Serio, M. A., 123
Sigaki, M., 351
Smagorinsky, J., 203
Smoot, L. D., 120
Sobiesiak, A., 61
Solomon, P. R., 123
Spalding, D. B., 175
Srivatsa, S. K., 185
Stabat, P., 146
Stein, S. E., 93
Sugiyama, S., 200
Sugiyama, T., 6, 9
Suzukawa, Y., 15

T

Tíen, C. L., 146, 194
Tamura, M., 155
Tanaka, K., 225
Tanaka, R., 7, 61, 71, 90, 131

Author Index

Tanigawa, T., 131
Thomas, S. D., 93
Truelove, J. S., 192

U

Ubhayakar, S. K., 123
Uede, M., 225
Ureshino, M., 352

V

Verlaan, A. L., 131, 134, 135, 140

W

Warnatz, J., 199, 201

Watanabe, Y., 90
Weber, R., 6, 46, 131
Weinberg, F. J., 3, 7, 9, 15, 131
Wen, C. Y., 353
Westmoreland, P. R., 93
Wunning, J. A., 6, 81
Wunning, J. G., 6, 81

Y

Yamamoto, T., 356
Yamamoto, Y., 124
Yoshikawa, K., 157, 343, 345, 346, 348

Z

Zhang, C., 200

For Product Safety Concerns and Information please contact our EU
representative GPSR@taylorandfrancis.com
Taylor & Francis Verlag GmbH, Kaufingerstraße 24, 80331 München, Germany

www.ingramcontent.com/pod-product-compliance
Ingram Content Group UK Ltd.
Pitfield, Milton Keynes, MK11 3LW, UK
UKHW021444080625
459435UK00011B/370